A Quantum Approach to Condensed Matter Physics

This textbook is a reader-friendly introduction to the theory underlying the many fascinating properties of solids. Assuming only an elementary knowledge of quantum mechanics, it describes the methods by which one can perform calculations and make predictions of some of the many complex phenomena that occur in solids and quantum liquids. The emphasis is on reaching important results by direct and intuitive methods, and avoiding unnecessary mathematical complexity. The authors lead the reader from an introduction to quasiparticles and collective excitations through to the more advanced concepts of skyrmions and composite fermions. The topics covered include electrons, phonons, and their interactions, density functional theory, superconductivity, transport theory, mesoscopic physics, the Kondo effect and heavy fermions, and the quantum Hall effect.

Designed as a self-contained text that starts at an elementary level and proceeds to more advanced topics, this book is aimed primarily at advanced undergraduate and graduate students in physics, materials science, and electrical engineering. Problem sets are included at the end of each chapter, with solutions available to lecturers on the internet. The coverage of some of the most recent developments in condensed matter physics will also appeal to experienced scientists in industry and academia working on the electrical properties of materials.

"*... recommended for reading because of the clarity and simplicity of presentation*"

Praise in *American Scientist* for Philip Taylor's *A Quantum Approach to the Solid State* (1970), on which this new book is based.

PHILIP TAYLOR received his Ph.D. in theoretical physics from the University of Cambridge in 1963 and subsequently moved to the United States, where he joined Case Western Reserve University in Cleveland, Ohio. Aside from periods spent as a visiting professor in various institutions worldwide, he has remained at CWRU, and in 1988 was named the Perkins Professor of Physics. Professor Taylor has published over 200 research papers on the theoretical physics of condensed matter and is the author of *A Quantum Approach to the Solid State* (1970).

OLLE HEINONEN received his doctorate from Case Western Reserve University in 1985 and spent the following two years working with Walter Kohn at the University of California, Santa Barbara. He returned to CWRU in 1987 and in 1989 joined the faculty of the University of Central Florida, where he became an Associate Professor in 1994. Since 1998 he has worked as a Staff Engineer with Seagate Technology. Dr Heinonen is also the co-author of *Many-Particle Theory* (1991) and the editor of *Composite Fermions* (1998).

A Quantum Approach to Condensed Matter Physics

PHILIP L. TAYLOR
Case Western Reserve University, Cleveland

OLLE HEINONEN
Seagate Technology, Seattle

PUBLISHED BY THE PRESS SYNDICATE OF THE UNIVERSITY OF CAMBRIDGE
The Pitt Building, Trumpington Street, Cambridge, United Kingdom

CAMBRIDGE UNIVERSITY PRESS
The Edinburgh Building, Cambridge CB2 2RU, UK
40 West 20th Street, New York, NY 10011-4211, USA
477 Williamstown Road, Port Melbourne, VIC 3207, Australia
Ruiz de Alarcón 13, 28014 Madrid, Spain
Dock House, The Waterfront, Cape Town 8001, South Africa

http://www.cambridge.org

© Cambridge University Press 2002

This book is in copyright. Subject to statutory exception
and to the provisions of relevant collective licensing agreements,
no reproduction of any part may take place without
the written permission of Cambridge University Press.

First published 2002

Printed in the United Kingdom at the University Press, Cambridge

Typeface *Times* System *3B2*

A catalogue record for this book is available from the British Library

Library of Congress Cataloguing in Publication data
Taylor, Philip L. (Philip Liddon)
A quantum approach to condensed matter physics / Philip L. Taylor and Olle Heinonen.
p. cm.
Includes bibliographical references and index.
ISBN 0 521 77103 X – ISBN 0 521 77827 1 (pc.)
1. Condensed matter. 2. Quantum theory. I. Heinonen, O. II. Title.

QC173.454 .T39 2002
530.4$'$1–dc21   2001037339

ISBN 0 521 77103 X hardback
ISBN 0 521 77827 1 paperback

Contents

Preface ix

Chapter 1
Semiclassical introduction 1

1.1 Elementary excitations 1
1.2 Phonons 4
1.3 Solitons 7
1.4 Magnons 10
1.5 Plasmons 12
1.6 Electron quasiparticles 15
1.7 The electron–phonon interaction 17
1.8 The quantum Hall effect 19
Problems 22

Chapter 2
Second quantization and the electron gas 26

2.1 A single electron 26
2.2 Occupation numbers 31
2.3 Second quantization for fermions 34
2.4 The electron gas and the Hartree–Fock approximation 42
2.5 Perturbation theory 50
2.6 The density operator 56
2.7 The random phase approximation and screening 60
2.8 Spin waves in the electron gas 71
Problems 75

Chapter 3
Boson systems 78

3.1 Second quantization for bosons 78
3.2 The harmonic oscillator 80
3.3 Quantum statistics at finite temperatures 82
3.4 Bogoliubov's theory of helium 88
3.5 Phonons in one dimension 93
3.6 Phonons in three dimensions 99
3.7 Acoustic and optical modes 102
3.8 Densities of states and the Debye model 104
3.9 Phonon interactions 107
3.10 Magnetic moments and spin 111
3.11 Magnons 117
Problems 122

Chapter 4
One-electron theory 125

4.1 Bloch electrons 125
4.2 Metals, insulators, and semiconductors 132
4.3 Nearly free electrons 135
4.4 Core states and the pseudopotential 143
4.5 Exact calculations, relativistic effects, and the structure factor 150
4.6 Dynamics of Bloch electrons 160
4.7 Scattering by impurities 170
4.8 Quasicrystals and glasses 174
Problems 179

Chapter 5
Density functional theory 182

5.1 The Hohenberg–Kohn theorem 182
5.2 The Kohn–Sham formulation 187
5.3 The local density approximation 191
5.4 Electronic structure calculations 195
5.5 The Generalized Gradient Approximation 198

5.6 More acronyms: TDDFT, CDFT, and EDFT   200
Problems   207

Chapter 6
Electron–phonon interactions   210

6.1 The Fröhlich Hamiltonian   210
6.2 Phonon frequencies and the Kohn anomaly   213
6.3 The Peierls transition   216
6.4 Polarons and mass enhancement   219
6.5 The attractive interaction between electrons   222
6.6 The Nakajima Hamiltonian   226
Problems   230

Chapter 7
Superconductivity   232

7.1 The superconducting state   232
7.2 The BCS Hamiltonian   235
7.3 The Bogoliubov–Valatin transformation   237
7.4 The ground-state wave function and the energy gap   243
7.5 The transition temperature   247
7.6 Ultrasonic attenuation   252
7.7 The Meissner effect   254
7.8 Tunneling experiments   258
7.9 Flux quantization and the Josephson effect   265
7.10 The Ginzburg–Landau equations   271
7.11 High-temperature superconductivity   278
Problems   282

Chapter 8
Semiclassical theory of conductivity in metals   285

8.1 The Boltzmann equation   285
8.2 Calculating the conductivity of metals   288
8.3 Effects in magnetic fields   295
8.4 Inelastic scattering and the temperature dependence of resistivity   299
8.5 Thermal conductivity in metals   304

8.6 Thermoelectric effects   308
Problems   313

Chapter 9
Mesoscopic physics   315

9.1 Conductance quantization in quantum point contacts   315
9.2 Multi-terminal devices: the Landauer–Büttiker formalism   324
9.3 Noise in two-terminal systems   329
9.4 Weak localization   332
9.5 Coulomb blockade   336
Problems   339

Chapter 10
The quantum Hall effect   342

10.1 Quantized resistance and dissipationless transport   342
10.2 Two-dimensional electron gas and the integer quantum Hall effect   344
10.3 Edge states   353
10.4 The fractional quantum Hall effect   357
10.5 Quasiparticle excitations from the Laughlin state   361
10.6 Collective excitations above the Laughlin state   367
10.7 Spins   370
10.8 Composite fermions   376
Problems   380

Chapter 11
The Kondo effect and heavy fermions   383

11.1 Metals and magnetic impurities   383
11.2 The resistance minimum and the Kondo effect   385
11.3 Low-temperature limit of the Kondo problem   391
11.4 Heavy fermions   397
Problems   403

Bibliography   405

Index   411

Preface

The aim of this book is to make the quantum theory of condensed matter *accessible*. To this end we have tried to produce a text that does not demand extensive prior knowledge of either condensed matter physics or quantum mechanics. Our hope is that both students and professional scientists will find it a user-friendly guide to some of the beautiful but subtle concepts that form the underpinning of the theory of the condensed state of matter.

The barriers to understanding these concepts are high, and so we do not try to vault them in a single leap. Instead we take a gentler path on which to reach our goal. We first introduce some of the topics from a semiclassical viewpoint before turning to the quantum-mechanical methods. When we encounter a new and unfamiliar problem to solve, we look for analogies with systems already studied. Often we are able to draw from our storehouse of techniques a familiar tool with which to cultivate the new terrain. We deal with BCS superconductivity in Chapter 7, for example, by adapting the canonical transformation that we used in studying liquid helium in Chapter 3. To find the energy of neutral collective excitations in the fractional quantum Hall effect in Chapter 10, we call on the approach used for the electron gas in the random phase approximation in Chapter 2. In studying heavy fermions in Chapter 11, we use the same technique that we found successful in treating the electron–phonon interaction in Chapter 6.

Experienced readers may recognize parts of this book. It is, in fact, an enlarged and updated version of an earlier text, *A Quantum Approach to the Solid State*. We have tried to preserve the tone of the previous book by emphasizing the overall structure of the subject rather than its details. We avoid the use of many of the formal methods of quantum field theory, and substitute a liberal amount of intuition in our effort to reach the goal of physical understanding with minimal mathematical complexity. For this we pay the penalty of losing some of the rigor that more complete analytical treatments can yield. The methods used to demonstrate results are typically simple and direct. They are expedient substitutes for the more thorough approaches to be found in some of the bulkier and more specialized texts cited in the Bibliography.

Some of the problems at the ends of the chapters are sufficiently challenging that it took the authors a longer time to solve them than it did to create them. Instructors using the text may therefore find it a time-saver to see our versions of the solutions. These are available by sending to *solutions@cambridge.org* an e-mail containing plausible evidence that the correspondent is in fact a busy instructor rather than a corner-cutting student pressed for time on a homework assignment.

The earlier version of this text owed much to Harold Hosack and Philip Nielsen for suggested improvements. The new version profits greatly from the comments of Harsh Mathur, Michael D. Johnson, Sankar Das Sarma, and Allan MacDonald. Any mistakes that remain are, of course, ours alone. We were probably not paying enough attention when our colleagues pointed them out to us.

Philip Taylor   Cleveland, Ohio
Olle Heinonen   Minneapolis, Minnesota

Chapter 1
Semiclassical introduction

1.1 Elementary excitations

The most fundamental question that one might be expected to answer is "why are there solids?" That is, if we were given a large number of atoms of copper, why should they form themselves into the regular array that we know as a crystal of metallic copper? Why should they not form an irregular structure like glass, or a superfluid liquid like helium?

We are ill-equipped to answer these questions in any other than a qualitative way, for they demand the solution of the many-body problem in one of its most difficult forms. We should have to consider the interactions between large numbers of identical copper nuclei – identical, that is, if we were fortunate enough to have an isotopically pure specimen – and even larger numbers of electrons. We should be able to omit neither the spins of the electrons nor the electric quadrupole moments of the nuclei. Provided we treated the problem with the methods of relativistic quantum mechanics, we could hope that the solution we obtained would be a good picture of the physical reality, and that we should then be able to predict all the properties of copper.

But, of course, such a task is impossible. Methods have not yet been developed that can find even the lowest-lying exact energy level of such a complex system. The best that we can do at present is to guess at the form the states will take, and then to try and calculate their energy. Thus, for instance, we might suppose that the copper atoms would either form a face-centered or body-centered cubic crystal. We should then estimate the relative energies of these two arrangements, taking into account all the interactions we could. If we found that the face-centered cubic structure had the lower energy we might be encouraged to go on and calculate the change in energy due to various small displacements of the atoms. But even though we found that all the small displacements that we tried only increased the energy of the system, that would still be no guarantee that we had found the lowest energy state. Fortunately we have tools, such as X-ray diffraction, with which we can satisfy ourselves that copper does indeed form a face-centered cubic crystal, so that calculations such as this do no more than test our assumptions and our mathematics. Accordingly, the philosophy of the quantum theory of condensed matter is often to accept the crystal structure as one of the given quantities of any problem. We then consider the wavefunctions of electrons in this structure, and the dynamics of the atoms as they undergo small displacements from it.

Unfortunately, we cannot always take this attitude towards the electronic structure of the crystal. Because we have fewer direct ways of investigating the electron wavefunction than we had for locating the nuclei, we must sometimes spend time questioning whether we have developed the most useful picture of the system. Before 1957, for example, people were unsuccessful in accounting for the properties of superconductors because they were starting from a ground state that was qualitatively different from what it is now thought to be. Occasionally, however, a new technique is introduced by means of which the symmetry of electronic states can be probed. An example is shown on the cover of this book. There the effect on the electronic structure of an impurity atom at the surface of a high-temperature superconductor is shown. The clover-leaf symmetry of the superconducting state is clearly seen in the scanning-tunneling-microscope image.

The interest of the experimentalist, however, is generally not directed towards the energy of the ground state of a substance, but more towards its response to the various stimuli that may be applied. One may measure its specific heat, for example, or its absorption of sound or microwaves. Such experiments generally involve raising the crystal from one of its low-lying states to an excited state of higher energy. It is thus the task of the theorist not only to make a reasonable guess at the ground state, but also to estimate the energies of excited states that are connected to the ground state in a simple way. Because the ground state may be of little further interest once its form has been postulated, it is convenient to forget about it altogether and to regard the process of raising the system to a higher state as one of creating something where nothing was before. The simplest such processes are known as the creation of *elementary excitations* of the system.

The usefulness of the concept of elementary excitations arises from a simple property that most many-body systems have in common. Suppose that there are two excited states, and that these have energies above the ground state of $\mathcal{E}_1$ and $\mathcal{E}_2$, respectively. Then it is frequently the case that there will also be one particular excited state whose energy, $\mathcal{E}_3$, is not far removed from $(\mathcal{E}_1 + \mathcal{E}_2)$. We should then say that in the state of energy $\mathcal{E}_3$ all the excitations that were present in the other two states are now present together. The difference $\Delta\mathcal{E}$ between $\mathcal{E}_3$ and $(\mathcal{E}_1 + \mathcal{E}_2)$ would be ascribed to an interaction between them (Fig. 1.1.1). If the states of energy $\mathcal{E}_1$ and $\mathcal{E}_2$ could not themselves be considered as collections of other excitations of lower energy then we say that these states represent elementary excitations of the system. As long as the interaction energy remains small we can with reasonable accuracy consider most of the excited states of a solid as collections of elementary excitations. This is clearly a very useful simplification of our original picture in which we just had a spectrum of energy levels which had no particular relationship to one another.

At this point it is useful to consider a division of the possible types of elementary excitations into two classes, known as *quasiparticle excitations* and *collective excitations*. The distinction between these is best illustrated by some simple examples. We know that if we have a gas of noninteracting particles, we can raise the energy of one of these particles without affecting the others at all. Thus if the gas were originally in its ground state we could describe this process as creating an elementary excitation. If we were now to raise the energy of another particle, the energies of the excitations would clearly add up to give the energy of the doubly excited system above its ground state. We should call these *particle excitations*. If now we include some interactions between the particles of the gas, we should expect these particle excitations to decay, since now the excited particle would scatter off the unexcited ones, and its energy and momentum would gradually be lost. However, if the particles obeyed the Pauli Exclusion Principle, and the energy of the excitation was very low, there would be very few empty states into which the particle could be scattered. We should expect the excitation to have a sufficiently long lifetime for the description in terms of particles to

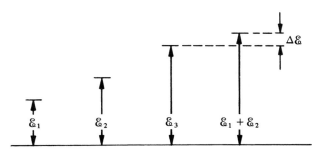

Figure 1.1.1. When two elementary excitations of energies $\mathcal{E}_1$ and $\mathcal{E}_2$ are present together the combined excitation has an energy $\mathcal{E}_3$ that is close to $\mathcal{E}_1 + \mathcal{E}_2$.

be a useful one. The energies of such excitations will differ from those for noninteracting particles because of the interactions. It is excitations such as these that we call *quasiparticles*.

A simple example of the other class of excitation is that of a sound wave in a solid. Because the interatomic forces in a solid are so strong, there is little profit in considering the motion of an atom in a crystal in terms of particle motion. Any momentum we might give to one atom is so quickly transmitted to its neighbors that after a very short time it would be difficult to tell which atom we had initially displaced. But we do know that a sound wave in the solid will exist for a much longer time before it is attenuated, and is therefore a much more useful picture of an excitation in the material. Since a sound wave is specified by giving the coordinates not of just one atom but of every atom in the solid, we call this a collective motion. The amplitude of such motion is quantized, a quantum unit of traveling sound wave being known as a *phonon*. A phonon is thus an example of a collective excitation in a solid.

We shall now consider semiclassically a few of the more important excitations that may occur in a solid. We shall postpone the more satisfying quantum-mechanical derivations until a later chapter. By that time the familiarity with the concepts that a semiclassical treatment gives may reduce somewhat the opacity of the quantum-mechanical procedures.

1.2 Phonons

The simplest example of collective motion that we can consider is that of a linear chain of equal masses connected by springs, as illustrated in Fig. 1.2.1. The vibrational modes of this system provide some insight into the atomic motion of a crystal lattice.

If the masses M are connected by springs of force constant K, and we call the displacement of the nth mass from its equilibrium position y_n, the equations

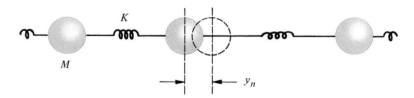

Figure 1.2.1. This chain of equal masses and springs supports collective motion in the form of traveling waves.

of motion of the system are

$$M \frac{d^2 y_n}{dt^2} = K[(y_{n+1} - y_n) - (y_n - y_{n-1})]$$
$$= K(y_{n+1} - 2y_n + y_{n-1}). \quad (1.2.1)$$

These equations are easily solved for any boundary conditions if we remember the recursion formula for cylindrical Bessel functions,

$$\frac{dJ_n}{dt} = -\frac{1}{2} [J_{n+1}(t) - J_{n-1}(t)],$$

from which

$$\frac{d^2 J_n}{dt^2} = \frac{1}{4} [J_{n+2}(t) - 2J_n(t) + J_{n-2}(t)].$$

The problem we considered in Section 1.1 was to find the motion of the masses if we displaced just one of them ($n = 0$, say) and then released it. The appropriate solution is then

$$y_n(t) = J_{2n}(\omega_m t)$$

where $\omega_m^2 = 4K/M$. This sort of behavior is illustrated in Fig. 1.2.2. The displacement of the zeroth mass, being given by $J_0(\omega_m t)$, is seen to exhibit oscillations which decay rapidly. After just a few oscillations $y_0(t)$ behaves as $t^{-1/2} \cos(\omega_m t)$. This shows that particle-like behavior, in which velocities are constant, has no relation to the motion of a component of such a system.

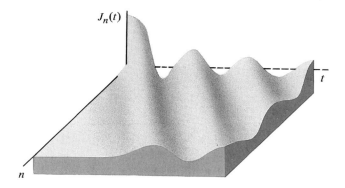

Figure 1.2.2. These Bessel functions are solutions of the equations of motion of the chain of masses and springs.

And this is quite apart from the fact that in a crystal whose atoms are vibrating we are not fortunate enough to know the boundary conditions of the problem. This direct approach is thus not very useful.

We find it more convenient to look for the normal modes of vibration of the system. We make the assumption that we can write

$$y_n \propto e^{i(\omega t + kna)}, \tag{1.2.2}$$

where ω is some function of the wavenumber k, and a is the spacing between masses. This satisfies the equations of motion if

$$-\omega^2 M = K(e^{ika} + e^{-ika} - 2),$$

that is, if

$$\omega = \pm \omega_m \sin\left(\tfrac{1}{2} ka\right).$$

The solution (1.2.2) represents traveling waves of frequency ω and wavenumber (defined for our purposes by $2\pi/\lambda$, where λ is the wavelength) equal to k. The group velocity v is given by $d\omega/dk$, the gradient of the curve shown in Fig. 1.2.3. We note that as ω approaches its maximum value, ω_m, the group velocity falls to zero. This explains why the Bessel function solution decayed to an oscillation of frequency ω_m after a short time, if we realize that the original equation for $y_n(t)$ can be considered as a superposition of waves of all wavenumbers. The waves of low frequency, having a large group velocity, travel quickly away from the zeroth site, leaving only the highest-frequency oscillations, whose group velocity is zero.

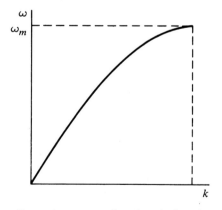

Figure 1.2.3. The dispersion curve for the chain of masses and springs.

It is formally straightforward enough to find the normal modes of vibration for systems more complicated than our linear chain of masses. The extension to three dimensions leads us to consider the *polarization* of the lattice waves, that is, the angle between **k**, which is now a vector, and the direction of displacement of the atoms. We can also introduce forces between atoms other than nearest neighbors. This makes the algebra of finding $\omega(\mathbf{k})$ more involved, but there are no difficulties of principle. Introduction of two or more different kinds of atom having different masses splits the graph of $\omega(\mathbf{k})$ into two or more branches, but as long as the restoring forces are all proportional to the displacement, then solutions like Eq. (1.2.2) can be found.

A *phonon* is the quantum-mechanical analog of the lattice wave described by Eq. (1.2.2). A single phonon of angular frequency ω carries energy $\hbar\omega$. A classical lattice wave of large amplitude corresponds to the quantum situation in which there are many phonons present in one mode. We shall see later that a collection of phonons bears some similarity to a gas of particles. When two particles collide we know that the total momentum is conserved in the collision. If we allow two phonons to interact we shall find that the total wavenumber is conserved in a certain sense. For this reason phonons are sometimes called quasiparticles, although we shall avoid this terminology here, keeping the distinction between collective and particle-like behavior.

1.3 Solitons

The chain of masses connected by Hookean springs that we considered in the previous section was a particularly easy problem to solve because the equations of motion (1.2.1) were linear in the displacements y_n. A real solid, on the other hand, consists of atoms or ions having hard, mutually repulsive cores. The equations of motion will now contain nonlinear (i.e., anharmonic) terms. How do these affect the type of excitation we may find?

If the amplitudes of the phonons are small then the effects of the anharmonic terms will be weak, and the problem can be treated as a system of interacting phonons. If the atomic displacements are large, on the other hand, then there arises a whole new family of elementary excitations known as solitary waves or *solitons*. In these excitations a localized wave of compression can travel through a solid, displacing the atoms momentarily but then leaving them as stationary as they were before the wave arrived.

The term soliton suggests by its word ending that it is a purely quantum-mechanical concept, but this is not the case. Solitary waves in classical systems had been observed as long ago as 1834, but it was only when their interactions were studied that it was found that in some cases two solitary waves could collide and then emerge from their collision with their shapes unchanged. This particle-like behavior led to the new terminology, which is now widely applied to solitary waves of all kinds.

We can begin to understand the relation between phonons in a harmonic solid and solitary waves in an anharmonic solid with the aid of an exactly soluble model due to Toda. We start by considering the simplest possible model that can support a soliton, namely a one-dimensional array of hard rods, as illustrated in Fig. 1.3.1. If we strike the rod at the left-hand end of the array, it will move and strike its neighbor, which in turn will strike another block. A solitary wave of compression will travel the length of the array leaving all but the final block at rest. The speed of this soliton will be determined entirely by the strength of the initial impact, and can take on any positive value. The wave is always localized to a single rod, in complete contrast to a sound wave in a harmonic solid, which is always completely delocalized.

Toda's achievement was to find a model that interpolated between these two systems. He suggested a chain in which the potential energy of interaction between adjacent masses was of the form

$$V(r) = ar + \frac{a}{b} e^{-br}. \tag{1.3.1}$$

In the limit where $b \to 0$ but where the product ab is equal to a finite constant c we regain the harmonic potential,

$$V(r) = \frac{a}{b} + \frac{1}{2} cr^2.$$

In the opposite limit, where $b \to \infty$ but $ab = c$, we find the hard-rod potential for which $V \to \infty$ if $r \leq 0$ and $V \to 0$ if $r > 0$.

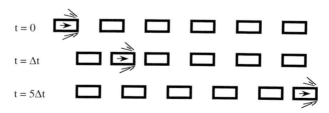

Figure 1.3.1. Through a series of elastic collisions, a solitary wave of compression propagates from left to right.

We construct a chain of equilibrium spacing d by having the potential $\Sigma_n V(R_n - R_{n-1} - d)$ act between masses located at R_n and R_{n-1}. In the notation where the displacement from equilibrium is $y_n = R_n - nd$, the equations of motion are

$$M \frac{d^2 y_n}{dt^2} = -a\left(e^{-b(y_{n+1}-y_n)} - e^{-b(y_n-y_{n-1})}\right).$$

If we now put $y_n - y_{n-1} \equiv r_n$ then we have

$$M \frac{d^2 r_n}{dt^2} = a\left(-e^{-br_{n+1}} + 2e^{-br_n} - e^{-br_{n-1}}\right).$$

One simple solution of this set of equations is the traveling wave for which

$$e^{-br_n} - 1 = \sinh^2 \mu \, \text{sech}^2(\mu n \pm \beta t) \tag{1.3.2}$$

with $\beta = \sqrt{ab/M} \sinh \mu$, and μ a number that determines both the amplitude of the wave and its spatial extent. Because the function $\text{sech}^2(\mu n \pm \beta t)$ becomes small unless its argument is small, we see that the width of the solitary wave is around d/μ. The speed v of the wave is $\beta d/\mu$, which on substitution of the expression for β becomes

$$v = d\sqrt{\frac{ab}{M}} \left(\frac{\sinh \mu}{\mu}\right).$$

For large-amplitude solitons the hard-rod feature of the potential dominates, and this speed becomes very large. For small-amplitude waves, on the other hand, $\sinh \mu/\mu \to 1$, and we recover the speed of sound, $d\sqrt{ab/M}$, of the harmonic chain.

The example of the Toda chain illustrates a number of points. It shows how the inclusion of nonlinearities may completely alter the qualitative nature of the elementary excitations of a system. The complete solution of the classical problem involves Jacobian elliptic functions, which shows how complicated even the simplest nonlinear model system can be. Finally, it also presents a formidable challenge to obtain solutions of the quantum-mechanical version of this model for a chain of more than a few particles.

1.4 Magnons

In a ferromagnet at a temperature below the Curie point, the magnetic moments associated with each lattice site l are lined up so that they all point in more or less the same direction. We call this the z-direction. In a simple model of the mechanism that leads to ferromagnetism, the torque acting on one of these moments is determined by the orientation of its nearest neighbors. Then the moment is subjected to an effective magnetic field, $\mathbf{H}_l$, given by

$$\mathbf{H}_l = A \sum_{l'} \boldsymbol{\mu}_{l'}$$

where A is a constant, $\boldsymbol{\mu}_{l'}$ is the moment at the site l', and the sum proceeds only over nearest neighbors. The torque acting on the moment at l is $\boldsymbol{\mu}_l \times \mathbf{H}_l$ and this must be equal to the rate of change of angular momentum. Since the magnetic moment of an atom is proportional to its angular momentum we have

$$\frac{d\boldsymbol{\mu}_l}{dt} \propto \boldsymbol{\mu}_l \times \mathbf{H}_l = A \sum_{l'} \boldsymbol{\mu}_l \times \boldsymbol{\mu}_{l'}. \tag{1.4.1}$$

As in the problem of the chain of masses and springs we look for a wave-like solution of these equations which will represent collective behavior. With the assumption that deviations of the $\boldsymbol{\mu}_l$ from the z-direction are small we write

$$\boldsymbol{\mu}_l = \boldsymbol{\mu}_z + \boldsymbol{\mu}_\perp e^{i(\omega t + \mathbf{k} \cdot \boldsymbol{l})},$$

where $\boldsymbol{\mu}_z$ points in the z-direction and where we have used the useful trick of writing the components in the x–y plane as a complex number, $\boldsymbol{\mu}_x + i\boldsymbol{\mu}_y$. That is, if $\boldsymbol{\mu}_\perp$ is in the x-direction, then $i\boldsymbol{\mu}_\perp$ is in the y-direction. On substitution in (1.4.1) we have, neglecting terms in $\boldsymbol{\mu}_\perp^2$,

$$i\omega\boldsymbol{\mu}_\perp \propto \boldsymbol{\mu}_z \times \boldsymbol{\mu}_\perp \sum_{l''} (e^{i\mathbf{k}\cdot\boldsymbol{l''}} - 1).$$

Here the l'' are the vectors joining the site l to its nearest neighbors. In a crystal with inversion symmetry the summation simplifies to

$$-2 \sum_{l''} \sin^2(\tfrac{1}{2}\mathbf{k}\cdot\boldsymbol{l''}).$$

This equation tells us that $\boldsymbol{\mu}_\perp$ rotates in the x–y plane with frequency

$$\omega \propto |\boldsymbol{\mu}_z| \sum_{l''} \sin^2(\tfrac{1}{2}\mathbf{k}\cdot\boldsymbol{l''}), \tag{1.4.2}$$

1.4 Magnons

the phase difference between atoms separated by a distance **r** being just **k** · **r**. This sort of situation is shown in Fig. 1.4.1, which indicates the direction in which **μ** points as one moves in the direction of **k** along a line of atoms. Because the magnetic moment involved is usually due to the spin of the electron, these waves are known as *spin waves*. The quantum unit of such a wave is known as a magnon.

The most important difference to note between phonons and magnons concerns the behavior of $\omega(\mathbf{k})$ for small **k** (Fig. 1.4.2). For phonons we found that the group velocity, $d\omega/dk$, tended to a constant as **k** tended to zero, this constant being of course the velocity of sound. For magnons,

Figure 1.4.1. The *k*-vector of this spin wave points to the left.

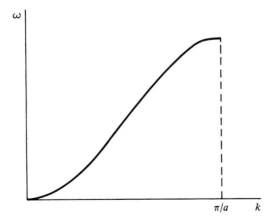

Figure 1.4.2. The dispersion curve for magnons is parabolic in shape for small wave numbers.

however, the group velocity tends to zero as **k** becomes small. This is of great importance in discussing the heat capacity and conductivity of solids at low temperatures.

In our simplified model we had to make some approximations in order to derive Eq. (1.4.2). This means that the spin waves we have postulated would eventually decay, even though our assumption about the effective field had been correct. In quantum-mechanical language we say that a crystal with two magnon excitations present is not an exact eigenstate of the system, and that magnon interactions are present even in our very simple model. This is not to say that a lattice containing phonons is an exact eigenstate of any physical system, for, of course, there are many factors we left out of consideration that limit the lifetime of such excitations in real crystals. Nevertheless, the fact that, in contrast to the phonon system, we cannot devise any useful model of a ferromagnet that can be solved exactly indicates how very difficult a problem magnetism is.

1.5 Plasmons

The model that we used to derive the classical analog of phonons was a system in which there were forces between nearest neighbors only. If we had included second and third nearest neighbors we might have found that the dispersion curve (the graph of ω against k) had a few extra maxima or minima, but ω would still be proportional to k for small values of k. That is, the velocity of sound would still be well defined. However, if we wanted to consider a three-dimensional crystal in which the atoms carried an electric charge e we would find some difficulties (Problem 1.3). Although the Coulomb force of electrostatic repulsion decays as r^{-2}, the number of neighbors at a distance of around r from an atom increases as r^2, and the equation for ω has to be treated very carefully. The result one finds for longitudinally polarized waves is that as **k** tends to zero ω now tends to a constant value Ω_p, known as the ion plasma frequency and given by

$$\Omega_p^2 = \frac{4\pi\rho_0 e^2}{M} \tag{1.5.1}$$

where e is the charge and M the mass of the particles, and ρ_0 the number of particles per unit volume of the crystal. We thus conclude that a Coulomb lattice does not support longitudinal sound waves in the usual sense, since Ω_p is no longer proportional to the wavenumber **k**.

1.5 Plasmons

This raises an interesting question about the collective excitations in metals. We think of a metal as composed of a lattice of positively charged ions embedded in a sea of nearly free conduction electrons. The ions interact by means of their mutual Coulomb repulsion, and so we might expect that the lattice would oscillate at the ion plasma frequency, Ω_p. Of course, we know from everyday experience that metals do carry longitudinal sound waves having a well defined velocity, and so the *effective* interaction between ions must be short-range in nature. It is clear then that the conduction electrons must play some part in this.

This leads to the concept of *screening*. We must suppose that in a sound wave in a metal the local variations in charge density due to the motion of the positively charged ions are cancelled out, or *screened*, by the motion of the conduction electrons. This influx of negative charge reduces the restoring force on the ions, and so the frequency of the oscillation is drastically reduced. That is to say, the ions and the electrons move in phase, and we should be able to calculate the velocity of sound by considering the motion of electrically neutral atoms interacting through short-range forces.

But if there is a mode of motion of the metallic lattice in which the electrons and ions move in phase, there should also be a mode in which they move *out of phase*. This is in fact the case, and it is these modes that are the true plasma oscillations of the system, since they do give rise to variations in charge density in the crystal. Their frequency, as we shall now show, is given for long wavelengths by Eq. (1.5.1), where now the ionic mass, M, is replaced by the electron mass, m. (In fact m should really be interpreted as the reduced mass of the electron in the center-of-mass coordinate system of an electron and an ion; however, since the mass of the ion is so many times greater than that of the electron this refinement is not necessary.)

We shall look for plasma oscillations by supposing that the density of electrons varies in a wave-like way, so that

$$\rho(\mathbf{r}) = \rho_0 + \rho_q \cos qx. \tag{1.5.2}$$

This density must be considered as an average over a distance that is large compared with the distance between an electron and its near neighbors, but small compared with q^{-1}. When the electrons are considered as point particles the density is really a set of delta-functions, but we take a local average of these to obtain $\rho(\mathbf{r})$. The electrostatic potential $\phi(\mathbf{r})$ will then be of the same form,

$$\phi(\mathbf{r}) = \phi_0 + \phi_q \cos qx, \tag{1.5.3}$$

and will be related to the density of electrons $\rho(\mathbf{r})$ and of ions $\rho_{\text{ion}}(\mathbf{r})$ by Poisson's equation,

$$\nabla^2 \phi(\mathbf{r}) = -4\pi e[\rho(\mathbf{r}) + \rho_{\text{ion}}(\mathbf{r})]. \tag{1.5.4}$$

If we take ρ_{ion} to be equal to ρ_0 we have on substitution of (1.5.2) and (1.5.3) in (1.5.4)

$$q^2 \phi_q = 4\pi e \rho_q.$$

The potential energy density is then

$$\frac{1}{2} e \rho \phi = \frac{1}{2} \frac{4\pi e^2 \rho_q^2}{q^2} \cos^2 qx.$$

The average kinetic energy density, $\frac{1}{2}m\overline{\rho v^2}$, is also altered by the presence of the plasma wave. The amplitude of the oscillation is $\rho_q/\rho_0 q$ and so an electron moving with the plasma suffers a velocity change of $(\dot{\rho}_q/\rho_0 q) \sin qx$ with $\dot{\rho}_q$ the time derivative of ρ_q. We must also take into account the heating of the plasma caused by adiabatic compression; since the fractional increase in density is $(\rho_q/\rho_0) \cos qx$ this effect will add to the velocity an amount of the order of $(v_0 \rho_q/\rho_0) \cos qx$. If we substitute these expressions into the classical Hamiltonian and take the spatial average we find an expression of the form

$$\bar{\mathcal{H}} = \frac{1}{4} \frac{4\pi e^2 \rho_q^2}{q^2} + \frac{1}{4} m \rho_0 \left[\left(\frac{\dot{\rho}_q}{\rho_0 q}\right)^2 + \left(\frac{\alpha v_0 \rho_q}{\rho_0}\right)^2 \right]$$

$$= \frac{m}{4\rho_0 q^2} \left[\dot{\rho}_q^2 + \rho_q^2 \left(\frac{4\pi \rho_0 e^2}{m} + \alpha^2 v_0^2 q^2\right) \right]$$

with α a constant of order unity. This is the Hamiltonian for a classical oscillator of frequency

$$\omega_q = (\omega_p^2 + \alpha^2 v_0^2 q^2)^{1/2},$$

where ω_p is the electron plasma frequency, $(4\pi \rho_0 e^2/m)^{1/2}$.

The important point to note about this approximate result is that ω_p is a very high frequency for electrons in metals, of the order of 10^{16} Hz, which corresponds to a quantum energy $\hbar \omega_p$ of several electron volts. Quanta of such oscillations are known as plasmons, and cannot be created thermally, since most metals melt at a thermal energy of the order of 0.1 eV. Thus the plasma oscillations represent degrees of freedom of the electron gas that are "frozen out." This accounts for the paradoxical result that the interaction between electrons is so strong that it may sometimes be ignored. One may contrast this situation with that of an atom in the solid considered in Section 1.2. There it was found that any attempt to give momentum to a single atom just resulted in the creation of a large number of collective excitations of low energy. An electron in the electron gas, on the other hand, retains its particle-like behavior much longer as it may not have the energy necessary to create a single plasmon.

1.6 Electron quasiparticles

Most of the phenomena we have considered so far have been collective motions. Our method of solving the equations of motion was to define a collective coordinate, y_k, which was a sum over the whole lattice of some factor times the particle coordinates, y_l. If we had had an infinite number of particles, then the coordinates of any one particle would only have played an infinitesimal role in the description of the motion. We now turn to the consideration of excitations in which the motion of one particle plays a finite role.

In Section 1.1 we have already briefly considered the problem of an assembly of particles that obey the Pauli Exclusion Principle. A gas of electrons is an example of such a system. As long as the electrons do not interact then the problem of classifying the energy levels is trivial. The momentum $\mathbf{p}$ of each of the electrons is separately conserved, and each has an energy $\mathcal{E} = \mathbf{p}^2/2m$. The spin of each electron may point either up or down, and no two electrons may have the same momentum $\mathbf{p}$ and spin s. If there are N electrons, the ground state of the whole system is that in which the N individual electron states of lowest energy are occupied and all others are empty. If the most energetic electron has momentum $\mathbf{p}_F$, then all states for which $|\mathbf{p}| < |\mathbf{p}_F|$ will be occupied. The spherical surface in momentum space defined by $|\mathbf{p}| = p_F$ is known as the *Fermi surface* (Fig. 1.6.1). The total energy of the system is then

$$\mathcal{E}_T = \sum_{s, |\mathbf{p}| < p_F} \frac{\mathbf{p}^2}{2m},$$

the sum being over states contained within the Fermi surface. We can write this another way by defining an occupation number, $n_{\mathbf{p},s}$, which is zero when

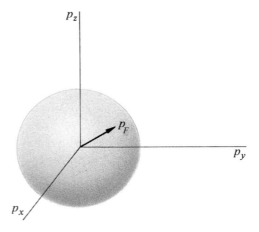

Figure 1.6.1. The Fermi sphere in momentum space contains all electron states with energy less than $\mathbf{p}_F^2/2m$.

the state with momentum **p** and spin s is empty and equal to 1 when it is occupied. Then

$$\mathcal{E}_T = \sum_{\text{all } s,\mathbf{p}} \frac{\mathbf{p}^2}{2m} n_{\mathbf{p},s}.$$

The usefulness of the concept of a quasiparticle rests on the fact that one may still discuss the occupancy of a state even when there are interactions between the particles. Although in the presence of interactions $n_{\mathbf{p},s}$ will no longer have to take on one of the two values 0 or 1, we can attach a meaning to it. We might, for instance, suppose that in with the electrons there is a positron at rest, and that it annihilates with one of the electrons. The total momentum of the gamma rays that would be emitted by the annihilating particles would be equal to their total momentum before annihilation. We could now ask what the probability is that this momentum be equal to **p**. Since for the noninteracting system this probability is proportional to $\sum_s n_{\mathbf{p},s}$, this provides an interpretation for $n_{\mathbf{p},s}$ in the interacting system.

In the noninteracting system we had a clear view of what constituted a particle excitation. The form of $n_{\mathbf{p},s}$ differed from that of the ground state in that one value of **p** less than p_F was unoccupied, and one greater than p_F was occupied (Fig. 1.6.2). We then consider the excited system as composed of the ground state plus an excitation comprising a particle and a "hole," the particle–hole pair having a well defined energy above that of the ground state. If we introduce interactions between the particles, and in particular if

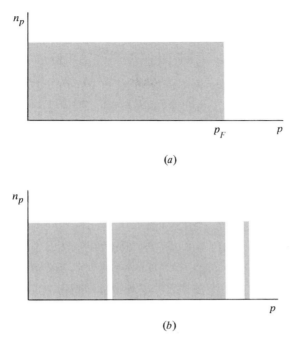

Figure 1.6.2. The excited state (b) is formed from the ground state (a) by the creation of a particle–hole pair.

we introduce the troublesome Coulomb interaction, it is hard to see whether the concept of a particle–hole excitation survives. It is, in fact, not only hard to see but also hard to calculate. One approach is to consider the effect of switching on the interactions between particles when the noninteracting system contains a particle–hole pair of energy $\mathcal{E}$. If the lifetime τ of the excitation is large compared with $\hbar/\mathcal{E}$, then it will still be useful to retain a similar picture of the excitations. Since now the interactions will have modified their energies, we refer to "quasielectrons" and "quasiholes."

1.7 The electron–phonon interaction

In Section 1.5 we discussed sound waves in a metal, and came to the conclusion that in these excitations the ions and electrons moved in phase. The long-range potential of the positively charged ions was thus screened, and the phonon frequency reduced from the ion plasma frequency Ω_p to some much smaller value. The way in which this occurs is illustrated in Fig. 1.7.1. We first imagine a vibration existing in the unscreened lattice of ions. We then

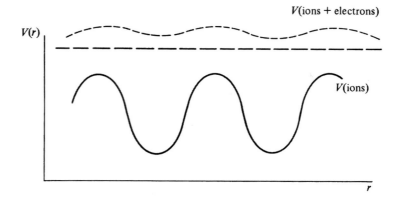

Figure 1.7.1. The deep potential due to the displacement of the ions by a phonon is screened by the flow of electrons.

suppose that the electron gas flows into the regions of compression and restores the electrical neutrality of the system on a macroscopic scale. There is, however, a difference between the motion of the ions and the electrons in that we assume the ions to be localized entities, while the electrons are described by wavefunctions that, in this case, will be small distortions of plane waves. When we increase the local density of electrons we must provide extra kinetic energy to take account of the Exclusion Principle. We might take the intuitive step of introducing the concept of a Fermi energy that is a function of position. We could then argue that the local kinetic energy density of the electron gas should be roughly equal to that of a uniform gas of free electrons, which happens to be $\frac{3}{5}\mathcal{E}_F \rho_0$. The sound wave in a metal is thus seen in this model as an interchange of kinetic energy between the ions and the electrons. We can calculate an order of magnitude for the velocity of sound by writing the classical Hamiltonian for the system in a similar approach to that of Section 1.5. The kinetic energy of the ions will be $M(\dot{\rho}_\mathbf{q})^2/2\rho_0\mathbf{q}^2$ when a wave of wavenumber $\mathbf{q}$ passes through a lattice of ions of mass M and average number density ρ_0. The total kinetic energy of the electrons is only changed to second order in $\rho_\mathbf{q}/\rho_0$, and so contributes an energy density of the order of $\mathcal{E}_F\rho_0(\rho_\mathbf{q}/\rho_0)^2$. Then

$$\mathcal{H} \simeq \frac{M}{2\rho_0\mathbf{q}^2}\left[(\dot{\rho}_\mathbf{q})^2 + \rho_\mathbf{q}^2\frac{\alpha\mathcal{E}_F\mathbf{q}^2}{M}\right],$$

where α is a constant of order unity. The frequency of the oscillator that this

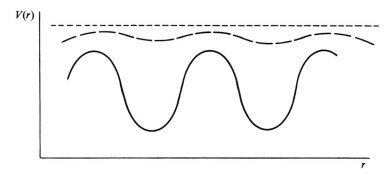

Figure 1.7.2. In a careful calculation the kinetic energy of the electrons is found to prevent a complete screening of the potential due to the displaced ions. The residual potential is shown as a dashed line.

Hamiltonian describes is

$$\omega = \sqrt{\frac{\alpha \mathcal{E}_F}{M}} q$$

which shows that the velocity of sound, v_s, can be written as

$$v_s \sim v_F \sqrt{\frac{m}{M}}$$

where v_F is the velocity of an electron with energy $\mathcal{E}_F$ and m/M is the mass ratio of electron to ion.

In a more careful treatment one would argue that the electron gas would not completely screen the electric field of the ions. Instead the electrons would flow until the sum of the electric potential energy and the kinetic energy of the electrons (the dotted line in Fig. 1.7.1) became uniform. There would then be a residual electric field (the dashed line in Fig. 1.7.2) tending to restore the ions to their equilibrium positions. It is the action of this residual electric field on the electrons that gives rise to the electron–phonon interaction which we shall study in Chapter 6.

1.8 The quantum Hall effect

We close this chapter with a first glimpse of a truly remarkable phenomenon. In the quantum Hall effect a current of electrons flowing along a surface gives rise to an electric field that is so precisely determined that it has become the basis for the legal standard of electrical resistance. This result is reproducible to better than one part in 10^8, even when one changes the material from which an experimental sample is made or alters the nature of the surface in which the current flows.

In the elementary theory of the *Hall effect* one argues that when electrons travel down a wire with average drift velocity $\mathbf{v}_d$ at right angles to an applied magnetic field $\mathbf{H}_0$ then they experience an average Lorentz force $e(\mathbf{v}_d/c) \times \mathbf{H}_0$ (Fig. 1.8.1). In order for the electrons to be undeflected in their motion this force must be counterbalanced by the *Hall field* $\mathbf{E}_H$, which arises from accumulations of charge on the surface of the wire. In the absence of applied electric fields we can then write

$$\mathbf{E} + \frac{\mathbf{v}_d}{c} \times \mathbf{H} = 0, \tag{1.8.1}$$

and since the current density j_y is given by $\rho_0 e v_d$, we have

$$E_x = \rho_H j_y, \tag{1.8.2}$$

with the Hall resistivity ρ_H equal to $H/\rho_0 ec$. The product $\rho_0 e$ can be interpreted as giving the density of charge carriers in a metal and also the sign of their effective charge (which may be positive or negative as a result of the effects of the lattice potential, as discussed in Chapter 4).

A special situation arises if the electrons are confined to a two-dimensional surface held perpendicular to the magnetic field. A semiclassical electron in the center of the sample will then travel in a circular orbit with the cyclotron frequency $\omega_c = eH/mc$. The x-component of this circular motion is reminiscent of a harmonic oscillator, and so it is no surprise to find that its energy

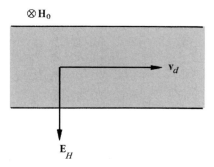

Figure 1.8.1. The Hall field $\mathbf{E}_H$ cancels the effect of the Lorentz force due to the applied magnetic field $\mathbf{H}_0$.

levels are quantized, with $\mathcal{E} = (n + \tfrac{1}{2})\hbar\omega_c$. These are known as Landau levels. Because its motion is circular, it does not contribute to any net current flowing through the sample, and so the question arises as to the origin of any such current.

The answer lies with the electrons at the edge of the sample. They cannot complete their little circles, as they keep bumping into a wall, bouncing off it, and then curving around to bump into it again (Fig. 1.8.2). In this way they can make their way down the length of the sample, and carry an electric current, the current of electrons at the top of the figure being to the right and the current at the bottom being to the left. To have a net current to the right, we must have more electrons at the top of the figure than at the bottom. The Fermi level must thus be higher there, and this translates into a higher electrical potential, and thus a Hall-effect voltage.

Suppose now that we gradually increase the density ρ_0 of electrons in the sample while keeping the Hall voltage constant. The number of circular orbits and edge states will increase proportionately, and the Hall resistance will decrease smoothly as $1/\rho_0$. This simple picture, however, is spoiled if there are impurities in the system. Then there will exist bound impurity states whose energies will lie between the Landau levels. Because these states carry no current, the Hall resistance will stop decreasing, and will remain constant until enough electrons have been added to raise the Fermi energy to lie in the next-highest Landau level.

This existence of *plateaus* in the Hall resistance as a function of number of electrons is known as the *integral quantum Hall effect*. In very pure samples plateaus can also be found when simple fractions (like 1/3 or 2/5) of the states in a Landau level are occupied. This occurs for a different reason, and is

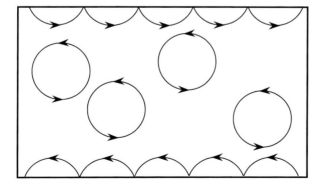

Figure 1.8.2. Only the "skipping orbits" at the edges can carry a current along the sample.

known as the *fractional quantum Hall effect*. The detailed origin of both these effects will be explored in Chapter 10.

Problems

1.1 The energy levels $\mathcal{E}_{ln}$ of a diatomic molecule are characterized by an angular momentum quantum number l and a vibrational one n. Assume a Hamiltonian of the form

$$\mathcal{H} = \frac{p_1^2}{2m} + \frac{p_2^2}{2m} + \frac{1}{4} K (d - d_0)^2$$

where d is the interatomic separation and $Kmd_0^4 \gg \hbar^2$. Calculate $\mathcal{E}_{10} - \mathcal{E}_{00}$ and $\mathcal{E}_{01} - \mathcal{E}_{00}$ (the energies of the two kinds of elementary excitation), and also the difference between the sum of these quantities and $\mathcal{E}_{11} - \mathcal{E}_{00}$. This difference represents the energy of interaction between the two excitations.

1.2 *The Mössbauer Effect* Suppose that an atom of ^{57}Fe emits a γ-ray of frequency ω_0 in the x-direction while it is moving in the same direction with velocity v. Then by the Doppler effect a stationary observer will see radiation of frequency approximately equal to $\omega_0 (1 + v/c)$. The spectrum of radiation emitted by a hot gas of iron atoms will thus be broadened by the thermal motion. Now suppose the iron atom to be bound in a solid, so that the x-component of its position is given by

$$x = \sum_q a_q \sin(\omega_q t + \phi_q),$$

where the phonon frequencies, ω_q, are much smaller than ω_0, and the phases, ϕ_q, are random. Derive an expression for the Fourier spectrum of the radiation intensity seen by a stationary observer, taking into account the frequency modulation caused by the motion of the atom.

[The fact that when the a_q are small a large proportion of the radiation has the unperturbed frequency, ω_0, is the basis of the *Mössbauer effect*. The emitted γ-ray may be resonantly absorbed by another iron atom in a process that is the converse of that described above.]

1.3 *Phonons in a Coulomb Lattice* If a particle at l carrying charge Ze is displaced a distance $\mathbf{y}_l$, the change in electric field experienced at distances **r** from *l* that are large compared with $\mathbf{y}_l$ is the field due to an electric dipole of moment $\boldsymbol{\mu}_l = Ze\mathbf{y}_l$, and is given by

$$\mathbf{E}_l(\mathbf{r}) = |\mathbf{r}|^{-5}[3(\boldsymbol{\mu}_l \cdot \mathbf{r})\mathbf{r} - r^2\boldsymbol{\mu}_l].$$

If the lattice is vibrating in a single longitudinal mode of wavenumber **q** then one may evaluate its frequency ω_q by calculating the total field $\sum_l \mathbf{E}_l$. For vanishingly small **q** the sum over lattice sites may be replaced by an integral [why?]. Evaluate $\omega(q \to 0)$ by (*i*) performing the integration over spherical polar coordinates ϕ, θ, and r, in that order, and (*ii*) performing the integral in cylindrical polar coordinates ϕ, r, and z, restricting the integration to points for which $r < R$, where R is some large distance. State in physical terms why these two results differ.

1.4 *Phonon Interactions* The velocity of sound in a solid depends, among other things, on the density. Since a sound wave is itself a density fluctuation we expect two sound waves to interact. In the situation shown in Fig. P1.1 a phonon of angular frequency ω' and wavenumber $\mathbf{q}'$ is incident upon a region of an otherwise homogeneous solid containing a line of density fluctuations due to another phonon of wavenumber **q**. By treating this as a moving diffraction grating obtain an expression for the wavenumber $\mathbf{q}''$ and frequency ω'' of the diffracted wave.

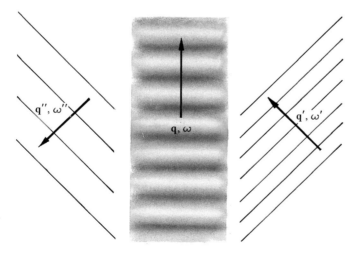

Figure P1.1. When two phonons are present simultaneously one of them may form an effective diffraction grating to scatter the other.

1.5 In a long chain, atoms of mass M interact through nearest-neighbor forces, and the potential energy is $V = \sum_n g(y_n - y_{n-1})^4$, where g is a constant and y_n is the displacement from equilibrium of the nth atom. A solitary wave travels down this chain with speed v. How does v vary with the amplitude of this wave?

1.6 Assume that the density of allowed states in momentum space for an electron is uniform, and that the only effect of an applied magnetic field $\mathbf{H}$ is to add to the energy of a particular momentum state the amount $\pm \mu_B H$, according to whether the electron spin is up or down (the only two possibilities). Derive an expression in terms of μ_B and the Fermi energy $\mathcal{E}_F$ for the magnetic field $\mathbf{H}$ that must be applied to increase the kinetic energy of an electron gas at 0 K by 5×10^{-8} of its original value.

1.7 In a certain model of ferromagnetism the energy of a free-electron gas has added to it an interaction term

$$\mathcal{E}_{int} = K(N_\uparrow^{4/3} + N_\downarrow^{4/3}),$$

where $N_\uparrow$ and $N_\downarrow$ are the total numbers of up- and down-spin electrons, respectively. By investigating the total energy of this system as a function of $N_\uparrow - N_\downarrow$ for constant $N(= N_\uparrow + N_\downarrow)$ decide for what range of K the magnetized state (in which $N_\uparrow \neq N_\downarrow$) will be (a) stable; (b) metastable; (c) unstable. Express your results for K in terms of N and $\mathcal{E}_F$, the Fermi energy in the absence of interactions.

1.8 In a classical antiferromagnet there are two oppositely magnetized sublattices, each of which is subject to a field

$$\mathbf{H}_l = -A \sum_{l'} \boldsymbol{\mu}_{l'},$$

the sum proceeding over sites l' that are nearest neighbors to l. Find the form of the spin-wave spectrum in a simple cubic crystal, and describe pictorially the motion of the spins at low and high frequencies.

1.9 In problem 1.6, you were asked to find the magnetic field that would increase the kinetic energy by a fraction 5×10^{-8}. Now redo this problem for the case where the *total* energy is *decreased* by a fraction 5×10^{-8}.

1.10 Calculate the energy of the soliton described by Eq. (1.3.2) in an infinite Toda chain. [Alternatively, as an easier problem just estimate this energy in the limits of small and large μ.]

Chapter 2

Second quantization and the electron gas

2.1 A single electron

We have taken a brief look from a semiclassical point of view at some of the kinds of behavior exhibited by many-particle systems, and have then used intuition to guess at how quantum mechanics might modify the properties we found. It is now time to adopt a more formal approach to these problems, and to see whether we can derive the previous results by solving the Schrödinger equation for the quantum-mechanical problem.

For a single electron we have the time-independent Schrödinger equation

$$\mathcal{H}\psi(\mathbf{r}) = \mathcal{E}\psi(\mathbf{r}), \tag{2.1.1}$$

where

$$\mathcal{H} = \frac{\mathbf{p}^2}{2m} + V(\mathbf{r})$$

and $\mathbf{p}$ is interpreted as the operator $-i\hbar\nabla$. This equation has physically meaningful solutions for an infinite number of energies $\mathcal{E}_\alpha$ ($\alpha = 1, 2, 3, \ldots$). The eigenfunctions $u_\alpha(\mathbf{r})$, for which

$$\mathcal{H}u_\alpha(\mathbf{r}) = \mathcal{E}_\alpha u_\alpha(\mathbf{r}),$$

form a complete set, meaning that any other function we are likely to need can be expanded in terms of them. The $u_\alpha(\mathbf{r})$ for different α are orthogonal, meaning that

$$\int u_\alpha^*(\mathbf{r})u_{\alpha'}(\mathbf{r})\,\mathbf{dr} = 0; \quad (\alpha \neq \alpha'), \tag{2.1.2}$$

where $\mathbf{dr}$ is an abbreviation for $dx\,dy\,dz$, u^* is the complex conjugate of u, and the integration is over all space. If the wavefunctions $u_\alpha(\mathbf{r})$ are normalized then the integral is equal to unity for $\alpha = \alpha'$.

It is convenient to adopt what is known as the *Dirac notation* to describe integrals of this kind. Because wavefunctions like $u_\alpha(\mathbf{r})$ have to be continuous, we can think of the integral in Eq. (2.1.2) as being equal to the limit of a sum like

$$\Delta\Omega \sum_i u_\alpha^*(\mathbf{r}_i) u_{\alpha'}(\mathbf{r}_i), \qquad (2.1.3)$$

where we have divided all space into a large number of cells centered on the points $\mathbf{r}_i$, and each of which encloses a volume equal to the vanishingly small quantity $\Delta\Omega$. If we look on $u_{\alpha'}(\mathbf{r})$ as being the *column vector* (i.e., the vertical array of numbers)

$$\begin{pmatrix} u_{\alpha'}(\mathbf{r}_1) \\ u_{\alpha'}(\mathbf{r}_2) \\ \cdot \\ \cdot \\ \cdot \end{pmatrix}$$

and $u_\alpha^*(\mathbf{r})$ as the *row vector*

$$(u_\alpha^*(\mathbf{r}_1), u_\alpha^*(\mathbf{r}_2), \ldots),$$

then the sum in expression (2.1.3) is just the matrix product of $u_\alpha^*(\mathbf{r})$ and $u_{\alpha'}(\mathbf{r})$. We adopt the notation of writing the row vector $u_\alpha^*(\mathbf{r})$ as $\langle\alpha|$ and of writing the column vector $u_{\alpha'}(\mathbf{r})$ as $|\alpha'\rangle$. Then we write the integral of Eq. (2.1.2) as $\langle\alpha|\alpha'\rangle$. For normalized wavefunctions Eq. (2.1.2) then becomes

$$\langle\alpha|\alpha'\rangle = \delta_{\alpha\alpha'} \qquad (2.1.4)$$

where $\delta_{\alpha\alpha'}$ is the Kronecker delta symbol, which is unity when $\alpha = \alpha'$ and zero otherwise.

An example of such a set of functions $u_\alpha(\mathbf{r})$ are the plane waves that are a solution of (2.1.1) when the potential $V(\mathbf{r}) = 0$. Then if $\psi(\mathbf{r}) \propto e^{i\mathbf{k}\cdot\mathbf{r}}$,

$$\mathcal{H}\psi = -\frac{\hbar^2}{2m}\nabla^2\psi = \frac{\hbar^2\mathbf{k}^2}{2m}\psi.$$

We can avoid the difficulty of normalizing such plane waves (which extend over all space) by only considering the wavefunction within a cubical box of volume $\Omega = L^3$, having corners at the points $(\pm L/2, \pm L/2, \pm L/2)$. Then we can take

$$u_\alpha(\mathbf{r}) = \Omega^{-1/2} e^{i\mathbf{k}_\alpha \cdot \mathbf{r}}.$$

We then impose *periodic boundary conditions*, by stipulating that the form of the wavefunction over any side of the box must be identical to its form over the opposite side. That is,

$$u_\alpha\left(-\frac{L}{2}, y, z\right) = u_\alpha\left(+\frac{L}{2}, y, z\right), \text{ etc.}$$

This means that $\mathbf{k}_\alpha$ can no longer be any vector, but is restricted to a discrete set of values such that

$$e^{i\mathbf{k}_\alpha \cdot (L,0,0)} = e^{i\mathbf{k}_\alpha \cdot (0,L,0)} = e^{i\mathbf{k}_\alpha \cdot (0,0,L)} = 1.$$

Hence

$$\mathbf{k}_\alpha = \left(\frac{2\pi m_x}{L}, \frac{2\pi m_y}{L}, \frac{2\pi m_z}{L}\right)$$

where m_x, m_y, and m_z are integers. Equation (2.1.4) is then obeyed. These allowed values of $\mathbf{k}_\alpha$ form a simple cubic lattice in k-space, the density of allowed points being $\Omega/(2\pi)^3$, which is independent of $\mathbf{k}$. Summations over α can then be interpreted as summations over allowed values of $\mathbf{k}$.

We can expand a function $\phi(\mathbf{r})$ in terms of the $u_\alpha(\mathbf{r})$ by writing

$$\phi(\mathbf{r}) = \sum_\alpha C_\alpha u_\alpha(\mathbf{r}) \tag{2.1.5}$$

and forming the integral

$$\int u_{\alpha'}^*(\mathbf{r}) \phi(\mathbf{r}) \, d\mathbf{r},$$

an integral that we would write in the notation of Eq. (2.1.4) as $\langle \alpha' | \phi \rangle$. On substituting from Eq. (2.1.5) we have

$$\langle \alpha' | \phi \rangle = \sum_\alpha C_\alpha \langle \alpha' | \alpha \rangle = \sum_\alpha C_\alpha \delta_{\alpha'\alpha} = C_{\alpha'},$$

so that

$$\phi(\mathbf{r}) = \sum_\alpha u_\alpha(\mathbf{r})\langle\alpha|\phi\rangle,$$

which in the Dirac notation becomes

$$|\phi\rangle = \sum_\alpha |\alpha\rangle\langle\alpha|\phi\rangle.$$

This allows us to consider the first part of the right-hand side of this equation as an operator that is identically equal to one;

$$\sum_\alpha |\alpha\rangle\langle\alpha| \equiv 1. \qquad (2.1.6)$$

The combination $|\alpha\rangle\langle\alpha|$ is the product of a column vector with a row vector; it is not a number but is an *operator*. When it acts on a wavefunction $|\phi\rangle$ it gives the combination $|\alpha\rangle\langle\alpha|\phi\rangle$, which is just the number $\langle\alpha|\phi\rangle$ multiplying the wavefunction $|\alpha\rangle$.

This notation can be extended by treating an integral like

$$\int u_\alpha^*(\mathbf{r})V(\mathbf{r})u_{\alpha'}(\mathbf{r})\,d\mathbf{r} \qquad (2.1.7)$$

in the same way that we handled Eq. (2.1.2). If we replace this integral by the sum

$$\Delta\Omega \sum_i u_\alpha^*(\mathbf{r}_i)V(\mathbf{r}_i)u_{\alpha'}(\mathbf{r}_i)$$

we find that we are considering $V(\mathbf{r})$ as the diagonal matrix

$$\begin{pmatrix} V(\mathbf{r}_1) & 0 & 0 & \cdots \\ 0 & V(\mathbf{r}_2) & 0 & \cdots \\ \cdots & \cdots & \cdots & \cdots \end{pmatrix}.$$

The matrix product that gives expression (2.1.7) would then be written as

$$\langle\alpha|V|\alpha'\rangle.$$

We then make use of the identity (2.1.6) to write

$$V(\mathbf{r}) = \left(\sum_{\alpha'} |\alpha'\rangle\langle\alpha'|\right) V \left(\sum_{\alpha''} |\alpha''\rangle\langle\alpha''|\right)$$
$$= \sum_{\alpha',\alpha''} \langle\alpha'|V|\alpha''\rangle |\alpha'\rangle\langle\alpha''|. \qquad (2.1.8)$$

The operator $|\alpha'\rangle\langle\alpha''|$ gives zero when it operates on any of the states $|\alpha\rangle$ except that for which $\alpha = \alpha''$, and then it gives the state $|\alpha'\rangle$. We can thus interpret $|\alpha'\rangle\langle\alpha''|$ as removing an electron from the state described by the wavefunction $u_{\alpha''}(\mathbf{r})$ and putting it into the state described by $u_{\alpha'}(\mathbf{r})$. In more dramatic language the operator *annihilates* an electron in the state $|\alpha''\rangle$ and *creates* one in the state $|\alpha'\rangle$. We can write this symbolically another way by introducing the rather difficult concept of the wavefunction $|0\rangle$ that denotes an empty box! That is, we define a column vector $|0\rangle$ that is normalized, so that

$$\langle 0|0\rangle = 1, \qquad (2.1.9)$$

but which is orthogonal to all the one-particle wavefunctions $u_\alpha(\mathbf{r})$, so that

$$\langle\alpha|0\rangle = 0$$

for all α. We call it the *vacuum state* and are careful not to confuse it with the number zero. This state $|0\rangle$ cannot be described by a wavefunction in the same sense as the $u_\alpha(\mathbf{r})$, in that we do not expect to be able to ask the same questions about probability densities that we might ask about an electron in the state $|\alpha\rangle$. It is instead just a useful device that allows us to insert Eq. (2.1.9) into Eq. (2.1.8) and write

$$|\alpha'\rangle\langle\alpha''| = |\alpha'\rangle\langle 0|0\rangle\langle\alpha''|$$
$$= (|\alpha'\rangle\langle 0|)(|0\rangle\langle\alpha''|).$$

We now define the combination $|0\rangle\langle\alpha''|$ as the operator $c_{\alpha''}$ that annihilates any electron it finds in the state $|\alpha''\rangle$, and the conjugate combination $|\alpha'\rangle\langle 0|$ as the creation operator $c^\dagger_{\alpha'}$ for the state $|\alpha'\rangle$. Then

$$|\alpha'\rangle\langle\alpha''| \equiv c^\dagger_{\alpha'} c_{\alpha''},$$

and

$$V(\mathbf{r}) \equiv \sum_{\alpha',\alpha''} \langle\alpha'|V|\alpha''\rangle c^\dagger_{\alpha'} c_{\alpha''}.$$

Similarly for the kinetic energy operator

$$\frac{\mathbf{p}^2}{2m} \equiv \frac{1}{2m} \sum_{\alpha',\alpha''} \langle \alpha' | \mathbf{p}^2 | \alpha'' \rangle c^\dagger_{\alpha'} c_{\alpha''}.$$

One might at this stage ask what use it serves to write the Hamiltonian for an electron in this form. As long as there is only one electron present then this way of representing operators is just a needless complication. It is when we have a large number of identical particles present that this language comes into its own.

2.2 Occupation numbers

Let us consider a collection of N identical free particles that do not interact. The Hamiltonian for this system is just the sum of the Hamiltonians for the individual particles,

$$\mathcal{H} = \sum_i \mathcal{H}_i = \frac{1}{2m} \sum_i \mathbf{p}_i^2.$$

We could form eigenfunctions of $\mathcal{H}$ simply by multiplying together the eigenfunctions of the individual $\mathcal{H}_i$. If

$$\mathcal{H}_i u_i(\mathbf{r}_i) = \mathcal{E}_i u_i(\mathbf{r}_i)$$

and

$$\Phi' = \prod_{i=1}^N u_i(\mathbf{r}_i) = u_1(\mathbf{r}_1) u_2(\mathbf{r}_2) \cdots u_N(\mathbf{r}_N)$$

then

$$\mathcal{H}\Phi' = \left(\sum_{i=1}^N \mathcal{E}_i \right) \Phi'.$$

The wavefunction Φ' however is not adequate as a physical solution of the Schrödinger equation because of its lack of symmetry. We know that if we interchange any two of the coordinates, then the wavefunction Φ must either remain unchanged or else be changed by this operation to $-\Phi$. That is, Φ must either be symmetric in the coordinates of the particles, which we then call *bosons*, or else be antisymmetric, in which case we refer to the particles as *fermions*. Electrons in particular are known to be fermions, and so Φ must be expressed as an antisymmetrized combination of the Φ'. We can achieve this by forming a determinant of the various $u_i(\mathbf{r}_j)$ and writing

$$\Phi = \frac{1}{\sqrt{N!}} \begin{vmatrix} u_1(\mathbf{r}_1) & u_1(\mathbf{r}_2) & \cdots & u_1(\mathbf{r}_N) \\ u_2(\mathbf{r}_1) & & & \cdots \\ \vdots & & & \vdots \\ u_N(\mathbf{r}_1) & \cdots & \cdots & u_N(\mathbf{r}_N) \end{vmatrix}.$$

This form of the wavefunction (known as a *Slater determinant*) is very cumbersome to handle, even in this simple case where the particles do not interact. When interactions are present Φ will no longer be an eigenfunction of $\mathcal{H}$, but since the various Φ form a complete set any N-electron wavefunction Ψ can be expanded in terms of N-electron Slater determinants. We could then write

$$\Psi = \sum_{\alpha_1 \cdots \alpha_N} C(\alpha_1 \cdots \alpha_N) \Phi(\alpha_1 \cdots \alpha_N)$$

where the α_i label the various states $u_{\alpha i}$ that occur in the determinant. If we knew the complex constants $C(\alpha_1 \ldots \alpha_N)$ then the wavefunction would be completely specified; it would, however, be grossly inconvenient in making calculations, each Φ alone being a sum of $N!$ terms.

The shorthand we use for describing the Φ is known as the *occupation number representation*, and cuts out all the redundant information contained in Φ. We know first of all that there are N particles, and that all the coordinates $\mathbf{r}_i$ come into the wavefunction on an equal footing. We also know that Φ is antisymmetrized, and so we do not need to write this explicitly every time we consider Φ. All we need specify are the states u_α that are occupied. We define an *occupation number*, n_α, that is equal to 1 if the state appears in the determinant describing Φ, and equal to 0 if it does not. Thus we specify the two-particle state

$$\Phi_{\alpha,\beta}(\mathbf{r}_1, \mathbf{r}_2) = \frac{1}{\sqrt{2}} \begin{vmatrix} u_\alpha(\mathbf{r}_1) & u_\alpha(\mathbf{r}_2) \\ u_\beta(\mathbf{r}_1) & u_\beta(\mathbf{r}_2) \end{vmatrix}$$

by writing

$$n_\alpha = n_\beta = 1; \text{ all other } n = 0.$$

2.2 Occupation numbers

We adopt the notation

$$|1, 1, 0, 0, \ldots\rangle$$

to signify the wavefunction in which

$$n_\alpha = 1; \quad n_\beta = 1; \quad n_\gamma = 0; \quad n_\delta = 0, \quad \text{etc.}$$

Clearly we can specify a Φ for any number of particles by means of this notation. We might further abbreviate by writing Φ as $|\{n_\mu\}\rangle$, where $\{n_\mu\}$ is understood to be the *set of occupation numbers* n_α, n_β, etc.

Now we know that for any two distinct Φ describing states containing the same number of particles

$$\int \Phi_\mu^* \Phi_{\mu'} \, \mathbf{dr}_1 \, \mathbf{dr}_2 \cdots \mathbf{dr}_N = 0. \tag{2.2.1}$$

We must also consider the Φ to be orthogonal when they specify different numbers of particles, since this integral is a product of the integrals over the various $\mathbf{r}_i$, and if Φ is a wavefunction for only $N - 1$ particles then it has zero component in the space defined by $\mathbf{r}_N$. (Note that we do *not* show this result by taking an integral of the form (2.2.1) when the Φ have different numbers of particles present, say N and $N - 1$. Rather we must convert the Φ for $N - 1$ particles to a function in N-particle space by multiplying by some function of $\mathbf{r}_N$. This function is of course zero, since there is no Nth particle!) We denote this orthogonality by writing

$$\langle \{n_\mu\} | \{n_{\mu'}\} \rangle = 0$$

for the cases where the sets $\{n_\mu\}$ and $\{n_{\mu'}\}$ are not identical. For normalized wavefunctions

$$\langle \{n_\mu\} | \{n_\mu\} \rangle = 1.$$

The many-particle wavefunctions that we describe with the notation $|\{n_\mu\}\rangle$ are a generalization of the single-particle wavefunctions $|\alpha\rangle$ of Section 2.1. The $|\{n_\mu\}\rangle$ form a complete set not just in one-particle space, as did the $|\alpha\rangle$, but also in the space known as *Fock space*, which can contain any number of particles. Corresponding to Eq. (2.1.6), which only referred to one-particle space, will be the relation

$$\sum_{\{n_\mu\}} |\{n_\mu\}\rangle \langle \{n_\mu\}| \equiv 1.$$

The sum here is over all possibilities for the set of numbers $\{n_\mu\}$. If V is an operator containing the coordinates of any number of particles we can write the identity

$$V \equiv \sum_{\{n_\mu\},\{n_{\mu'}\}} |\{n_\mu\}\rangle\langle\{n_\mu\}|V|\{n_{\mu'}\}\rangle\langle\{n_{\mu'}\}|. \tag{2.2.2}$$

If we further abbreviate $\langle\{n_\mu\}|V|\{n_{\mu'}\}\rangle$ by $V_{\mu\mu'}$ this becomes

$$V \equiv \sum_{\{n_\mu\},\{n_{\mu'}\}} V_{\mu\mu'} |\{n_\mu\}\rangle\langle\{n_{\mu'}\}|.$$

Our next task is to interpret the many-particle operators $|\{n_\mu\}\rangle\langle\{n_{\mu'}\}|$ in terms of the annihilation and creation operators that were introduced at the end of Section 2.1.

2.3 Second quantization for fermions

The notation $|\{n_\mu\}\rangle\langle\{n_{\mu'}\}|$ that we adopted in the last section was an abbreviation for

$$|n_1, n_2, \ldots\rangle\langle n_1', n_2', \ldots|. \tag{2.3.1}$$

In the case where our identical particles are fermions the individual occupation numbers, n_i, can only take on the values 0 and 1. Because each wavefunction $|n_1, n_2, \ldots\rangle$ symbolizes a determinant of the wavefunctions u_i for which $n_i = 1$, we must be sure that we always take the various u_i in the same order. If we took them in a different order we would be doing the equivalent of interchanging some columns of the determinant defining Φ, and we might end up with $-\Phi$ instead.

The simplest operator of the form (2.3.1) is one in which just one of the n_i' is different from n_i. That is, we consider

$$|n_1, n_2, \ldots, n_p = 0, \ldots\rangle\langle n_1, n_2, \ldots, n_p = 1, \ldots|. \tag{2.3.2}$$

This clearly has something in common with the annihilation operator of Section 2.1, since it acts only on a wavefunction that has the pth one-particle state occupied, and gives a wavefunction with the pth state empty. If we want an operator that acts on *any* wavefunction for which $n_p = 1$, we should have to sum expression (2.3.2) over all the possibilities for the other n_i.

2.3 Second quantization for fermions

This would give

$$\sum_{\{n_i\}(i\neq p)} |n_1, \ldots, n_p = 0, \ldots\rangle\langle n_1, \ldots, n_p = 1, \ldots|.$$

Finally, we need to keep track of the sign of the new wavefunction that this operator gives, as mentioned above. This can be achieved by defining a number

$$N_p = \sum_{j=1}^{p-1} n_j$$

and multiplying the operator by $(-1)^{N_p}$. We define the result of this as the *annihilation operator* for the fermion system.

$$c_p = \sum_{\{n_i\}(i\neq p)} (-1)^{N_p} |\ldots, n_i, \ldots, n_p = 0, \ldots\rangle\langle\ldots, n_i, \ldots, n_p = 1, \ldots|. \quad (2.3.3)$$

Now consider the effect of c_p upon a wavefunction in which the pth state is empty. Since

$$\langle\ldots n_p = 1, \ldots | \ldots n_p = 0, \ldots\rangle = 0$$

we have

$$c_p |\ldots n_p = 0, \ldots\rangle = 0.$$

If the pth state is occupied, however, there will be one term in the summation that will not be orthogonal to the wavefunction, and

$$c_p |n_1, n_2, \ldots, n_p = 1, \ldots\rangle = (-1)^{N_p} |n_1, n_2, \ldots, n_p = 0, \ldots\rangle.$$

To operate twice with c_p would be to try and destroy two particles from the same state, and so

$$c_p^2 = 0,$$

as may also be seen from the definition of c_p.

The *creation operator* is the operator conjugate to c_p. It is defined by

$$c_p^\dagger = \sum_{\{n_i\}(i\neq p)} (-1)^{N_p} |\ldots n_p = 1, \ldots\rangle\langle\ldots n_p = 0, \ldots|,$$

and has the effect of introducing a particle into the formerly empty pth state.

$$c_p^\dagger|\ldots n_p = 0, \ldots\rangle = (-1)^{N_p}|\ldots n_p = 1, \ldots\rangle$$
$$c_p^\dagger|\ldots n_p = 1, \ldots\rangle = 0.$$

Any more complicated operator of the form $|\{n_\mu\}\rangle\langle\{n_{\mu'}\}|$ can be expressed in terms of the various c_p and $c_p^\dagger$, for we can always write

$$|\{n_\mu\}\rangle\langle\{n_{\mu'}\}| = |\{n_\mu\}\rangle\langle\{n_{\mu''}\}|\{n_{\mu''}\}\rangle\langle\{n_{\mu'''}\}|\cdots\rangle\langle\{n_{\mu'}\}|,$$

choosing $|\{n_{\mu''}\}\rangle$ to differ from $|\{n_\mu\}\rangle$ in only one occupation number, and so on. Thus we could write

$$\sum_{\{n_i\}(i\neq p,q)} |\ldots n_p = 0, n_q = 0, \ldots\rangle\langle\ldots n_p = 1, n_q = 1, \ldots|$$

$$= \sum_{\{n_i\}(i\neq p,q)} |\ldots n_p = 0, n_q = 0\rangle\langle\ldots n_p = 0, n_q = 1, \ldots|$$

$$|\ldots n_p = 0, n_q = 1, \ldots\rangle\langle\ldots n_p = 1, n_q = 1, \ldots|$$

$$= (-1)^{N_q}c_q(-1)^{N_p}c_p. \qquad (2.3.4)$$

Alternatively we might have inserted a term

$$\langle\ldots n_p = 1, n_q = 0, \ldots|\ldots n_p = 1, n_q = 0\ldots\rangle$$

into the operator. Then we would have found it to be equal to

$$(-1)^{N_p}c_p(-1)^{N_q}c_q. \qquad (2.3.5)$$

Now c_p destroys the particle in the pth state, and so the value of N_q depends on whether we evaluate it before or after operating with c_p if we assume $q > p$. We then find that

$$(-1)^{N_q}c_p = c_p(-1)^{N_q+1}. \qquad (2.3.6)$$

N_p, on the other hand, does not depend on n_q. We thus find, combining Eqs. (2.3.4), (2.3.5), and (2.3.6),

$$c_p c_q + c_q c_p = 0 \quad (p \neq q). \qquad (2.3.7)$$

2.3 Second quantization for fermions

We say that c_p and c_q *anticommute*, and abbreviate Eq. (2.3.7) by writing

$$\{c_p, c_q\} = 0 \quad (p \neq q). \tag{2.3.8}$$

By similar arguments it can also be shown that

$$\{c_p^\dagger, c_q^\dagger\} = \{c_p^\dagger, c_q\} = 0 \quad (p \neq q). \tag{2.3.9}$$

The combination $c_p^\dagger c_p$, in which c_p first operates followed by $c_p^\dagger$, is particularly important. It is easy to see that it has eigenvalue zero when it operates on a state for which n_p is zero, and has eigenvalue unity when it operates on a state that has $n_p = 1$. We can consequently identify the operator $c_p^\dagger c_p$ as the *number operator*,

$$c_p^\dagger c_p = n_p.$$

Now if we first operate with $c_p^\dagger$ upon a state with $n_p = 0$, we obtain the state with $n_p = 1$. Thus when the combination $c_p c_p^\dagger$ operates upon a wavefunction in which the pth state is empty it gives unity. Similarly we know that

$$c_p c_p^\dagger | \ldots n_p = 1, \ldots \rangle = 0$$

since $c_p^\dagger$ cannot create another particle in an already occupied state. We are thus led to the conclusion that

$$c_p c_p^\dagger = 1 - n_p,$$

from which

$$c_p^\dagger c_p + c_p c_p^\dagger = 1.$$

Thus in the notation of Eq. (2.3.8) we have

$$\{c_p^\dagger, c_p\} = 1.$$

In summary the commutation relations for c_p are

$$\{c_p^\dagger, c_q^\dagger\} = \{c_p, c_q\} = 0$$
$$\{c_p, c_q^\dagger\} = \delta_{pq}. \tag{2.3.10}$$

We are now in a position to use Eq. (2.2.2) to write any operator in terms of annihilation and creation operators. Let us first consider a *single-particle operator*, such as an ordinary potential, $V(\mathbf{r})$. This will enter the many-body Hamiltonian in the form

$$\sum_{i=1}^{N} V(\mathbf{r}_i)$$

since it acts equally on all the particles. We then find that the only matrix elements, given by

$$\langle \{n_\mu\}|V|\{n_{\mu'}\}\rangle = \int \Phi^*(\alpha_1 \ldots \alpha_N) V(\mathbf{r}_i) \Phi(\alpha'_1 \ldots \alpha'_N) \, \mathbf{dr}_1 \cdots \mathbf{dr}_N, \quad (2.3.11)$$

that do not vanish are those in which not more than one of the α'_i is different from α_i. Then the integral reduces to

$$\int u_i^*(\mathbf{r}) V(\mathbf{r}) u_{i'}(\mathbf{r}) \, \mathbf{dr}.$$

In the occupation number representation we should write this as

$$V_{ii'} \equiv \langle n_1, \ldots, n_{\alpha_i} = 1, n_{\alpha'_i} = 0, \ldots |V| n_1, \ldots, n_{\alpha_i} = 0, n_{\alpha'_i} = 1, \ldots \rangle.$$

Then from Eq. (2.2.2)

$$V \equiv \sum_{\substack{\alpha_i, \alpha'_i \\ \{n_j\}(j \neq \alpha_i, \alpha'_i)}} V_{ii'} | \ldots n_{\alpha_i} = 1, n_{\alpha'_i} = 0, \ldots \rangle \langle \ldots n_{\alpha_i} = 0, n_{\alpha'_i} = 1, \ldots |$$

$$= \sum_{\alpha_i, \alpha'_i} V_{ii'} c^\dagger_{\alpha_i} c_{\alpha'_i}.$$

We could do the same with the kinetic energy part, T, of the Hamiltonian, and write the Hamiltonian as a sum of T and V. It is usually most convenient to choose a set of functions u_α such that T only has diagonal matrix elements. The plane wave representation of Section 2.1 satisfies this criterion, so that if we choose

$$u_\alpha(\mathbf{r}) = \Omega^{-1/2} e^{i\mathbf{k}_\alpha \cdot \mathbf{r}} \quad (2.3.12)$$

as the states out of which to build our determinants Φ, we shall have

$$T_{\mathbf{kk}'} = \Omega^{-1} \int e^{-i\mathbf{k}\cdot\mathbf{r}} \left(\frac{-\hbar^2}{2m}\nabla^2\right) e^{i\mathbf{k}'\cdot\mathbf{r}}\, \mathbf{dr} = \frac{\hbar^2 \mathbf{k}^2}{2m} \delta_{\mathbf{kk}'},$$

and

$$V_{\mathbf{kk}'} = \Omega^{-1} \int e^{i(\mathbf{k}'-\mathbf{k})\cdot\mathbf{r}} V(\mathbf{r})\, \mathbf{dr}.$$

The Hamiltonian then becomes

$$\mathcal{H} = \sum_{\mathbf{k}} \mathcal{E}_{\mathbf{k}} c_{\mathbf{k}}^\dagger c_{\mathbf{k}} + \sum_{\mathbf{k},\mathbf{k}'} V_{\mathbf{kk}'} c_{\mathbf{k}}^\dagger c_{\mathbf{k}'}$$

where we have written $\mathcal{E}_{\mathbf{k}}$ for $T_{\mathbf{kk}}$. Note that the annihilation and creation operators always appear in pairs; a potential cannot remove a particle from a state without putting it back in some other state.

As long as there are no interactions between the particles – that is, as long as the Hamiltonian can be split up into a sum of parts each of which refers to only one particle – there is little profit in rephrasing the problem in this notation, which is, for irrelevant reasons, known as *second quantization*. When, however, we introduce interactions between particles this formalism provides the only workable approach. Consider, for example, an interaction between particles that is expressible as a simple potential. That is, we add to the single-particle Hamiltonian terms of the form

$$V = \tfrac{1}{2} \sum_{i,j} V(\mathbf{r}_i - \mathbf{r}_j); \quad i \neq j$$

(the factor of $\tfrac{1}{2}$ prevents us from counting interactions twice). Then

$$\langle \{n_\mu\} | V | \{n_{\mu'}\} \rangle$$
$$= \sum_{i,j} \tfrac{1}{2} \int \Phi^*(\alpha_1 \cdots \alpha_N) V(\mathbf{r}_i - \mathbf{r}_j) \Phi(\alpha_1' \cdots \alpha_N')\, \mathbf{dr}_1 \cdots \mathbf{dr}_N.$$

As in the case of Eq. (2.3.11), this integral may be simplified. The determinants are sums of products of the functions u_α, and so the integral is a sum of terms of the form

$$\int u_a^*(\mathbf{r}_1) u_b^*(\mathbf{r}_2) \cdots u_z^*(\mathbf{r}_N) V(\mathbf{r}_i - \mathbf{r}_j) u_{a'}(\mathbf{r}_1) \cdots u_{z'}(\mathbf{r}_N)\, \mathbf{dr}_1 \cdots \mathbf{dr}_N.$$

This can be separated into a product of integrals over one coordinate,

$$\int u_a^*(\mathbf{r}_1) u_{a'}(\mathbf{r}_1) \, d\mathbf{r}_1 \int u_b^*(\mathbf{r}_2) u_{b'}(\mathbf{r}_2) \, d\mathbf{r}_2 \int \cdots, \qquad (2.3.13)$$

with the exception of the integration over $d\mathbf{r}_i$ and $d\mathbf{r}_j$, which cannot be separated, and gives a term of the form

$$V_{\alpha\beta\gamma\delta} = \int u_\alpha^*(\mathbf{r}_i) u_\beta^*(\mathbf{r}_j) V(\mathbf{r}_i - \mathbf{r}_j) u_\gamma(\mathbf{r}_j) u_\delta(\mathbf{r}_i) \, d\mathbf{r}_i \, d\mathbf{r}_j. \qquad (2.3.14)$$

Because the $u(\mathbf{r})$ are orthogonal, the integrals (2.3.13) vanish unless $a = a', b = b'$, etc. We are thus left with the fact that V can only alter the occupation of the states α, β, γ and δ. Thus V may be expressed as

$$V \equiv \tfrac{1}{2} \sum_{\alpha,\beta,\gamma,\delta} V_{\alpha\beta\gamma\delta} c_\alpha^\dagger c_\beta^\dagger c_\gamma c_\delta. \qquad (2.3.15)$$

It is not obvious that we have the correct numerical factor and the correct order of the operators c_γ and c_δ in this expression. Because these operators anticommute, incorrect ordering would describe the interaction $(-V)$. We can check the validity of the expression by considering the simplest case, where $V = 1$. Then the contribution of V to the Hamiltonian is

$$\tfrac{1}{2} \sum_{i \neq j} 1 = \tfrac{1}{2} N(N-1), \qquad (2.3.16)$$

with N the total number of particles present. From (2.3.14) we have

$$V_{\alpha\beta\gamma\delta} = \delta_{\alpha\delta} \delta_{\beta\gamma}$$

and so from Eq. (2.3.15)

$$V = \tfrac{1}{2} \sum_{\alpha,\beta} c_\alpha^\dagger c_\beta^\dagger c_\beta c_\alpha$$

$$= \tfrac{1}{2} \sum_{\alpha,\beta} c_\alpha^\dagger (c_\alpha c_\beta^\dagger - \delta_{\alpha\beta}) c_\beta$$

$$= \tfrac{1}{2} (N^2 - N)$$

in agreement with Eq. (2.3.16).

2.3 Second quantization for fermions

The matrix elements of V take on a particularly simple form when the $u(\mathbf{r})$ are the plane waves defined in (2.3.12). Then

$$V_{\mathbf{k},\mathbf{k}',\mathbf{k}'',\mathbf{k}'''} = \Omega^{-2} \int e^{i(\mathbf{k}'''-\mathbf{k})\cdot\mathbf{r}_1} e^{i(\mathbf{k}''-\mathbf{k}')\cdot\mathbf{r}_2} V(\mathbf{r}_1 - \mathbf{r}_2) \, \mathbf{dr}_1 \, \mathbf{dr}_2$$

$$= \Omega^{-2} \int e^{\frac{1}{2}i(\mathbf{k}'''-\mathbf{k}+\mathbf{k}''-\mathbf{k}')\cdot(\mathbf{r}_1+\mathbf{r}_2)} e^{\frac{1}{2}i(\mathbf{k}'''-\mathbf{k}-\mathbf{k}''+\mathbf{k}')\cdot(\mathbf{r}_1-\mathbf{r}_2)}$$

$$\times V(\mathbf{r}_1 - \mathbf{r}_2) \, \mathbf{dr}_1 \, \mathbf{dr}_2.$$

If we change to relative coordinates by writing $\mathbf{r} = \mathbf{r}_1 - \mathbf{r}_2$ and $\mathbf{R} = \frac{1}{2}(\mathbf{r}_1 + \mathbf{r}_2)$ we have

$$V_{\mathbf{k},\mathbf{k}',\mathbf{k}'',\mathbf{k}'''} = \Omega^{-2} \int e^{i(\mathbf{k}'''-\mathbf{k}+\mathbf{k}''-\mathbf{k}')\cdot\mathbf{R}} \, \mathbf{dR} \int e^{\frac{1}{2}i(\mathbf{k}'''-\mathbf{k}-\mathbf{k}''+\mathbf{k}')\cdot\mathbf{r}} V(\mathbf{r}) \, \mathbf{dr}.$$

The integral over $\mathbf{dR}$ vanishes unless $\mathbf{k}''' - \mathbf{k} + \mathbf{k}'' - \mathbf{k}' = 0$, in which case it gives Ω, so that

$$V_{\mathbf{k},\mathbf{k}',\mathbf{k}'',\mathbf{k}'''} = \delta_{\mathbf{k}+\mathbf{k}',\mathbf{k}'''+\mathbf{k}''} \Omega^{-1} \int e^{i(\mathbf{k}'-\mathbf{k}'')\cdot\mathbf{r}} V(\mathbf{r}) \, \mathbf{dr}.$$

The δ-function is no more than an expression of the conservation of momentum in a scattering process, since for a plane wave $\mathbf{p} = \hbar\mathbf{k}$. The integral over $\mathbf{dr}$ is the Fourier transform of the interparticle potential; we shall call it $V_{\mathbf{k}'-\mathbf{k}''}$. Then

$$V \equiv \frac{1}{2} \sum_{\mathbf{k},\mathbf{k}',\mathbf{k}''} V_{\mathbf{k}'-\mathbf{k}''} c^\dagger_\mathbf{k} c^\dagger_{\mathbf{k}'} c_{\mathbf{k}''} c_{\mathbf{k}+\mathbf{k}'-\mathbf{k}''}.$$

We can make it clearer that momentum is being transferred by defining $\mathbf{k}' - \mathbf{k}'' = \mathbf{q}$ and renaming the other variables. This gives

$$V \equiv \frac{1}{2} \sum_{\mathbf{k},\mathbf{k}',\mathbf{q}} V_\mathbf{q} c^\dagger_{\mathbf{k}-\mathbf{q}} c^\dagger_{\mathbf{k}'+\mathbf{q}} c_{\mathbf{k}'} c_\mathbf{k}.$$

It is often useful to interpret this product of operators pictorially. Particles in the states $\mathbf{k}$ and $\mathbf{k}'$ are destroyed, while particles in the states $\mathbf{k} - \mathbf{q}$ and $\mathbf{k}' + \mathbf{q}$ are created. This can be seen as a scattering of one particle by the other – a process in which an amount of momentum equal to $\hbar\mathbf{q}$ is transferred (Fig. 2.3.1).

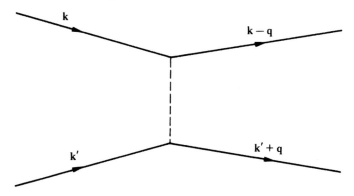

Figure 2.3.1. This diagram shows electrons in states $\mathbf{k}$ and $\mathbf{k}'$ being scattered into states $\mathbf{k} - \mathbf{q}$ and $\mathbf{k}' + \mathbf{q}$, and represents the product of operators $c^\dagger_{\mathbf{k}-\mathbf{q}} c^\dagger_{\mathbf{k}'+\mathbf{q}} c_{\mathbf{k}'} c_{\mathbf{k}}$.

2.4 The electron gas and the Hartree–Fock approximation

Our principal motivation in studying the theory of systems of interacting fermions is the hope that we might in this way better understand the behavior of the conduction electrons in a metal. Accordingly the first system to which we apply this formalism is that of unit volume of a gas of N spinless electrons, interacting by means of the Coulomb electrostatic repulsion. Then

$$V(\mathbf{r}_i - \mathbf{r}_j) = \frac{e^2}{|\mathbf{r}_i - \mathbf{r}_j|},$$

and

$$\begin{aligned} V_\mathbf{q} &= \Omega^{-1} \int e^{i\mathbf{q}\cdot\mathbf{r}} \frac{e^2}{|\mathbf{r}|} \, d\mathbf{r} \\ &= \frac{2\pi e^2}{\Omega} \int_0^\infty dr \int_0^\pi d\theta \, e^{iqr\cos\theta} r \, \sin\theta \\ &= \frac{4\pi e^2}{\Omega q} \int_0^\infty \sin qr \, dr. \end{aligned}$$

This integral does not converge, owing to the long range of the Coulomb potential, and so one uses the trick of supposing that the potential does not vary merely as r^{-1}, but as $r^{-1} e^{-\alpha r}$ and then takes the limit as α tends to zero.

2.4 The electron gas and the Hartree–Fock approximation

One finds

$$V_{\mathbf{q}} = \frac{4\pi e^2}{\Omega q^2} \lim_{\alpha \to 0} \int_0^\infty e^{-\alpha y/q} \sin y \, dy$$

$$= \frac{4\pi e^2}{\Omega q^2},$$

provided $q \neq 0$. The divergence of $V_{\mathbf{q}}$ for $q = 0$ is a difficulty that we might have anticipated had we looked more closely at the physics of the situation we are considering. Because the electrons are described by traveling waves the box may be considered as a conductor carrying a charge Ne, and it is an elementary result of electrostatics that charge always resides at the outer surface of a conductor.

It is now clear that our model of a metal as a mere gas of electrons was too simplified to be useful. We must take into account the presence of the positively charged ions that maintain the overall electrical neutrality of the metal. This, however, makes the problem a very difficult one indeed. Even for a single electron it is not trivial to solve the Schrödinger equation for a periodic lattice potential, as we shall see in Chapter 4, and so it is necessary to keep as simple a model as possible. This is achieved by replacing the lattice of positively charged ions by a fixed uniform distribution of positive charge, and investigating the interaction of the electrons in the presence of this background charge. This simplified model of a metal is sometimes known as *jellium*. The positive charge background adds to the Hamiltonian an extra one-particle potential term V^+ which, as we saw in the previous section, can be written

$$V^+ \equiv \sum_{\mathbf{k},\mathbf{q}} V_{\mathbf{q}}^+ c_{\mathbf{k}-\mathbf{q}}^\dagger c_{\mathbf{k}}.$$

As this charge density is uniform, the Fourier transform $V_{\mathbf{q}}^+$ of the potential due to it vanishes unless $q = 0$. Thus

$$V^+ = \sum_{\mathbf{k}} V_0^+ n_{\mathbf{k}} = N V_0^+$$

and no scattering is caused by this term, which is just a constant. Now if we look back to our transcription of the electron interaction potential in second-quantized form, we see that the troublesome coefficient V_0 also occurs in terms that cause no scattering. The part of $\mathcal{H}$ containing V_0 was

$$\tfrac{1}{2}\sum_{\mathbf{k},\mathbf{k}'} V_0 c_{\mathbf{k}}^\dagger c_{\mathbf{k}'}^\dagger c_{\mathbf{k}'} c_{\mathbf{k}} = \tfrac{1}{2}\sum_{\mathbf{k},\mathbf{k}'} V_0 c_{\mathbf{k}}^\dagger (c_{\mathbf{k}} c_{\mathbf{k}'}^\dagger - \delta_{\mathbf{k}\mathbf{k}'}) c_{\mathbf{k}'}$$

$$= \tfrac{1}{2}\sum_{\mathbf{k},\mathbf{k}'} V_0 (n_{\mathbf{k}} n_{\mathbf{k}'} - n_{\mathbf{k}} \delta_{\mathbf{k}\mathbf{k}'})$$

$$= \tfrac{1}{2} N(N-1) V_0,$$

which is again a constant. It is then reasonable to suppose that we can choose the density of the positive charge background in such a way that these terms cancel. We write

$$\tfrac{1}{2} N(N-1) V_0 + N V_0^+ + W_0^+ = 0,$$

where the energy W_0^+ of interaction of the positive charge with itself has been included in the sum of divergent terms that must cancel. The details of this cancellation are left as an exercise, and it suffices for us to know that the divergence can be removed. We shall assume that this has been done in what follows, and always omit the term for $q = 0$. The Hamiltonian then becomes

$$\mathcal{H} = \sum_{\mathbf{k}} \mathcal{E}_{\mathbf{k}} c_{\mathbf{k}}^\dagger c_{\mathbf{k}} + \sum_{\mathbf{k},\mathbf{k}',\mathbf{q}} \frac{2\pi e^2}{\Omega \mathbf{q}^2} c_{\mathbf{k}-\mathbf{q}}^\dagger c_{\mathbf{k}'+\mathbf{q}}^\dagger c_{\mathbf{k}'} c_{\mathbf{k}}, \qquad (2.4.1)$$

where once again we have written $\mathcal{E}_{\mathbf{k}}$ for $\hbar^2 \mathbf{k}^2/2m$. We are interested in finding the eigenfunctions of this Hamiltonian and the corresponding eigenvalues. Such a task proves to be immensely difficult, as many of the techniques that are used for single-particle problems fail in this instance. We shall not go too deeply into the complicated procedures that can be developed to get around these difficulties, but instead shall just examine some simple approximation methods.

A zero-order solution can be arrived at by neglecting the interaction term altogether. Then

$$\mathcal{H}_0 = \sum_{\mathbf{k}} \mathcal{E}_{\mathbf{k}} c_{\mathbf{k}}^\dagger c_{\mathbf{k}} = \sum_{\mathbf{k}} \mathcal{E}_{\mathbf{k}} n_{\mathbf{k}}.$$

We can easily construct eigenfunctions of the operator $n_{\mathbf{k}}$. First let us start with the state $|0\rangle$ in which there are no particles at all. This is just the vacuum state defined in Section 2.1. For all $\mathbf{k}$

$$n_{\mathbf{k}}|0\rangle = 0.$$

2.4 The electron gas and the Hartree–Fock approximation

We can next create one particle in the state $\mathbf{k}'$ by operating upon the vacuum state with $c^\dagger_{\mathbf{k}'}$,

$$|\mathbf{k}'\rangle = c^\dagger_{\mathbf{k}'}|0\rangle.$$

Then

$$\begin{aligned}n_{\mathbf{k}}|\mathbf{k}'\rangle &= c^\dagger_{\mathbf{k}} c_{\mathbf{k}} c^\dagger_{\mathbf{k}'}|0\rangle \\ &= c^\dagger_{\mathbf{k}}(\delta_{\mathbf{k}\mathbf{k}'} - c^\dagger_{\mathbf{k}'} c_{\mathbf{k}})|0\rangle\end{aligned}$$

from the commutation relations. But we know that $c_{\mathbf{k}}$ operating upon the vacuum state gives zero, since there is no particle there to destroy, and so

$$n_{\mathbf{k}}|\mathbf{k}'\rangle = \delta_{\mathbf{k}\mathbf{k}'}|\mathbf{k}'\rangle.$$

We can go on to build up a wavefunction $|\Phi\rangle$ containing N particles by repeatedly operating upon the vacuum state with different $c^\dagger_{\mathbf{k}'}$, so that

$$|\Phi\rangle = \left(\prod_{i=1}^{N} c^\dagger_{\mathbf{k}_i}\right)|0\rangle.$$

We can similarly show that

$$n_{\mathbf{k}}|\Phi\rangle = \sum_{i=1}^{N} \delta_{\mathbf{k}\mathbf{k}_i}|\Phi\rangle.$$

Then

$$\begin{aligned}\mathcal{H}_0|\Phi\rangle &= \sum_{\mathbf{k}} \mathcal{E}_{\mathbf{k}} n_{\mathbf{k}}|\Phi\rangle \\ &= \sum_{i=1}^{N} \mathcal{E}_{\mathbf{k}_i}|\Phi\rangle.\end{aligned}$$

Thus $|\Phi\rangle$ is an eigenfunction of the Hamiltonian with eigenvalue $\sum_i \mathcal{E}_{\mathbf{k}_i}$. The solution of this kind having the lowest energy is clearly that in which the wavenumbers $\mathbf{k}_i$ represent the N single-particle states for which the individual energies $\mathcal{E}_{\mathbf{k}_i}$ are the lowest. Since

$$\mathcal{E}_{\mathbf{k}_i} = \frac{\hbar^2 \mathbf{k}_i^2}{2m},$$

this is no more than our picture of a Fermi surface in momentum space. If there are N states for which $|\mathbf{k}| < k_F$, then all such states are filled to give the ground state of a gas of noninteracting fermions. This picture is sometimes known as a *Sommerfeld gas* (Fig. 2.4.1).

Now that we have found the eigenfunctions of $\mathcal{H}_0$ we can calculate an approximation to the energy of the interacting system. The exact energy we know to be $\langle \Psi | \mathcal{H} | \Psi \rangle$, where Ψ is the exact wavefunction. If we assume Ψ and Φ to be not too dissimilar we can calculate an approximate energy, $\mathcal{E}_{HF}$, by forming $\langle \Phi | \mathcal{H} | \Phi \rangle$. Using Eq. (2.4.1) we find

$$\mathcal{E}_{HF} = \langle \Phi | \mathcal{H}_0 + V | \Phi \rangle$$
$$= \mathcal{E}_0 + \langle \Phi | \sum_{\mathbf{k},\mathbf{k}',\mathbf{q}} \frac{2\pi e^2}{\Omega \mathbf{q}^2} c^\dagger_{\mathbf{k}-\mathbf{q}} c^\dagger_{\mathbf{k}'+\mathbf{q}} c_{\mathbf{k}'} c_{\mathbf{k}} | \Phi \rangle,$$

where we call the energy of the noninteracting system $\mathcal{E}_0$.

Now the effect of the potential is to take particles out of the states $\mathbf{k}$ and $\mathbf{k}'$ and put them into the states $\mathbf{k} - \mathbf{q}$ and $\mathbf{k}' + \mathbf{q}$. If these states are different from the original ones, then the wavefunction formed in this way will be orthogonal to Φ, and the matrix element will be zero. This means that the only terms that do not vanish will be those that *do not change the occupation numbers of* $|\Phi\rangle$. Then either

$$\mathbf{k}' = \mathbf{k}' + \mathbf{q} \quad \text{and} \quad \mathbf{k} = \mathbf{k} - \mathbf{q}$$

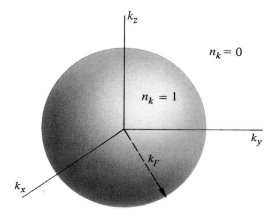

Figure 2.4.1. The Fermi surface of the Sommerfeld gas separates k-states for which $n_\mathbf{k} = 1$ from those for which $n_\mathbf{k} = 0$.

2.4 The electron gas and the Hartree–Fock approximation

or else

$$\mathbf{k}' = \mathbf{k} - \mathbf{q} \quad \text{and} \quad \mathbf{k} = \mathbf{k}' + \mathbf{q}.$$

The first possibility implies that $q = 0$, and we have agreed to omit this term, which does not correspond to any scattering at all (Fig. 2.4.2(a)). This leaves us with the second possibility, Fig. 2.4.2(b), known as *exchange scattering*, in which the particle that was in the state $\mathbf{k}$ is scattered into the state $\mathbf{k}'$, and vice versa. The correction this gives to $\mathcal{E}_0$ is known as the *exchange energy*. We find

$$\mathcal{E}_{HF} = \mathcal{E}_0 + \sum_{\mathbf{k},\mathbf{k}'} \langle \Phi | \frac{2\pi e^2}{\Omega |\mathbf{k} - \mathbf{k}'|^2} c^\dagger_{\mathbf{k}'} c^\dagger_{\mathbf{k}} c_{\mathbf{k}'} c_{\mathbf{k}} | \Phi \rangle$$

$$= \mathcal{E}_0 + \sum_{\mathbf{k},\mathbf{k}'} \langle \Phi | \frac{2\pi e^2}{\Omega |\mathbf{k} - \mathbf{k}'|^2} (-n_{\mathbf{k}'} n_{\mathbf{k}}) | \Phi \rangle. \quad (2.4.2)$$

This we can write as

$$\mathcal{E}_{HF} = \mathcal{E}_0 - \sum_{\mathbf{k},\mathbf{k}' \text{ occupied}} \frac{2\pi e^2}{\Omega |\mathbf{k} - \mathbf{k}'|^2}$$

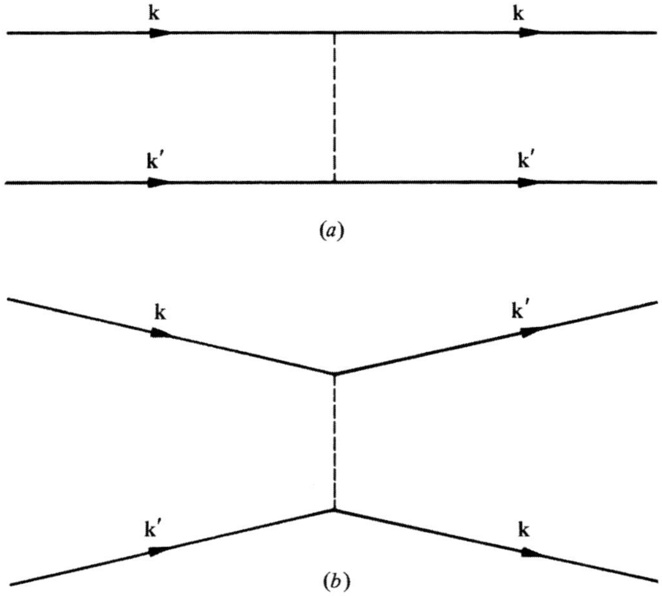

Figure 2.4.2. In the Hartree–Fock approximation only direct scattering (a) and exchange scattering (b) between identical states can occur.

where now the summation is over states **k** and **k**′ that are *both occupied*. This method of treating the electron gas is a special case of an approach known as the *Hartree–Fock approximation*, and is the simplest way in which we can take the interactions into account. Even this approach can become very complicated, however, if in addition to the interactions there is some additional one-particle potential applied to the system.

The interesting point to note is that in this approximation the energy is reduced *below* that of the Sommerfeld gas, $\mathcal{E}_0$. It is not paradoxical that the repulsive interaction should decrease the energy of the system, for we must not forget that we have also effectively added a uniform background of positive charge when we eliminated the interaction $V_{q=0}$. The reduction in energy comes from the fact that the particles are kept apart by the antisymmetrization of the wavefunction, and so are acted upon more by the positive charge background than by their neighbors.

So far we have considered only a system of spinless particles. In fact we know that the electron has a spin angular momentum of $\frac{1}{2}\hbar$, which means that a single electron can occupy a state **k** in two ways, either with spin up or with spin down. We denote this by naming the states **k**↑ and **k**↓. This modifies the energy we find for the Sommerfeld gas, since now the Fermi wavenumber k_F is determined by the condition that there be only $\frac{1}{2}N$ values of **k** for which $|\mathbf{k}| < k_F$. The inclusion of the spin of the electron in our model must necessarily introduce a number of other complications, for the spin will be accompanied by a magnetic moment, and the electrons will interact by virtue of their magnetic fields. However, we ignore these effects, retaining only the Coulomb interaction, and ask how the Hartree–Fock energy is modified by the inclusion of spin.

The first point we note is that since **k**↑ and **k**↓ denote separate states all anticommutators for states of opposite spin vanish. Thus for example

$$\{c_{\mathbf{k}\uparrow}, c_{\mathbf{k}\downarrow}^\dagger\} = 0.$$

Next we see that since the Coulomb interaction does not contain the spin coordinates it cannot cause an exchange of particles with opposite spin. The integration over $\mathbf{r}_j$, for example, in Eq. (2.3.14) demands that the spin of state u_β be the same as that of u_γ. The Hamiltonian thus becomes

$$\mathcal{H} = \sum_{\mathbf{k},s} \mathcal{E}_\mathbf{k} c_{\mathbf{k}s}^\dagger c_{\mathbf{k}s} + \frac{1}{2} \sum_{\mathbf{k},\mathbf{k}',\mathbf{q},s,s'} V_\mathbf{q} c_{\mathbf{k}-\mathbf{q},s}^\dagger c_{\mathbf{k}'+\mathbf{q},s'}^\dagger c_{\mathbf{k}',s'} c_{\mathbf{k},s}$$

and only terms of the form $n_{\mathbf{k}'\uparrow} n_{\mathbf{k}\uparrow}$ and $n_{\mathbf{k}'\downarrow} n_{\mathbf{k}\downarrow}$ will come into the expression

2.4 The electron gas and the Hartree–Fock approximation

for the exchange energy. We again find

$$\mathcal{E}_{HF} = \mathcal{E}_0 - \sum \frac{2\pi e^2}{\Omega|\mathbf{k}-\mathbf{k}'|^2}$$

but now the summation proceeds only over states $\mathbf{k}$ and $\mathbf{k}'$ that are occupied and that have the same spin. This reduces the amount by which the Hartree–Fock energy $\mathcal{E}_{HF}$ is less than $\mathcal{E}_0$. One may express this by saying that now the antisymmetrization only keeps apart the electrons having parallel spins. We can guess that in the exact solution to the problem *all* the electrons will try to keep apart from one another regardless of their spins; they will reduce the potential energy of the system by doing so. We say that their motions will be correlated, and that the difference between the exact energy and $\mathcal{E}_{HF}$ is the *correlation energy*.

One might now be tempted to ask how the energy of a single electron is altered by the interaction. A little reflection shows this question to be meaningless, since the energy of interaction of two particles cannot be associated with either one, and it makes no sense to share out the interaction energy between the particles in some arbitrary way. One can only talk about the total energy of the system. However, as long as we are within the Hartree–Fock approximation we can ask how the total energy changes when we change the approximate wavefunction Φ. In particular we might ask how $\mathcal{E}_{HF}$ changes when we remove an electron. If we consider $\langle\Phi|n_\mathbf{p}|\Phi\rangle$ as the number $\langle n_\mathbf{p}\rangle$ (as opposed to the operator $n_\mathbf{p} = c_\mathbf{p}^\dagger c_\mathbf{p}$), then to take away an electron in the $\mathbf{p}$th state is to reduce $\langle n_\mathbf{p}\rangle$ by 1. The energy change is consequently $\partial\mathcal{E}/\partial\langle n_\mathbf{p}\rangle$. From Eq. (2.4.2) this gives for the example where we neglected spin

$$\frac{\partial\mathcal{E}_{HF}}{\partial\langle n_\mathbf{p}\rangle} = \frac{\partial}{\partial\langle n_\mathbf{p}\rangle}\left\{\sum_\mathbf{k}\mathcal{E}_\mathbf{k}\langle n_\mathbf{k}\rangle - \sum_{\mathbf{k},\mathbf{k}'}\frac{2\pi e^2}{\Omega|\mathbf{k}-\mathbf{k}'|^2}\langle n_\mathbf{k} n_{\mathbf{k}'}\rangle\right\}$$

$$= \mathcal{E}_\mathbf{p} - \sum_\mathbf{k}\frac{4\pi e^2}{\Omega|\mathbf{k}-\mathbf{p}|^2}\langle n_\mathbf{k}\rangle. \qquad (2.4.3)$$

We might then put this electron back into another state $\mathbf{p}'$. The work required to take the electron from $\mathbf{p}$ to $\mathbf{p}'$ would then be

$$\Delta W = \frac{\partial\mathcal{E}_{HF}}{\partial\langle n_{\mathbf{p}'}\rangle} - \frac{\partial\mathcal{E}_{HF}}{\partial\langle n_\mathbf{p}\rangle}.$$

If the states were separated by a very small momentum difference, so that

$\mathbf{p}' = \mathbf{p} + \boldsymbol{\delta}$ we should have

$$\Delta W = \boldsymbol{\delta} \cdot \frac{\partial}{\partial \mathbf{p}} \left(\frac{\partial \mathcal{E}_{HF}}{\partial \langle n_\mathbf{p} \rangle} \right). \tag{2.4.4}$$

Because of the spherical symmetry in momentum space of the interactions and of the function $\langle n_\mathbf{p} \rangle$, we know that the derivative with respect to $\mathbf{p}$ (by which we mean the gradient in p-space) must be in the direction of $\mathbf{p}$, so that only the component, δ_p, of $\boldsymbol{\delta}$ that is parallel to $\mathbf{p}$ enters Eq. (2.4.4). In the limit that $\delta_p \to 0$ we have

$$\frac{\Delta W}{\delta_p} \to \frac{dW}{dp} = \frac{\partial}{\partial p} \left(\frac{\partial \mathcal{E}_{HF}}{\partial \langle n_\mathbf{p} \rangle} \right),$$

which from (2.4.3) is

$$\frac{dW}{dp} = \frac{\partial \mathcal{E}_\mathbf{p}}{\partial p} - \frac{\partial}{\partial p} \sum_\mathbf{k} \frac{4\pi e^2}{\Omega |\mathbf{k} - \mathbf{p}|^2} \langle n_\mathbf{k} \rangle. \tag{2.4.5}$$

This is of interest because it tells us the energy of the lowest-lying group of excitations of the system within the Hartree–Fock approximation. When we took an electron below the Fermi surface of the noninteracting system and put it in a higher energy state, we said that we had created a particle–hole pair, which was an elementary excitation. In the limiting case that we had taken an electron from a vanishingly small distance $\delta/2$ below the Fermi surface, and put it in a state $\delta/2$ above the surface, the excitation would have had energy $\delta \times (\partial \mathcal{E}/\partial p)_{p=p_F}$, corresponding to the first term in (2.4.5). In the Hartree–Fock approximation the elementary excitation has the energy $\delta \times (dW/dp)_{p=p_F}$, which includes the second term in (2.4.5). However, inspection of (2.4.5) shows that as $p \to p_F$, dW/dp becomes infinite. This has consequences that are at variance with the experimentally determined properties of the electron gas as found in metals, and is a first indication that the Hartree–Fock approximation may be inadequate where Coulomb forces are involved.

2.5 Perturbation theory

The only system whose wavefunctions we have studied has been the gas of noninteracting fermions described by the Hamiltonian

$$\mathcal{H}_0 = \sum_\mathbf{k} \mathcal{E}_\mathbf{k} n_\mathbf{k}.$$

2.5 Perturbation theory

In the Hartree–Fock approximation we merely took the wavefunctions $|\Phi\rangle$ of the noninteracting system, and worked out the expectation value in the state $|\Phi\rangle$ of a Hamiltonian containing interactions. While we could look upon this as a variational approach – we guess that the wavefunction might be like $|\Phi\rangle$ and we work out the energy it would give – we could also consider it as the first term in a perturbation expansion. Let us now quickly look at the methods of perturbation theory, and see how they apply to many-body systems.

We start by stating a solution of a simple problem,

$$\mathcal{H}_0|\phi\rangle = \mathcal{E}_0|\phi\rangle, \quad (2.5.1)$$

and consider the solutions of

$$(\mathcal{H}_0 + V)|\psi\rangle = \mathcal{E}|\psi\rangle. \quad (2.5.2)$$

From the perturbed Eq. (2.5.2) we have that

$$\langle\phi|(\mathcal{H}_0 + V)|\psi\rangle = \langle\phi|\mathcal{E}|\psi\rangle$$

and so if we normalize $|\psi\rangle$ with the condition

$$\langle\phi|\psi\rangle = 1,$$

then because $\langle\phi|\mathcal{H}_0 = \langle\phi|\mathcal{E}_0$ we have

$$\mathcal{E} - \mathcal{E}_0 = \langle\phi|V|\psi\rangle. \quad (2.5.3)$$

To proceed further we need to define what we mean by a *function of an operator*. From (2.5.1), for instance, we can find by operating with $\mathcal{H}_0$ on both sides that

$$\mathcal{H}_0\mathcal{H}_0|\phi\rangle = \mathcal{H}_0^2|\phi\rangle = \mathcal{E}_0^2|\phi\rangle,$$

and in general that

$$(\mathcal{H}_0)^n|\phi\rangle = (\mathcal{E}_0)^n|\phi\rangle.$$

Thus if we interpret $f(\mathcal{H}_0)$ as a power series expansion in $\mathcal{H}_0$ we should have

$$f(\mathcal{H}_0)|\phi\rangle = f(\mathcal{E}_0)|\phi\rangle. \quad (2.5.4)$$

We generalize this definition to include functions like $\mathcal{H}^{1/2}$ that have no power series expansions. In particular, the operator function $(z - \mathcal{H}_0)^{-1}$ is the operator whose eigenfunctions are $|\phi\rangle$ and whose corresponding eigenvalues are $(z - \mathcal{E}_0)^{-1}$, provided $z \neq \mathcal{E}_0$.

There are various ways in which we can write the solution of (2.5.2) using the expression for the perturbed energy (2.5.3). We could write

$$|\psi\rangle = |\phi\rangle + (\mathcal{E}_0 - \mathcal{H}_0)^{-1}(1 - |\psi\rangle\langle\phi|)V|\psi\rangle, \qquad (2.5.5)$$

which is the starting point of *Rayleigh–Schrödinger* perturbation theory. Another possibility is

$$|\psi\rangle = |\phi\rangle + (\mathcal{E} - \mathcal{H}_0)^{-1}(1 - |\phi\rangle\langle\phi|)V|\psi\rangle, \qquad (2.5.6)$$

which is the starting point for *Brillouin–Wigner* perturbation theory. These equations can be verified by operating upon them with $(\mathcal{E}_0 - \mathcal{H}_0)$ and $(\mathcal{E} - \mathcal{H}_0)$, respectively. Because the right-hand sides still contain the unknown $|\psi\rangle$, one iterates these by using the equation itself to substitute for $|\psi\rangle$. Thus if we write

$$(\mathcal{E} - \mathcal{H}_0)^{-1}(1 - |\phi\rangle\langle\phi|) = G_0$$

then the Brillouin–Wigner formula becomes

$$|\psi\rangle = |\phi\rangle + G_0 V |\phi\rangle + G_0 V G_0 V |\phi\rangle + \cdots.$$

The Rayleigh–Schrödinger expansion, for example, allows us to write the energy, to second order in V, as

$$\mathcal{E} = \mathcal{E}_0 + \langle\phi|V|\phi\rangle + \langle\phi|V(\mathcal{E}_0 - \mathcal{H}_0)^{-1}(1 - |\phi\rangle\langle\phi|)V|\phi\rangle.$$

This sort of expression is evaluated by remembering that

$$\sum_{p'} |\phi_{p'}\rangle\langle\phi_{p'}| \equiv 1$$

when the $|\phi_{p'}\rangle$ form a complete set. Then for the energy of the state $|\psi_p\rangle$ corresponding to the unperturbed state $|\phi_p\rangle$ of energy $\mathcal{E}_p$ one finds

$$\mathcal{E} = \mathcal{E}_p + \langle\phi_p|V|\phi_p\rangle + \sum_{p'(p'\neq p)} \langle\phi_p|V|\phi_{p'}\rangle(\mathcal{E}_p - \mathcal{E}_{p'})^{-1}\langle\phi_{p'}|V|\phi_p\rangle. \qquad (2.5.7)$$

2.5 Perturbation theory

Again, this expression is valid only to second order in V. The term $p' = p$ is excluded from the summation by the projection operator, $1 - |\phi_p\rangle\langle\phi_p|$. This removes one term for which the energy denominator would vanish.

We can interpret the final term in expression (2.5.7) in the following way. First the interaction V causes the system to make a transition from the state $|\phi_p\rangle$ to the state $|\phi_{p'}\rangle$. This process does not conserve energy, and so the system can only remain in the intermediate state a time of the order of $\hbar/(\mathcal{E}_p - \mathcal{E}_{p'})$, which is all that the Uncertainty Principle allows. It must then make a transition back to the original state, again by means of the perturbation V. The higher-order expansions become very complicated, especially in the Rayleigh–Schrödinger formula, where $|\psi\rangle$ occurs twice. In fact, it turns out that the Brillouin–Wigner expression is less well suited to many-body systems than the Rayleigh–Schrödinger one from the point of view of convergence. Although there are a number of elegant methods that mitigate the awkwardness of keeping track of the terms in the Rayleigh–Schrödinger expansion the approach remains a difficult one, and we shall for the most part leave this approach to the more specialized texts.

In the case of the electron gas in particular, one runs into difficulty even in the second-order expansion for the energy, as this turns out to diverge. We noticed the danger signals flying in the first-order term, which was the Hartree–Fock approximation. There we found in Eq. (2.4.5) that although the energy W required to add another electron to the system was finite, its derivative, dW/dp, was infinite. This was due to the fact that for the Coulomb interaction $V_q \propto q^{-2}$, and reflects the long-range nature of the Coulomb force. If we try to take this to second order in perturbation theory by using (2.5.7) we find that the total energy itself diverges logarithmically. To see this we note from Eq. (2.5.7) that the contribution to the energy that is of second order in V takes on the form

$$\mathcal{E}^{(2)} = \sum_{\beta(\neq\alpha)} \langle\Phi_\alpha|V|\Phi_\beta\rangle(\mathcal{E}_\alpha - \mathcal{E}_\beta)^{-1}\langle\Phi_\beta|V|\Phi_\alpha\rangle,$$

where Φ_α and Φ_β are the initial and intermediate wavefunctions describing the N independent electrons. Now

$$V = \tfrac{1}{2}\sum_{\mathbf{k},\mathbf{k'},\mathbf{q}} V_\mathbf{q} c^\dagger_{\mathbf{k}-\mathbf{q}} c^\dagger_{\mathbf{k'}+\mathbf{q}} c_{\mathbf{k'}} c_\mathbf{k}$$

and so Φ_β differs from Φ_α in having electrons removed from states $\mathbf{k}$ and $\mathbf{k'}$ and put back in states $\mathbf{k'} + \mathbf{q}$ and $\mathbf{k} - \mathbf{q}$. Now the second time that V appears in $\mathcal{E}^{(2)}$ it must transform Φ_β back into Φ_α. If we now write

$$V = \tfrac{1}{2} \sum_{\mathbf{p},\mathbf{p}',\mathbf{q}'} V_{\mathbf{q}'} c^\dagger_{\mathbf{p}-\mathbf{q}'} c^\dagger_{\mathbf{p}'+\mathbf{q}'} c_{\mathbf{p}'} c_{\mathbf{p}}$$

then there are only two possible ways in which this can happen. These are illustrated in Fig. 2.5.1. We must either have $\mathbf{p} = \mathbf{k}' + \mathbf{q}$ and $\mathbf{p}' = \mathbf{k} - \mathbf{q}$ as in Fig. 2.5.1(a) or else $\mathbf{p} = \mathbf{k} - \mathbf{q}$ and $\mathbf{p}' = \mathbf{k}' + \mathbf{q}$ as in Fig. 2.5.1(b). In the former case the operators $c_{\mathbf{p}'}$ and $c_{\mathbf{p}}$ annihilate the electrons that were scattered into states $\mathbf{k} - \mathbf{q}$ and $\mathbf{k}' + \mathbf{q}$, respectively, and electrons are created in states $\mathbf{k} - \mathbf{q} + \mathbf{q}'$ and $\mathbf{k}' + \mathbf{q} - \mathbf{q}'$. In order for the net result of all this to be the original state Φ_α we must either have $\mathbf{q} = \mathbf{q}'$ or else $\mathbf{k}' + \mathbf{q} - \mathbf{q}' = \mathbf{k}$. We call the first possibility a "direct" term and the second an "exchange" term. We shall investigate only the contribution to $\mathcal{E}^{(2)}$ of the direct term, as this turns out to be the more important one. If we follow up the alternative

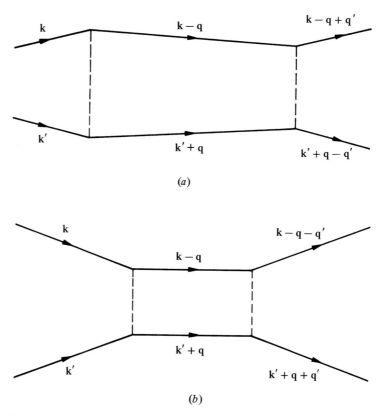

Figure 2.5.1. The two possible second-order scattering processes.

2.5 Perturbation theory

possibility for the choice of **p** and **p**′, we find that this also leads to a direct term and an exchange scattering term of the same magnitude as before. The total contribution of direct terms to $\mathcal{E}^{(2)}$ is thus

$$\mathcal{E}^{(2)}_{\text{direct}} = 2 \sum_{\mathbf{k},\mathbf{k}',\mathbf{q}} \left(\frac{1}{2}V_{\mathbf{q}}\right)^2 \langle \Phi_\alpha | c^\dagger_{\mathbf{k}'} c^\dagger_{\mathbf{k}} c_{\mathbf{k}-\mathbf{q}} c_{\mathbf{k}'+\mathbf{q}}$$

$$\times \frac{1}{\mathcal{E}_{\mathbf{k}} + \mathcal{E}_{\mathbf{k}'} - \mathcal{E}_{\mathbf{k}-\mathbf{q}} - \mathcal{E}_{\mathbf{k}'+\mathbf{q}}} c^\dagger_{\mathbf{k}-\mathbf{q}} c^\dagger_{\mathbf{k}'+\mathbf{q}} c_{\mathbf{k}'} c_{\mathbf{k}} |\Phi_\alpha\rangle.$$

We may use the fact that $\mathcal{E}_{\mathbf{k}} = \hbar^2 k^2/2m$ to simplify the energy denominator and then commute the c-operators into pairs that form number operators. We then have

$$\mathcal{E}^{(2)}_{\text{direct}} = \frac{m}{2\hbar^2} \sum_{\mathbf{k},\mathbf{k}',\mathbf{q}} V_{\mathbf{q}}^2 \langle \Phi_\alpha | \frac{(1 - n_{\mathbf{k}-\mathbf{q}})(1 - n_{\mathbf{k}'+\mathbf{q}}) n_{\mathbf{k}'} n_{\mathbf{k}}}{-\mathbf{q}\cdot(\mathbf{q} + \mathbf{k}' - \mathbf{k})} |\Phi_\alpha\rangle, \tag{2.5.8}$$

and we now take Φ_α to be the noninteracting ground state. If we include the spin of the electron we should have to multiply this result by a factor of 4, since both states **k** and **k**′ can have spin either up or down.

The difficulty with expression (2.5.8) lies in the terms for which q is small. The factor of $(1 - n_{\mathbf{k}-\mathbf{q}}) n_{\mathbf{k}}$ then restricts the summation over **k** to a thin layer of states of thickness q on one side of the Fermi surface, as indicated in Fig. 2.5.2. The summation over **k**′ is similarly restricted to a layer on the opposite side. The two summations thus contribute a factor of order q^2. The volume element for the summation over **q** will be $4\pi q^2 \, dq$, and $V_{\mathbf{q}}$ is proportional to

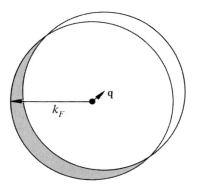

Figure 2.5.2. The product $(1 - n_{\mathbf{k}-\mathbf{q}}) n_{\mathbf{k}}$ vanishes everywhere outside the shaded volume.

q^{-2}. The net result is that $\mathcal{E}^{(2)}_{\text{direct}}$ contains a factor $\int_0 q^{-1} \, dq$, which diverges logarithmically.

We might also have guessed that perturbation theory could not be applied in any straightforward way from our semiclassical approach to the electron gas. There we found that a form of collective motion – the plasma oscillation – played an important role in the dynamics of the system. It is clear that we could not arrive at a description of collective motion by just taking a couple of terms of a perturbation expansion starting with a scheme of independent particles. As it turns out, rather sophisticated methods have been devised whereby one can sum an infinite number of terms selected from the perturbation expansion, and arrive at a picture of collective behavior. We, however, shall first take a simpler approach. We know that plasma oscillations represent density fluctuations, and so we shall deliberately search for a solution of the Schrödinger equation that describes these.

2.6 The density operator

So far we have used the occupation number representation to define operators that create particles in various states, u_α. When these are plane wave states the probability of finding the particle is constant at all points in space. We now ask whether it is possible to define operators that will create or destroy particles at one particular point in space.

For the plane wave states contained in a cubical box of volume Ω, we found in Section 2.1 that the allowed values of $\mathbf{k}$ were given by

$$\mathbf{k} = \left(\frac{2\pi m_x}{L}, \frac{2\pi m_y}{L}, \frac{2\pi m_z}{L} \right),$$

where the m were integers. We now first show the important relation

$$\sum_{\mathbf{k}} e^{i\mathbf{k}\cdot\mathbf{r}} = \Omega\,\delta(\mathbf{r}) \tag{2.6.1}$$

where $\delta(\mathbf{r})$ is the three-dimensional Dirac delta-function, which is zero for $r \neq 0$ and for which $\int_\Omega f(\mathbf{r})\,\delta(\mathbf{r})\,d\mathbf{r} = f(0)$ when $r = 0$ is within Ω. To see this we first substitute for $\mathbf{k}$ to find

$$\sum_{\mathbf{k}} e^{i\mathbf{k}\cdot\mathbf{r}} = \sum_{m_x=-\infty}^{\infty} \exp\left(\frac{2\pi i m_x x}{L}\right) \sum_{m_y=-\infty}^{\infty} \exp\left(\frac{2\pi i m_y y}{L}\right) \sum_{m_z=-\infty}^{\infty} \exp\left(\frac{2\pi i m_z z}{L}\right).$$

2.6 The density operator

Now

$$\sum_{m=-M}^{M} \exp\left(\frac{2\pi i m x}{L}\right) = \frac{\sin\left[\frac{(2M+1)\pi x}{L}\right]}{\sin\left(\frac{\pi x}{L}\right)}$$

so that

$$\int \sum_{m=-M}^{M} \exp\left(\frac{2\pi i m x}{L}\right) f(x)\, dx = \int \frac{L}{\pi} \frac{\sin[(2M+1)\zeta]}{\sin \zeta} f\left(\frac{L\zeta}{\pi}\right) d\zeta.$$

As $M \to \infty$ the term $\sin[(2M+1)\zeta]$ begins to oscillate so rapidly that only the region near $\zeta = 0$ gives any contribution to the integral. Then we can replace $\sin \zeta$ by ζ and $f(L\zeta/\pi)$ by $f(0)$ so that the integral becomes

$$\frac{L}{\pi} f(0) \int_{M\zeta=-\infty}^{\infty} \frac{\sin[(2M+1)\zeta]}{M\zeta} d(M\zeta) = Lf(0).$$

Thus

$$\int f(\mathbf{r}) \sum_{\mathbf{k}} e^{i\mathbf{k}\cdot\mathbf{r}}\, d\mathbf{r} = L^3 f(0),$$

which proves (2.6.1).

This relation suggests that we define an operator $\psi^\dagger(\mathbf{r})$ by

$$\psi^\dagger(\mathbf{r}) = \Omega^{-1/2} \sum_{\mathbf{k}} e^{-i\mathbf{k}\cdot\mathbf{r}} c_{\mathbf{k}}^\dagger. \qquad (2.6.2)$$

This operator, not to be confused with a one-particle wavefunction, is known as a *fermion field operator*. Its conjugate is

$$\psi(\mathbf{r}) = \Omega^{-1/2} \sum_{\mathbf{k}} e^{i\mathbf{k}\cdot\mathbf{r}} c_{\mathbf{k}}. \qquad (2.6.3)$$

Then

$$\{\psi(\mathbf{r}), \psi^\dagger(\mathbf{r}')\} = \Omega^{-1} \sum_{\mathbf{k},\mathbf{k}'} e^{i(\mathbf{k}\cdot\mathbf{r}-\mathbf{k}'\cdot\mathbf{r}')} \{c_\mathbf{k}, c_{\mathbf{k}'}^\dagger\}$$

$$= \Omega^{-1} \sum_{\mathbf{k}} e^{i\mathbf{k}\cdot(\mathbf{r}-\mathbf{r}')}$$

$$= \delta(\mathbf{r}-\mathbf{r}'). \qquad (2.6.4)$$

The fact that the field operators have these anticommutation relations shows that they are the operators we are looking for. If $\psi(\mathbf{r})$ does indeed annihilate a particle at $\mathbf{r}$ we should expect that

$$\langle 0|\psi^\dagger(\mathbf{r}')\psi(\mathbf{r})|0\rangle = 0$$

and

$$\langle 0|\psi(\mathbf{r})\psi^\dagger(\mathbf{r}')|0\rangle = \delta(\mathbf{r}-\mathbf{r}'),$$

for $\psi(\mathbf{r})$ will always give zero when operating on the vacuum state $|0\rangle$ unless we first operate with $\psi^\dagger(\mathbf{r})$. This is compatible with (2.6.4).

The operators that we expressed in terms of the $c_\mathbf{k}^\dagger$ and $c_\mathbf{k}$ may equally well be expressed in terms of the $\psi^\dagger(\mathbf{r})$ and $\psi(\mathbf{r})$. For instance,

$$V(\mathbf{r}) \equiv \int \psi^\dagger(\mathbf{r}) V(\mathbf{r}) \psi(\mathbf{r})\, d\mathbf{r},$$

and

$$V(\mathbf{r}, \mathbf{r}') \equiv \int \psi^\dagger(\mathbf{r})\psi^\dagger(\mathbf{r}') V(\mathbf{r},\mathbf{r}') \psi(\mathbf{r}')\psi(\mathbf{r})\, d\mathbf{r}\, d\mathbf{r}',$$

as may be verified by substitution from the definitions (2.6.2) and (2.6.3) and use of (2.6.1).

In particular, we can use the field operators to represent the density of particles, $\rho(\mathbf{r})$, at the point $\mathbf{r}$. The density is defined as the sum over particle coordinates, $\mathbf{r}_i$,

$$\rho(\mathbf{r}) = \sum_i \delta(\mathbf{r}-\mathbf{r}_i),$$

2.6 The density operator

and in terms of the field operators this becomes

$$\rho(\mathbf{r}) \equiv \int \psi^\dagger(\mathbf{r}')\, \delta(\mathbf{r}-\mathbf{r}')\psi(\mathbf{r}')\, d\mathbf{r}'$$
$$= \psi^\dagger(\mathbf{r})\psi(\mathbf{r}). \qquad (2.6.5)$$

The Fourier transform of the particle density is

$$\rho_\mathbf{q} = \Omega^{-1} \int e^{-i\mathbf{q}\cdot\mathbf{r}} \rho(\mathbf{r})\, d\mathbf{r}$$
$$= \Omega^{-2} \int e^{-i\mathbf{q}\cdot\mathbf{r}} \sum_{\mathbf{k},\mathbf{k}'} e^{i(\mathbf{k}'-\mathbf{k})\cdot\mathbf{r}} c^\dagger_\mathbf{k} c_{\mathbf{k}'}\, d\mathbf{r}$$
$$= \Omega^{-1} \sum_{\mathbf{k},\mathbf{k}'} c^\dagger_\mathbf{k} c_{\mathbf{k}'}\, \delta_{\mathbf{k}'-\mathbf{k},\mathbf{q}}$$
$$= \Omega^{-1} \sum_\mathbf{k} c^\dagger_\mathbf{k} c_{\mathbf{k}+\mathbf{q}}.$$

Because $\rho(\mathbf{r}) = \rho^\dagger(\mathbf{r})$ it follows that $\rho^\dagger_\mathbf{q} = \rho_{-\mathbf{q}}$.

As an example of the usefulness of the density operator, we show how the Hamiltonian for the electron gas can be expressed in terms of the number and density operators. From Eq. (2.4.1) we have

$$\mathcal{H} = \sum_\mathbf{k} \varepsilon_\mathbf{k} n_\mathbf{k} + \sum_{\mathbf{k},\mathbf{k}',\mathbf{q}} \frac{2\pi e^2}{\Omega q^2} c^\dagger_{\mathbf{k}-\mathbf{q}} c^\dagger_{\mathbf{k}'+\mathbf{q}} c_{\mathbf{k}'} c_\mathbf{k}.$$

Now

$$\sum_{\mathbf{k},\mathbf{k}',\mathbf{q}} c^\dagger_{\mathbf{k}-\mathbf{q}} c^\dagger_{\mathbf{k}'+\mathbf{q}} c_{\mathbf{k}'} c_\mathbf{k} = -\sum_{\mathbf{k},\mathbf{k}',\mathbf{q}} c^\dagger_{\mathbf{k}-\mathbf{q}} c^\dagger_{\mathbf{k}'+\mathbf{q}} c_\mathbf{k} c_{\mathbf{k}'}$$
$$= -\sum_{\mathbf{k},\mathbf{k}',\mathbf{q}} c^\dagger_{\mathbf{k}-\mathbf{q}} (\delta_{\mathbf{k},\mathbf{k}'+\mathbf{q}} - c_\mathbf{k} c^\dagger_{\mathbf{k}'+\mathbf{q}}) c_{\mathbf{k}'}$$
$$= -\sum_{\mathbf{k}',\mathbf{q}} c^\dagger_{\mathbf{k}'} c_{\mathbf{k}'} + \sum_\mathbf{q} \left(\sum_\mathbf{k} c^\dagger_{\mathbf{k}-\mathbf{q}} c_\mathbf{k}\right)\left(\sum_{\mathbf{k}'} c^\dagger_{\mathbf{k}'+\mathbf{q}} c_{\mathbf{k}'}\right)$$
$$= -\sum_\mathbf{q}\left(\sum_{\mathbf{k}'} n_{\mathbf{k}'}\right) + \Omega^2 \sum_\mathbf{q} \rho_\mathbf{q} \rho_{-\mathbf{q}}.$$

But since $\sum_{\mathbf{k}'} n_{\mathbf{k}'} = N$, the total number of particles, we have

$$\mathcal{H} = \sum_{\mathbf{k}} \mathcal{E}_{\mathbf{k}} n_{\mathbf{k}} + \sum_{\mathbf{q}} \frac{2\pi e^2}{\Omega \mathbf{q}^2} (\Omega^2 \rho_{\mathbf{q}}^\dagger \rho_{\mathbf{q}} - N).$$

Note that the summation over $\mathbf{q}$ cannot be performed for each component of the interaction separately, as these would not converge.

2.7 The random phase approximation and screening

Let us suppose that there is an operator $B^\dagger$ that creates an excitation of a many-body system. If the ground-state wavefunction is $|\Psi\rangle$ then we should have

$$\mathcal{H}|\Psi\rangle = \mathcal{E}_0|\Psi\rangle$$

and

$$\mathcal{H} B^\dagger |\Psi\rangle = (\mathcal{E}_0 + \mathcal{E}_b) B^\dagger |\Psi\rangle,$$

since the excited state of the system, $B^\dagger|\Psi\rangle$, is also an eigenfunction of $\mathcal{H}$, having an energy that is greater than the ground state by an amount $\mathcal{E}_b$, the excitation energy. Thus

$$\mathcal{H} B^\dagger |\Psi\rangle - B^\dagger \mathcal{H} |\Psi\rangle = \mathcal{E}_b B^\dagger |\Psi\rangle.$$

If in particular

$$\mathcal{H} B^\dagger - B^\dagger \mathcal{H} = \mathcal{E}_b B^\dagger$$

then we can say that $B^\dagger$ creates an excitation of energy $\mathcal{E}_b$ *irrespective of which eigenstate of $\mathcal{H}$ it operates on* (provided $B^\dagger|\Psi\rangle \neq 0$). We abbreviate this condition by

$$[\mathcal{H}, B^\dagger] = \mathcal{E}_b B^\dagger,$$

where $[\mathcal{H}, B^\dagger]$ is known as the *commutator* of $\mathcal{H}$ and $B^\dagger$, in distinction to the anticommutator, in which the minus sign is replaced by a plus.

Let us now examine the case where

$$B^\dagger = c_{\mathbf{p+q}}^\dagger c_{\mathbf{p}},$$

2.7 The random phase approximation and screening

and see under what conditions this would create an excitation in the electron gas. In this case

$$\mathcal{H} = \sum_{\mathbf{k}} \mathcal{E}_{\mathbf{k}} c^{\dagger}_{\mathbf{k}} c_{\mathbf{k}} + \tfrac{1}{2} \sum_{\mathbf{k},\mathbf{k}',\mathbf{q}'} V_{\mathbf{q}'} c^{\dagger}_{\mathbf{k}-\mathbf{q}'} c^{\dagger}_{\mathbf{k}'+\mathbf{q}'} c_{\mathbf{k}'} c_{\mathbf{k}},$$

and it takes but a tedious half-hour to show that

$$[\mathcal{H}, c^{\dagger}_{\mathbf{p}+\mathbf{q}} c_{\mathbf{p}}] = (\mathcal{E}_{\mathbf{p}+\mathbf{q}} - \mathcal{E}_{\mathbf{p}}) c^{\dagger}_{\mathbf{p}+\mathbf{q}} c_{\mathbf{p}}$$
$$+ \sum_{\mathbf{q}'} \frac{V_{\mathbf{q}'}}{2} [(c^{\dagger}_{\mathbf{p}+\mathbf{q}-\mathbf{q}'} c_{\mathbf{p}} - c^{\dagger}_{\mathbf{p}+\mathbf{q}} c_{\mathbf{p}+\mathbf{q}'}) \rho^{\dagger}_{\mathbf{q}'} \Omega$$
$$+ \Omega \rho_{\mathbf{q}'} (c^{\dagger}_{\mathbf{p}+\mathbf{q}'+\mathbf{q}} c_{\mathbf{p}} - c^{\dagger}_{\mathbf{p}+\mathbf{q}} c_{\mathbf{p}-\mathbf{q}'})]. \qquad (2.7.1)$$

We recall that for the noninteracting system the operator $c^{\dagger}_{\mathbf{p}+\mathbf{q}} c_{\mathbf{p}}$ creates a particle–hole pair when it operates on the ground state of a system in which the **p**th state is occupied and the (**p** + **q**)th state is empty. We do not know what the ground state of the interacting system is, but we can see under what conditions the operator $c^{\dagger}_{\mathbf{p}+\mathbf{q}} c_{\mathbf{p}}$ will create an excitation in it. Let us consider this operator when |**q**| is much greater than the Fermi radius, k_F (Fig. 2.7.1). Then for any occupied **p** we can choose **q** such that $\mathcal{E}_{\mathbf{p}+\mathbf{q}} - \mathcal{E}_{\mathbf{p}}$ is as large as we like. Then we can ignore the second term in (2.7.1), and to a good approximation the commutator of $\mathcal{H}$ and $c^{\dagger}_{\mathbf{p}+\mathbf{q}} c_{\mathbf{p}}$ is a number times this operator itself. This means that where *large* momentum transfers between particles are concerned, we are justified in considering quasiparticle excitations of the system. But now let us look at the case where **q** is small. Then the argument that the first term will dominate is no longer valid, indicating that the quasiparticle picture may not be appropriate where small momentum transfers are

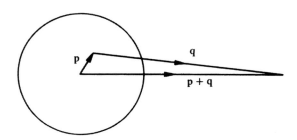

Figure 2.7.1. For large **q** the change in kinetic energy on scattering from **p** to **p** + **q** is generally large compared with the matrix element of the potential for this process.

concerned. Accordingly we go on to ask whether $\rho_{\mathbf{q}}^{\dagger}$, the creation operator for a density fluctuation, might not be the one we are after.

We can form the commutator of $\rho_{\mathbf{q}}^{\dagger}$ with the Hamiltonian simply by summing (2.7.1) over all $\mathbf{p}$. The interaction terms cancel exactly, since

$$\sum_{\mathbf{p}} c_{\mathbf{p}+\mathbf{q}-\mathbf{q}'}^{\dagger} c_{\mathbf{p}} = \sum_{\mathbf{p}} c_{\mathbf{p}+\mathbf{q}}^{\dagger} c_{\mathbf{p}+\mathbf{q}'} = \Omega \rho_{\mathbf{q}-\mathbf{q}'}^{\dagger},$$

which leaves us with

$$[\mathcal{H}, \rho_{\mathbf{q}}^{\dagger}] = \Omega^{-1} \sum_{\mathbf{p}} (\mathcal{E}_{\mathbf{p}+\mathbf{q}} - \mathcal{E}_{\mathbf{p}}) c_{\mathbf{p}+\mathbf{q}}^{\dagger} c_{\mathbf{p}}.$$

Although this does not appear to be proportional to $\rho_q^{\dagger}$, we cannot be sure that it will not have a similar effect when acting on the wavefunction of the interacting system. Accordingly we optimistically persevere, and once again take the commutator with the Hamiltonian $\mathcal{H}$. If $[\mathcal{H}, \rho_q^{\dagger}]$ is effectively equivalent to $\hbar \omega \rho_q^{\dagger}$ then we should have

$$[\mathcal{H}, [\mathcal{H}, \rho_q^{\dagger}]] = (\hbar \omega)^2 \rho_q^{\dagger}.$$

In fact, our tenacity will be rewarded. We find

$$[\mathcal{H}, [\mathcal{H}, \rho_q^{\dagger}]] = \Omega^{-1} \left[\mathcal{H}, \sum_{\mathbf{p}} (\mathcal{E}_{\mathbf{p}+\mathbf{q}} - \mathcal{E}_{\mathbf{p}}) c_{\mathbf{p}+\mathbf{q}}^{\dagger} c_{\mathbf{p}} \right]$$

$$= \Omega^{-1} \sum_{\mathbf{p}} (\mathcal{E}_{\mathbf{p}+\mathbf{q}} - \mathcal{E}_{\mathbf{p}}) [\mathcal{H}, c_{\mathbf{p}+\mathbf{q}}^{\dagger} c_{\mathbf{p}}]$$

$$= \Omega^{-1} \sum_{\mathbf{p}} (\mathcal{E}_{\mathbf{p}+\mathbf{q}} - \mathcal{E}_{\mathbf{p}})^2 c_{\mathbf{p}+\mathbf{q}}^{\dagger} c_{\mathbf{p}} + \sum_{\mathbf{p},\mathbf{q}'} (\mathcal{E}_{\mathbf{p}+\mathbf{q}} - \mathcal{E}_{\mathbf{p}}) \frac{V_{\mathbf{q}'}}{2}$$

$$\times [(c_{\mathbf{p}+\mathbf{q}-\mathbf{q}'}^{\dagger} c_{\mathbf{p}} - c_{\mathbf{p}+\mathbf{q}}^{\dagger} c_{\mathbf{p}+\mathbf{q}'}) \rho_{\mathbf{q}'}^{\dagger} + \rho_{\mathbf{q}'} (c_{\mathbf{p}+\mathbf{q}'+\mathbf{q}}^{\dagger} c_{\mathbf{p}} - c_{\mathbf{p}+\mathbf{q}}^{\dagger} c_{\mathbf{p}-\mathbf{q}'})].$$

This simplifies when we remember that

$$\mathcal{E}_{\mathbf{p}+\mathbf{q}} = \frac{\hbar^2}{2m} (\mathbf{p}+\mathbf{q})^2; \quad \mathcal{E}_{\mathbf{p}} = \frac{\hbar^2 \mathbf{p}^2}{2m}; \quad \mathcal{E}_{\mathbf{p}+\mathbf{q}} - \mathcal{E}_{\mathbf{p}} = \frac{\hbar^2}{2m} (2\mathbf{p} \cdot \mathbf{q} + \mathbf{q}^2).$$

2.7 The random phase approximation and screening

This means that

$$\sum_{\mathbf{p}} (\mathcal{E}_{\mathbf{p+q}} - \mathcal{E}_{\mathbf{p}})(c^{\dagger}_{\mathbf{p+q-q'}}c_{\mathbf{p}} - c^{\dagger}_{\mathbf{p+q}}c_{\mathbf{p+q'}})$$

$$= \frac{\hbar^2}{2m} \sum_{\mathbf{p}} (2\mathbf{p}\cdot\mathbf{q} + q^2) c^{\dagger}_{\mathbf{p+q-q'}}c_{\mathbf{p}}$$

$$- \frac{\hbar^2}{2m} \sum_{\mathbf{p'}} [2(\mathbf{p'}-\mathbf{q'})\cdot\mathbf{q} + q^2] c^{\dagger}_{\mathbf{p'+q-q'}}c_{\mathbf{p'}},$$

where we have written $\mathbf{p'}$ for $\mathbf{p}+\mathbf{q'}$. This reduces to

$$\frac{\hbar^2}{2m} \sum_{\mathbf{p'}} 2\mathbf{q'}\cdot\mathbf{q} c^{\dagger}_{\mathbf{p'+q-q'}}c_{\mathbf{p'}} = \frac{\hbar^2 \mathbf{q'}\cdot\mathbf{q}}{m} \rho^{\dagger}_{\mathbf{q-q'}},$$

so that

$$(\hbar\omega)^2 \rho^{\dagger}_{\mathbf{q}} = \sum_{\mathbf{p}} \left[\frac{\hbar^2}{2m}(2\mathbf{p}\cdot\mathbf{q}+q^2) \right]^2 c^{\dagger}_{\mathbf{p+q}}c_{\mathbf{p}}/\Omega$$

$$+ \sum_{\mathbf{q'}} \frac{V_{\mathbf{q'}}}{2} \frac{\hbar^2 \mathbf{q}\cdot\mathbf{q'}}{m} (\rho_{\mathbf{q'-q}}\rho^{\dagger}_{\mathbf{q'}} - \rho^{\dagger}_{-\mathbf{q'}}\rho_{-\mathbf{q-q'}})\Omega. \quad (2.7.2)$$

Now the zero Fourier component of the density, ρ_0, plays a very different role from all the other components. It is just the average density of particles in the system. Consider, for instance, a box of electrons of average density ρ_0, the box being of length L (Fig. 2.7.2). The first nonzero Fourier component

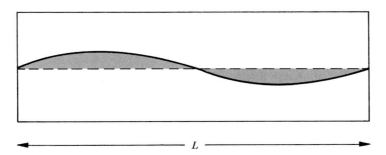

Figure 2.7.2. When $q = 2\pi/L$ the operator ρ_q measures a quantity approximately equal to the difference in the number of particles in the two halves of the container.

of ρ will be approximately equal to the difference between the number of particles in the left- and right-hand sides of the box, since it is just

$$\Omega^{-1} \int \rho(\mathbf{r}) \exp\left(\frac{2i\pi x}{L}\right) d\mathbf{r}.$$

When the number of particles present is large the difference between numbers in the two halves will be very small compared with the total number, and so ρ_0 will be the most important term in the summation over $\mathbf{q}'$ in (2.7.2). Because the term with $\mathbf{q}' = 0$ is omitted, it is only when $\mathbf{q}' = \pm\mathbf{q}$ that such a term will appear. The neglect of all terms for which $\mathbf{q}' \neq \pm\mathbf{q}$ is known as the *random phase approximation*, or RPA. This, combined with neglect of the summation over $\mathbf{p}$, which is small when $\mathbf{q}$ is small, leaves us with

$$(\hbar\omega)^2 \rho_\mathbf{q}^\dagger = \frac{4\pi e^2}{2\mathbf{q}^2} \frac{\hbar^2 \mathbf{q}^2}{m} (\rho_0 \rho_\mathbf{q}^\dagger + \rho_\mathbf{q}^\dagger \rho_0),$$

and since all the $\rho_\mathbf{q}$ commute (Problem 2.7) we have

$$\omega^2 = \frac{4\pi e^2 \rho_0}{m},$$

which gives just the classical plasma frequency, ω_p.

We thus see that the relevant excitations of low wavenumber are not particle–hole pairs, but collective motions of the electron gas. Bohm and Pines argue that we should consider the electrons as interacting through a matrix element $V_\mathbf{q}$ that is equal to $4\pi e^2/\Omega \mathbf{q}^2$ only when q is greater than some characteristic value q_c. Below q_c the interactions contribute only to the plasma oscillations and can be left out of the particle interaction terms in the Hamiltonian. We expect q_c^{-1} to be of the order of the average interparticle distance, since plasma waves can only exist when their wavelength is greater than this value. Thus we put $V_\mathbf{q} = 0$ for $q < q_c$. This means that the interaction potential in configuration space, which is the Fourier transform of $V_\mathbf{q}$, will be

$$V(\mathbf{r}) = \sum_{q>q_c} e^{i\mathbf{q}\cdot\mathbf{r}} \frac{4\pi e^2}{\Omega \mathbf{q}^2}.$$

This gives a function rather like the Yukawa potential,

$$(e^2/r) \exp(-q_c r),$$

2.7 The random phase approximation and screening

which is an example of a *screened Coulomb potential*. An electron at a point tends to repel all the others from its vicinity, which effectively gives a region of net positive charge surrounding each electron. This partially cancels (or *screens*) the mutual repulsion of the electrons at large distances.

We can understand the concept of screening within the framework of perturbation theory by considering the effect of a weak sinusoidal potential applied to the electron gas. The total Hamiltonian would then be

$$\mathcal{H} = \mathcal{H}_0 + V + U,$$

with V the Coulomb interaction of the electrons, and U the externally applied potential, being given by

$$U = 2U_\mathbf{q} \cos \mathbf{q} \cdot \mathbf{r},$$

which in the notation of second quantization is

$$U = U_\mathbf{q} \sum_\mathbf{p} (c^\dagger_{\mathbf{p}+\mathbf{q}} c_\mathbf{p} + c^\dagger_{\mathbf{p}-\mathbf{q}} c_\mathbf{p}). \tag{2.7.3}$$

Perturbation theory can then be used to express the wavefunction and energy as a power series in $(U + V)$, which we can then rearrange in the form of a power series in U. In the Rayleigh–Schrödinger expansion for the wavefunction, for example, we have

$$|\Psi\rangle = |\Phi\rangle + (\mathcal{E}_0 - \mathcal{H}_0)^{-1}(1 - |\Phi\rangle\langle\Phi|)(U + V)|\Phi\rangle + \cdots$$
$$= [|\Phi\rangle + (\mathcal{E}_0 - \mathcal{H}_0)^{-1}(1 - |\Phi\rangle\langle\Phi|)V|\Phi\rangle + \cdots]$$
$$+ (\mathcal{E}_0 - \mathcal{H}_0)^{-1}(1 - |\Phi\rangle\langle\Phi|)U|\Phi\rangle + (\mathcal{E}_0 - \mathcal{H}_0)^{-1}$$
$$\times (1 - |\Phi\rangle\langle\Phi|)V(\mathcal{E}_0 - \mathcal{H}_0)^{-1}(1 - |\Phi\rangle\langle\Phi|)U|\Phi\rangle + \cdots$$
$$+ [\text{terms of order } U^2] + \cdots.$$

We investigate the response of the system to weak applied fields by examining those terms that are linear in U. We notice that we could write the sum of these contributions in the form

$$(\mathcal{E}_0 - \mathcal{H}_0)^{-1}(1 - |\Phi\rangle\langle\Phi|)U_{\text{eff}}|\Phi\rangle$$

66   Second quantization and the electron gas

if we were to define an *effective potential* U_{eff} by the equation

$$U_{\text{eff}} = U + V(\mathcal{E}_0 - \mathcal{H}_0)^{-1}(1 - |\Phi\rangle\langle\Phi|)U$$
$$+ U(\mathcal{E}_0 - \mathcal{H}_0)^{-1}(1 - |\Phi\rangle\langle\Phi|)V + \cdots. \quad (2.7.4)$$

Let us now substitute for V, the Coulomb interaction, and for U, and simplify U by considering only the first part of the summand in expression (2.7.3). Then the second term on the right-hand side of Eq. (2.7.4), for example, becomes

$$\tfrac{1}{2}\sum_{\mathbf{k},\mathbf{k}',\mathbf{q}'} V_{\mathbf{q}'} c^\dagger_{\mathbf{k}-\mathbf{q}'} c^\dagger_{\mathbf{k}'+\mathbf{q}'} c_{\mathbf{k}'} c_{\mathbf{k}} (\mathcal{E}_0 - \mathcal{H}_0)^{-1}(1 - |\Phi\rangle\langle\Phi|) \sum_{\mathbf{p}} U_{\mathbf{q}} c^\dagger_{\mathbf{p}+\mathbf{q}} c_{\mathbf{p}}. \quad (2.7.5)$$

This component of U_{eff} is thus a sum of terms that annihilate the electrons in states $\mathbf{p}, \mathbf{k}$, and $\mathbf{k}'$, and create them again in states $\mathbf{p}+\mathbf{q}, \mathbf{k}'+\mathbf{q}'$, and $\mathbf{k}-\mathbf{q}'$. Such complicated processes could be represented by diagrams like Fig. 2.7.3, and are not easily interpreted in physical terms.

There are, however, some terms from this sum that contribute in a special way to U_{eff}, and whose effect has a simple interpretation. Let us look, for example, at the term in which $\mathbf{q} = \mathbf{q}'$ and $\mathbf{p}+\mathbf{q} = \mathbf{k}$. Then we can join together the two parts of Fig. 2.7.3 and represent the scattering in the form shown in Fig. 2.7.4. We note the interesting fact that the net result of

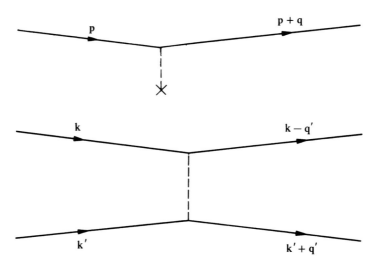

Figure 2.7.3. In this diagram an electron is scattered by the externally applied potential U, and then two other electrons interact through their Coulomb repulsion V.

2.7 The random phase approximation and screening

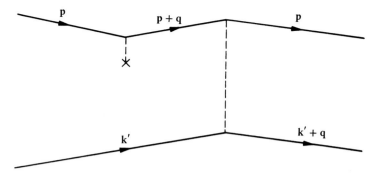

Figure 2.7.4. In this special case of the preceding diagram the same electron participates in both scattering processes.

these interactions is that an electron is scattered from $\mathbf{k}'$ to $\mathbf{k}'+\mathbf{q}$. The physical interpretation of this is that the externally applied potential U causes a density fluctuation in the electron gas, and it is this density fluctuation that scatters the electron originally in the state $\mathbf{k}'$. The contribution to expression (2.7.5) from these processes is

$$\tfrac{1}{2}\sum_{\mathbf{p}} V_{\mathbf{q}} n_{\mathbf{p}} (1 - n_{\mathbf{p}+\mathbf{q}})(\mathcal{E}_{\mathbf{p}} - \mathcal{E}_{\mathbf{p}+\mathbf{q}})^{-1} \sum_{\mathbf{k}'} U_{\mathbf{q}} c^{\dagger}_{\mathbf{k}'+\mathbf{q}} c_{\mathbf{k}'},$$

the energy denominator $(\mathcal{E}_{\mathbf{p}} - \mathcal{E}_{\mathbf{p}+\mathbf{q}})^{-1}$ coming from the effect of U on the state Φ. There is also a set of terms for which $\mathbf{q} = -\mathbf{q}'$ and $\mathbf{p}+\mathbf{q} = \mathbf{k}'$, which contribute an equal amount again. To these must then be added a set of terms from the third component of the right-hand side of expression (2.7.4) in which V acts first, followed by U. From these we select the terms shown in Fig. 2.7.5, which contribute an amount

$$\sum_{\mathbf{p}} V_{\mathbf{q}} n_{\mathbf{p}+\mathbf{q}} (1 - n_{\mathbf{p}})(\mathcal{E}_{\mathbf{p}+\mathbf{q}} - \mathcal{E}_{\mathbf{p}})^{-1} \sum_{\mathbf{k}} U_{\mathbf{q}} c^{\dagger}_{\mathbf{k}+\mathbf{q}} c_{\mathbf{k}}.$$

We identify the sums over $\mathbf{k}$ and $\mathbf{k}'$ as just being equal to U itself, and so our approximation for Eq. (2.7.4) becomes

$$U_{\text{eff}} \simeq U + V_{\mathbf{q}} \left(\sum_{\mathbf{p}} \frac{n_{\mathbf{p}+\mathbf{q}} - n_{\mathbf{p}}}{\mathcal{E}_{\mathbf{p}+\mathbf{q}} - \mathcal{E}_{\mathbf{p}}} \right) U + \cdots.$$

If we make similar approximations for the terms of higher order in this series we shall have contributions of the form shown in Fig. 2.7.6, which can be

68   *Second quantization and the electron gas*

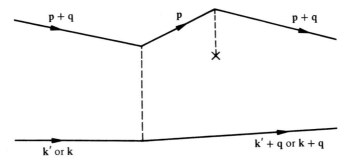

Figure 2.7.5. Here an electron is first scattered by another electron and then by the applied potential U.

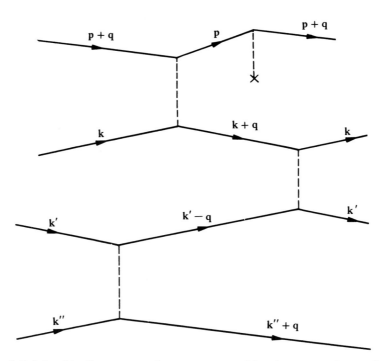

Figure 2.7.6. In this diagram an electron scattered by the externally applied potential passes its extra momentum to another electron through a chain of Coulomb interactions.

given a simpler aspect if we think of the scattering of the electron from $\mathbf{p}+\mathbf{q}$ to $\mathbf{p}$ as the creation of a particle–hole pair, and represent the hole of wavenumber $\mathbf{p}+\mathbf{q}$ by an arrow pointing backwards. Figure 2.7.6 can then be redrawn as in Fig. 2.7.7. All these complicated diagrams will have the net effect of scattering just one electron and increasing its wavenumber by an

2.7 The random phase approximation and screening

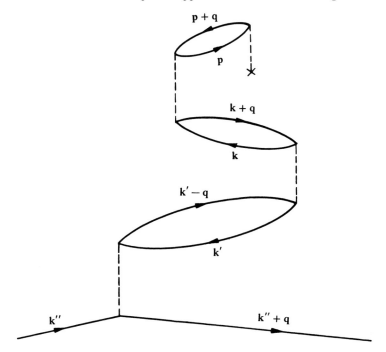

Figure 2.7.7. This redrawing of Fig. 2.7.6. depicts the absence of an electron in a given k-state as a line pointing backwards.

amount **q**. They in fact form a geometric series, which allows us to write

$$U_{\text{eff}} \simeq \left[1 + V_q\left(\sum_p \frac{n_{p+q} - n_p}{\mathcal{E}_{p+q} - \mathcal{E}_p}\right) + V_q^2\left(\sum_p \frac{n_{p+q} - n_p}{\mathcal{E}_{p+q} - \mathcal{E}_p}\right)^2 + \cdots\right]U$$

$$= \frac{U}{\epsilon(\mathbf{q})},$$

where

$$\epsilon(\mathbf{q}) = 1 - V_q \sum_p \frac{n_{p+q} - n_p}{\mathcal{E}_{p+q} - \mathcal{E}_p}. \tag{2.7.6}$$

Defined in this way, $\epsilon(\mathbf{q})$ plays the role of a *dielectric constant* in that it is the factor by which the applied field, which may be likened to the electric displacement **D**, exceeds the actual field **E** within the electron gas. Because a conductor like the electron gas cannot support a steady uniform electric field we expect $\epsilon(\mathbf{q})$ to become infinite as $q \to 0$. This does indeed occur, since $V_\mathbf{q}$ varies as q^{-2} while the summation over $\mathbf{p}$ remains finite.

Because any potential can be analyzed into its Fourier components, this theory gives us an approximate result for the modification by the electron gas of a potential of any shape. If, for example, we put a charged impurity into the electron gas the potential U would be $-Ze^2/r$. This is the sum of Fourier components $-4\pi Ze^2/\Omega q^2$, each of which would be screened in our linear approximation by the dielectric constant $\epsilon(\mathbf{q})$. The result would be a screened potential of Fourier transform

$$U_{\text{eff}}(\mathbf{q}) \simeq \frac{-4\pi Ze^2/\Omega \mathbf{q}^2}{1 - (4\pi e^2/\Omega \mathbf{q}^2) \sum_\mathbf{p} [(n_{\mathbf{p}+\mathbf{q}} - n_\mathbf{p})/(\mathcal{E}_{\mathbf{p}+\mathbf{q}} - \mathcal{E}_\mathbf{p})]}.$$

This expression remains finite as $q \to 0$, and thus represents a potential that again has some similarity to the Yukawa potential.

Improvements on this theory are fairly arduous, even in the linear approximation. The most obvious correction would be to include exchange scattering in our analysis by considering processes of the type shown in Fig. 2.7.8 in addition to those of Fig. 2.7.4. In higher orders, however, these processes do not reduce to simple products that can be summed as geometric series, and their investigation lies beyond the scope of this book.

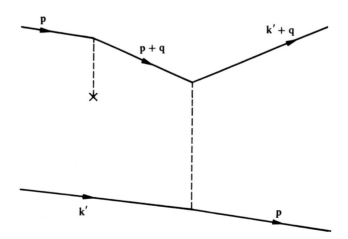

Figure 2.7.8. Exchange processes such as this one, in which the final k-state of one electron is identical to the initial state of another electron, are neglected in deriving Eq. (2.7.6).

2.8 Spin waves in the electron gas

An interesting application of the random phase approximation occurs in the theory of metallic ferromagnets. We saw in Section 2.4 that in the Hartree–Fock approximation the exchange energy is negative. It is illustrated in Problem 2.3 that at low electron densities this exchange energy becomes large enough in comparison to the kinetic energy that a magnetized phase, in which all the electron spins are pointing in the same direction, appears the most stable. While we are aware of the failings of the Hartree–Fock approximation and should not accept its predictions unquestioningly, we are led to the conclusion that in a metal such as nickel it is the presence of some effective electron interaction that gives rise to ferromagnetism. We cannot accept an alternative model of the type we assumed in Section 1.4, in which each spin is localized at a lattice site, because measurements show there to be a nonintegral number of spins per atom in this metal. We thus assume a Hamiltonian of the form

$$\mathcal{H} = \sum_{\mathbf{k},s} \mathcal{E}_\mathbf{k} c^\dagger_{\mathbf{k},s} c_{\mathbf{k},s}$$
$$+ \frac{1}{2} \sum_{\mathbf{k},\mathbf{k}',\mathbf{q}',s,s'} V_{\mathbf{q}'} c^\dagger_{\mathbf{k}-\mathbf{q}',s} c^\dagger_{\mathbf{k}'+\mathbf{q}',s'} c_{\mathbf{k}',s'} c_{\mathbf{k},s},$$

which is identical to our previous form for the Hamiltonian of the electron gas except that we shall take $V_{\mathbf{q}'}$ to be an effective interaction, and not necessarily the pure Coulomb interaction. We make the assumption that the ground state of this system is magnetized, so that $N_\downarrow$, the total number of electrons with spin down, is greater than $N_\uparrow$.

We now look for collective excitations of this system that can be interpreted as spin waves. We first consider the commutator of the Hamiltonian with the operator

$$B^\dagger_\mathbf{p} = c^\dagger_{\mathbf{p}+\mathbf{q}\uparrow} c_{\mathbf{p}\downarrow}. \tag{2.8.1}$$

This differs from the operator considered in the previous section in that it reverses the spin of the electron on which it acts, and can thus change the magnetization of the electron gas. We find

$$[\mathcal{H}, B^\dagger_\mathbf{p}] = (\mathcal{E}_{\mathbf{p}+\mathbf{q}} - \mathcal{E}_\mathbf{p}) B^\dagger_\mathbf{p} + \sum_{\mathbf{q}',\mathbf{k},s} V_{\mathbf{q}'} (c^\dagger_{\mathbf{p}+\mathbf{q}-\mathbf{q}'\uparrow} c^\dagger_{\mathbf{k}+\mathbf{q}',s} c_{\mathbf{k},s} c_{\mathbf{p}\downarrow}$$
$$+ c^\dagger_{\mathbf{p}+\mathbf{q}\uparrow} c^\dagger_{\mathbf{k}-\mathbf{q}',s} c_{\mathbf{p}-\mathbf{q}'\downarrow} c_{\mathbf{k},s}). \tag{2.8.2}$$

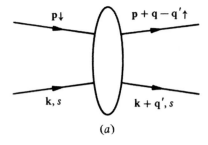

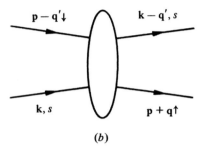

Figure 2.8.1. Of these scattering processes we retain only those in which one of the final electron states is the same as one of the initial ones.

The summation is thus over interactions of the form shown in Fig. 2.8.1.

We now make the random phase approximation by retaining only those processes in which one electron leaves in a state identical to one of the original states. We thus select from Fig. 2.8.1(a) only those processes for which $\mathbf{k} + \mathbf{q}', s = \mathbf{p}\downarrow$ or for which $\mathbf{k}, s = \mathbf{p} + \mathbf{q} - \mathbf{q}'\uparrow$. With a similar selection from the processes of Fig. 2.8.1(b) we find that Eq. (2.8.2) becomes

$$[\mathcal{H}, B_{\mathbf{p}}^{\dagger}] \simeq (\mathcal{E}_{\mathbf{p}+\mathbf{q}} - \mathcal{E}_{\mathbf{p}})B_{\mathbf{p}}^{\dagger} + \sum_{\mathbf{q}'} V_{\mathbf{q}'}[(n_{\mathbf{p}-\mathbf{q}'\downarrow} - n_{\mathbf{p}-\mathbf{q}'+\mathbf{q}\uparrow})B_{\mathbf{p}}^{\dagger}$$
$$+ (n_{\mathbf{p}+\mathbf{q}\uparrow} - n_{\mathbf{p}\downarrow})B_{\mathbf{p}-\mathbf{q}'}^{\dagger}], \qquad (2.8.3)$$

where

$$B_{\mathbf{p}-\mathbf{q}'}^{\dagger} = c_{\mathbf{p}-\mathbf{q}'+\mathbf{q}\uparrow}^{\dagger} c_{\mathbf{p}-\mathbf{q}'\downarrow}.$$

This can be characterized as a random phase approximation because it retains only those terms involving the number operators, and it is the sum

2.8 Spin waves in the electron gas

of the number operators that gives the zeroth Fourier component of the density.

The fact that the right-hand side of Eq. (2.8.3) involves terms in $B^\dagger_{\mathbf{p}-\mathbf{q}'}$ shows that $B^\dagger_{\mathbf{p}}$ does not create eigenstates of $\mathcal{H}$ when acting on the ground state. It does, however, suggest that we once again form a linear combination of these operators by writing

$$B^\dagger = \sum_{\mathbf{p}} \alpha_{\mathbf{p}} B^\dagger_{\mathbf{p}},$$

where the $\alpha_{\mathbf{p}}$ are constants. If this operator does indeed create spin waves of energy $\hbar\omega_{\mathbf{q}}$ we shall find

$$\sum_{\mathbf{p}} \alpha_{\mathbf{p}} [\mathcal{H}, B^\dagger_{\mathbf{p}}] = \hbar\omega_{\mathbf{q}} \sum_{\mathbf{p}} \alpha_{\mathbf{p}} B^\dagger_{\mathbf{p}}.$$

We substitute in this relation from Eq. (2.8.3) and equate the coefficients of $B^\dagger_{\mathbf{p}}$ to find

$$(\hbar\omega_{\mathbf{q}} - \mathcal{E}_{\mathbf{p}+\mathbf{q}} + \mathcal{E}_{\mathbf{p}})\alpha_{\mathbf{p}} = \sum_{\mathbf{q}'} V_{\mathbf{q}'} [(n_{\mathbf{p}-\mathbf{q}'\downarrow} - n_{\mathbf{p}-\mathbf{q}'+\mathbf{q}\uparrow})\alpha_{\mathbf{p}}$$
$$+ (n_{\mathbf{p}+\mathbf{q}+\mathbf{q}'\uparrow} - n_{\mathbf{p}+\mathbf{q}'\downarrow})\alpha_{\mathbf{p}+\mathbf{q}'}].$$

At this point we simplify the problem by assuming that $V_{\mathbf{q}'}$ can be taken as a positive constant V. We can then write

$$(\hbar\omega_{\mathbf{q}} - \mathcal{E}_{\mathbf{p}+\mathbf{q}} + \mathcal{E}_{\mathbf{p}} - VN_{\downarrow} + VN_{\uparrow})\alpha_{\mathbf{p}} = V \sum_{\mathbf{p}'} (n_{\mathbf{p}'+\mathbf{q}\uparrow} - n_{\mathbf{p}'\downarrow})\alpha_{\mathbf{p}'},$$

where we have written $\mathbf{p}'$ for $\mathbf{p}+\mathbf{q}'$. This can be solved by noting that the right-hand side is independent of $\mathbf{p}$. We can thus multiply both sides of this equation by a factor

$$\frac{n_{\mathbf{p}+\mathbf{q}\uparrow} - n_{\mathbf{p}\downarrow}}{\hbar\omega_{\mathbf{q}} - \mathcal{E}_{\mathbf{p}+\mathbf{q}} + \mathcal{E}_{\mathbf{p}} - VN_{\downarrow} + VN_{\uparrow}},$$

sum over $\mathbf{p}$, and find

$$V \sum_{\mathbf{p}} \frac{n_{\mathbf{p}+\mathbf{q}\uparrow} - n_{\mathbf{p}\downarrow}}{\hbar\omega_{\mathbf{q}} - \mathcal{E}_{\mathbf{p}+\mathbf{q}} + \mathcal{E}_{\mathbf{p}} + V(N_{\uparrow} - N_{\downarrow})} = 1.$$

This equation determines $\omega_\mathbf{q}$. For small $\mathbf{q}$, which is the regime in which the random phase approximation is best justified, we can expand the left-hand side binomially to find

$$\frac{1}{N_\uparrow - N_\downarrow} \sum_\mathbf{p} (n_{\mathbf{p}+\mathbf{q}\uparrow} - n_{\mathbf{p}\downarrow}) \left[1 - \frac{\hbar\omega_\mathbf{q} - \mathcal{E}_{\mathbf{p}+\mathbf{q}} + \mathcal{E}_\mathbf{p}}{V(N_\uparrow - N_\downarrow)} \right.$$

$$\left. + \left(\frac{\hbar\omega_\mathbf{q} - \mathcal{E}_{\mathbf{p}+\mathbf{q}} + \mathcal{E}_\mathbf{p}}{V(N_\uparrow - N_\downarrow)} \right)^2 - \cdots \right] = 1.$$

If $\mathcal{E}_\mathbf{p}$ is just the free-electron energy, $\hbar^2 \mathbf{p}^2/2m$, and we retain only terms of order $\mathbf{q}^2$ or greater, then this reduces to

$$\hbar\omega_\mathbf{q} = \sum_\mathbf{p} \left(\frac{n_{\mathbf{p}+\mathbf{q}\uparrow} - n_{\mathbf{p}\downarrow}}{N_\uparrow - N_\downarrow} \right) \left[\frac{\hbar^2}{2m}(2\mathbf{p}\cdot\mathbf{q} + \mathbf{q}^2) + \frac{\hbar^4}{m^2} \frac{(\mathbf{p}\cdot\mathbf{q})^2}{V(N_\uparrow - N_\downarrow)} \right]$$

$$= \frac{\hbar^2 \mathbf{q}^2}{2m}(\alpha + \beta), \qquad (2.8.4)$$

where α is a constant of order unity and independent of V, while β is inversely proportional to V. These constants are most simply evaluated by considering the ground state of the system to consist of two filled Fermi spheres in momentum space – a large one for the down-spin electrons and a small one for the up-spin electrons. The form of the magnetic excitation spectrum is then as shown in Fig. 2.8.2, and consists of two branches. The spin waves have an energy $\hbar\omega_\mathbf{q}$ that increases as q^2 for small q, and they represent the collective motion of the system. There are, however, also the quasiparticle excitations of energy around $V(N_\downarrow - N_\uparrow)$ that are represented by the diagonal terms in Eq. (2.8.3).

This calculation presents a very much oversimplified picture of magnons in a metal, and should not be taken too seriously. It has the disadvantage, for instance, that states of the system in which spin waves are excited are eigenstates of $S_\uparrow$, the total spin in the up direction, but not of S^2, the square of the total spin angular momentum. The model suffices to show, however, the possibility of the existence of a type of magnon quite dissimilar to that introduced in the localized model of Section 1.4. It is also interesting to note that there are some materials, such as palladium, in which the interactions are not quite strong enough to lead to ferromagnetism, but are strong enough to allow spin fluctuations to be transmitted an appreciable distance before decaying. Such critically damped spin waves are known as *paramagnons*.

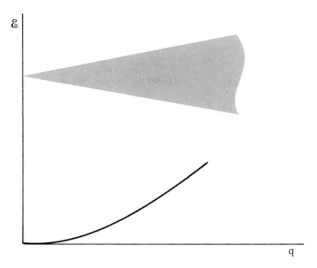

Figure 2.8.2. The spectrum of elementary excitations of the ferromagnetic electron gas. The lower branch shows the magnons while the upper band represents quasi-particle excitations.

Problems

2.1 Using the definitions of $c_\mathbf{p}$ and $c_\mathbf{p}^\dagger$ given, verify that
$$\{c_\mathbf{p}, c_{\mathbf{p}'}^\dagger\} = \delta_{\mathbf{p}\mathbf{p}'}; \quad \{c_\mathbf{p}, c_{\mathbf{p}'}\} = \{c_\mathbf{p}^\dagger, c_{\mathbf{p}'}^\dagger\} = 0.$$

2.2 Verify the statement that dW/dp as defined by Eq. (2.4.5) becomes infinite as $p \to p_F$, the radius of the Fermi surface.

2.3 In the Hartree–Fock approximation the energy of the electron gas is composed of kinetic and exchange energies. In a certain set of units the kinetic energy per electron is 2.21 rydbergs and the exchange energy -0.916 rydbergs when the gas is at unit density and zero temperature, and the up- and down-spin levels are equally populated. Estimate the density at which a magnetic phase, in which all spins are pointing up, becomes the more stable one.

2.4 The operators $\gamma_{\mathbf{k}0}$ and $\gamma_{\mathbf{k}1}$ are defined in terms of electron annihilation and creation operators by the relations
$$\gamma_{\mathbf{k}0} = u_\mathbf{k} c_{\mathbf{k}\uparrow} - v_\mathbf{k} c_{-\mathbf{k}\downarrow}^\dagger; \quad \gamma_{\mathbf{k}1} = v_\mathbf{k} c_{\mathbf{k}\uparrow}^\dagger + u_\mathbf{k} c_{-\mathbf{k}\downarrow},$$
where $c_{\mathbf{k}\uparrow}^\dagger$, for instance, creates an electron of wavenumber $\mathbf{k}$ with spin

76   Second quantization and the electron gas

up, and $u_\mathbf{k}$ and $v_\mathbf{k}$ are real constants such that $u_\mathbf{k}^2 + v_\mathbf{k}^2 = 1$. What are the various anticommutation relations of the γ and $\gamma^\dagger$?

2.5 Verify that
$$\left[\sum_\mathbf{k} n_\mathbf{k}, \mathcal{H}\right] = 0$$
for the electron gas.

2.6 Verify Eq. (2.7.1).

2.7 Verify that $[\rho_\mathbf{k}, \rho_{\mathbf{k}'}] = 0$.

2.8 Calculate the contribution to U_eff as defined in Eq. (2.7.4) of the exchange scattering processes shown in Fig. 2.7.8.

2.9 The theory of the dielectric constant of the electron gas can be generalized to include the responses to applied fields that vary with time. If a potential $U(\mathbf{r})e^{-i\omega t}$ is applied then scattering of an electron occurs by absorption of a photon of energy $\hbar\omega$, and the energy denominator of Eq. (2.7.6) is modified to give

$$\epsilon(q, \omega) = 1 - V_\mathbf{q} \sum_\mathbf{p} \frac{n_{\mathbf{p+q}} - n_\mathbf{p}}{\mathcal{E}_{\mathbf{p+q}} - \mathcal{E}_\mathbf{p} + \hbar\omega}.$$

Show that for vanishingly small q the dielectric constant itself vanishes when ω is the plasma frequency ω_p.

2.10 Evaluate the constants α and β of Eq. (2.8.4).

2.11 If the sum of coefficients $\alpha + \beta$ in Eq. (2.8.4) becomes negative, then the magnetic system will be unstable. Use your answer to Problem 2.10 to find the minimum value that $V(N_\downarrow - N_\uparrow)/\mathcal{E}_F$ must have to ensure that the magnet is stable. (Here $\mathcal{E}_F$ is the Fermi energy of the unmagnetized system.) Does your result agree qualitatively with the semiclassical argument that $N_\downarrow - N_\uparrow$ should be equal to the difference between the integrated densities of states $\mathcal{N}(\mathcal{E}_{F\downarrow}) - \mathcal{N}(\mathcal{E}_{F\uparrow})$?

2.12 Sketch a contour map of $\epsilon(q, \omega)$ as determined from the expression given in Problem 2.9. That is, estimate the sign and magnitude of $\epsilon(q, \omega)$ for various q and ω, and plot lines of constant ϵ in the q–ω plane.

2.13 Consider a system consisting of a large number N of spinless interacting fermions in a large one-dimensional box of length L. There are periodic boundary conditions. The particles interact via a delta-function potential, and so the Hamiltonian is

$$\mathcal{H} = \sum_k Ak^2 c_k^\dagger c_k + (V/2L) \sum_{k,k',q} c_{k-q}^\dagger c_{k'+q}^\dagger c_{k'} c_k$$

with A and V constants. The sums proceed over all permitted values of k, k', and q. That is, the terms with $q = 0$ are not excluded from the sum.

(a) Calculate the energy of the ground state of the noninteracting system.
(b) Calculate the energy of the ground state of the interacting system in the Hartree–Fock approximation.
(c) State in physical terms why the answer you obtained to part (b) must be an exact solution of the problem.

Chapter 3
Boson systems

3.1 Second quantization for bosons

In the formalism that we developed for dealing with fermions the number operator, n_p, played an important role, as we found that the Hamiltonian for the noninteracting system could be expressed in terms of it to give

$$\mathcal{H}_0 = \sum_{\mathbf{k}} \mathcal{E}_{\mathbf{k}} n_{\mathbf{k}}.$$

Now we turn to the consideration of systems in which we can allow more than one particle to occupy the same state. This time we shall need to define a number operator that has not only the eigenvalues 0 and 1, but all the nonnegative integers. The wavefunctions Φ that describe the noninteracting system will no longer be determinants of one-particle states, but will be symmetrized products of them, such that Φ remains unaltered by the interchange of any two particles.

In analogy with the fermion case we define annihilation and creation operators for boson systems

$$a_p = \sum_{\{n_i\}} \sqrt{n_p}\, |n_1, \ldots (n_p - 1), \ldots\rangle\langle n_1, \ldots n_p, \ldots| \qquad (3.1.1)$$

and

$$a_p^\dagger = \sum_{\{n_i\}} \sqrt{n_p}\, |n_1, \ldots n_p, \ldots\rangle\langle n_1, \ldots (n_p - 1), \ldots|. \qquad (3.1.2)$$

The summation is understood to be over all possible sets of numbers n_i, including n_p, with the sole condition that $n_p > 0$. These operators reduce or increase by one the number of particles in the pth state. The factor of $\sqrt{n_p}$ is included so that the combination $a_p^\dagger a_p$ will correspond to the number operator and have eigenvalues n_p. We do, in fact, find that with these definitions

$$a_p^\dagger a_p = \sum_{\{n_i\}} n_p |\ldots n_p, \ldots\rangle\langle \ldots n_p, \ldots |,$$

so that

$$a_p^\dagger a_p |\ldots n_p \ldots\rangle = n_p |\ldots n_p \ldots\rangle.$$

On the other hand,

$$a_p a_p^\dagger |\ldots n_p \ldots\rangle = (n_p + 1) |\ldots n_p \ldots\rangle$$

and so

$$a_p a_p^\dagger - a_p^\dagger a_p = 1.$$

We write this as

$$[a_p, a_p^\dagger] = 1,$$

which says that the commutator of a_p and $a_p^\dagger$ is equal to unity. We can further show that

$$[a_p^\dagger, a_{p'}^\dagger] = [a_p, a_{p'}] = 0$$

$$[a_p, a_{p'}^\dagger] = \delta_{pp'}.$$

These results are in terms of commutators rather than anticommutators because of the fact that we form the same wavefunction irrespective of the order in which we create the particles.

We can then write the Hamiltonian for a noninteracting system of bosons in the form

$$\mathcal{H}_0 = \sum_{\mathbf{k}} \mathcal{E}_{\mathbf{k}} a_{\mathbf{k}}^\dagger a_{\mathbf{k}} = \sum_{\mathbf{k}} \mathcal{E}_{\mathbf{k}} n_{\mathbf{k}}.$$

An important difference between the fermion and boson systems that we consider is that while for the Fermi systems the total number of particles, N, is constant in time, this is not generally so for true Bose systems, where N may be determined by thermodynamic considerations; for example, the total number of phonons present in a solid may be increased by raising the temperature of the system. On the other hand, there are also systems such as atoms of ^{4}He for which N is conserved but which behave as *pseudobosons* in that their behavior is *approximately* described by boson commutation relations. This distinction will be made clearer in Section 3.3.

3.2 The harmonic oscillator

The simplest example of a system of noninteracting bosons is provided by the case of the three-dimensional harmonic oscillator, where a particle of mass m is imagined to be in a potential $\frac{1}{2}m\omega^2\mathbf{r}^2$. The Hamiltonian is

$$\mathcal{H} = \sum_{i=1}^{3} \frac{1}{2m}[p_i^2 + (m\omega x_i)^2],$$

where the three components of momentum and position obey the commutation relations

$$[x_i, p_j] = i\hbar\,\delta_{ij}.$$

We then define

$$a_i = \sqrt{\frac{1}{2m\omega\hbar}}(m\omega x_i + ip_i); \quad a_i^\dagger = \sqrt{\frac{1}{2m\omega\hbar}}(m\omega x_i - ip_i), \qquad (3.2.1)$$

and are not in the least surprised to find that

$$[a_i, a_j^\dagger] = \delta_{ij}$$

and that

$$\mathcal{H} = \sum_i \tfrac{1}{2}\hbar\omega(a_i^\dagger a_i + a_i a_i^\dagger)$$
$$= \sum_i (n_i + \tfrac{1}{2})\hbar\omega,$$

where $n_i = a_i^\dagger a_i$.

In Section 3.1 we started our discussion of boson systems with the *assumption* that there was a number operator whose eigenvalues were the positive integers and zero, and deduced the commutation relations for the a and $a^\dagger$. What we could have done in this section is to proceed in the reverse direction,

3.2 The harmonic oscillator

starting with the commutation relations and hence deducing that the eigenvalues of n_i are the natural numbers. Our solution to the harmonic oscillator problem is then complete. The eigenfunctions are constructed from the ground state by creating excitations with the $a_i^\dagger$,

$$|\Phi\rangle = A \prod_{i=1}^{3} (a_i^\dagger)^{n_i} |0\rangle,$$

the energy eigenvalues of these states being simply $\sum_i (n_i + \frac{1}{2})\hbar\omega$. (Here A is some normalizing constant.)

As an exercise in using boson annihilation and creation operators we shall now consider a simple example – the anharmonic oscillator in one dimension. To the oscillator Hamiltonian,

$$\mathcal{H}_0 = \frac{1}{2m}[p^2 + (m\omega x)^2],$$

we add a perturbation

$$V = \beta x^3.$$

Since

$$x = \sqrt{\frac{\hbar}{2m\omega}} (a^\dagger + a), \qquad (3.2.2)$$

then

$$\beta x^3 = \beta \left(\frac{\hbar}{2m\omega}\right)^{3/2} (a^\dagger + a)^3$$

and

$$\mathcal{H} = \hbar\omega \left(a^\dagger a + \frac{1}{2}\right) + \beta \left(\frac{\hbar}{2m\omega}\right)^{3/2} (a^\dagger + a)^3.$$

We try to find the energy levels of the anharmonic oscillator by using perturbation theory. For the unperturbed state $|\Phi\rangle$, which we can write as $|n\rangle$ (since it is characterized solely by its energy $(n + \frac{1}{2})\hbar\omega$), the perturbed energy to second order in V will be

$$\mathcal{E} = \left(n + \frac{1}{2}\right)\hbar\omega + \langle n|V|n\rangle + \langle n|V \frac{1}{\mathcal{E}_n - \mathcal{H}_0} V|n\rangle. \qquad (3.2.3)$$

In this particular case the first-order energy change, $\langle n|V|n\rangle$, will be zero, since V is a product of an odd number of annihilation or creation operators, and cannot recreate the same state when it operates upon $|n\rangle$. In second order, however, we shall find terms like

$$\beta^2\left(\frac{\hbar}{2m\omega}\right)^3 \langle n|a^\dagger aa \frac{1}{\mathcal{E}_n - \mathcal{H}_0} a^\dagger aa^\dagger|n\rangle, \tag{3.2.4}$$

which will give a contribution. From the definitions (3.1.1) and (3.1.2) we have

$$a^\dagger aa^\dagger|n\rangle = (n+1)^{3/2}|n+1\rangle,$$

so that

$$(\mathcal{E}_n - \mathcal{H}_0)^{-1} a^\dagger aa^\dagger|n\rangle = [n\hbar\omega - (n+1)\hbar\omega]^{-1}(n+1)^{3/2}|n+1\rangle.$$

Thus the expression (3.2.4) is equal to

$$-\beta^2\left(\frac{\hbar}{2m\omega}\right)^3 \frac{n(n+1)^2}{\hbar\omega} = -n(n+1)^2 \frac{\hbar^2\beta^2}{8m^3\omega^4}.$$

The energy to second order will be the sum of a handful of terms similar to this, and is easily enough evaluated. Note that had we been using wavefunctions $\Phi_n(x)$ instead of the occupation-number representation $|n\rangle$ we would have had to calculate the energy shift by forming integrals of the form

$$\int \Phi_n^*(x) x^3 \Phi_{n'}(x)\, dx,$$

which would have required knowledge of the properties of integrals of Hermite polynomials. Note also that in this particular case the perturbation series must eventually diverge, because the potential βx^3 becomes indefinitely large and negative for large negative x. This does not detract from the usefulness of the theory for small x.

3.3 Quantum statistics at finite temperatures

In the last section we saw that the excited energy levels of a harmonic oscillator could be regarded as an assembly of noninteracting bosons. It is clear that for such systems the total number of bosons present is not constant,

3.3 Quantum statistics at finite temperatures

since exciting the oscillator to a higher level is equivalent to increasing the number of bosons present. While we were considering the electron gas we always had a Hamiltonian that conserved the total number of particles, so that we were then able to write the equation

$$\left[\mathcal{H}, \sum_{\mathbf{k}} n_{\mathbf{k}}\right] = 0.$$

This was because any term in $\mathcal{H}$ that contained annihilation operators always contained an equal number of creation operators. In the boson case this was not so, as it is easily verified that

$$[(a^\dagger + a)^3, a^\dagger a] \neq 0,$$

and so the interactions in the anharmonic oscillator change the total number of particles. This leads us to the consideration of systems at temperatures different from zero, for if the system of noninteracting bosons is at zero temperature, then there are no bosons present, and we have nothing left to study. At a finite temperature the system will not be in its ground state, but will have a wavefunction in which the various k-states are occupied according to the rules of statistical mechanics. This contrasts with the system of non-interacting fermions, where at zero temperature

$$|\Phi\rangle = |1, 1, \ldots 1, 0, 0, \ldots\rangle.$$

If the density of fermions is reasonably large, as in the case of electrons in a metal, the average energy per particle is large compared with thermal energies. Thermal excitation is then only of secondary importance in determining the total energy of the system. In the case of bosons, however, the thermal energy is of primary interest, and so we now turn briefly to a consideration of the form we expect $|\Phi\rangle$ to take at a finite temperature T.

A fundamental result of statistical mechanics is that the probability of a system being in a state $|i\rangle$ of energy $\mathcal{E}_i$ is proportional to $e^{-\beta \mathcal{E}_i}$ where $\beta = 1/kT$ and k is Boltzmann's constant. Thus the average value of a quantity A that has values A_i in the states $|i\rangle$ is given by

$$\bar{A} = \frac{\sum_i A_i e^{-\beta \mathcal{E}_i}}{\sum_i e^{-\beta \mathcal{E}_i}}.$$

This can be expressed as

$$\bar{A} = \frac{\text{Tr}(Ae^{-\beta\mathcal{H}})}{\text{Tr}(e^{-\beta\mathcal{H}})} \qquad (3.3.1)$$

where the operation of taking the trace is defined by

$$\text{Tr}(Ae^{-\beta\mathcal{H}}) = \sum_j \langle j|Ae^{-\beta\mathcal{H}}|j\rangle.$$

We show this by using the identity operator (2.1.6) to write

$$\begin{aligned}\text{Tr}(Ae^{-\beta\mathcal{H}}) &= \sum_{i,j} \langle j|A|i\rangle\langle i|e^{-\beta\mathcal{H}}|j\rangle \\ &= \sum_{i,j} \langle j|A|i\rangle e^{-\beta\mathcal{E}_j}\delta_{ij} \\ &= \sum_i A_i e^{-\beta\mathcal{E}_i}.\end{aligned}$$

Fortunately, a trace is always independent of the choice of basis functions, and so here we have chosen the most convenient set, $|j\rangle$, the eigenfunctions of the Hamiltonian.

When the Hamiltonian refers to a system of interacting bosons whose total number N is not conserved, the operation of taking the trace must include summing over all possible values of N, as these are all valid states of the system. But while it is the case that for true bosons N may not be constant, there are some systems in which the total number of particles is conserved, and whose commutation relations are very similar to those for bosons. We shall see in the theory of superconductivity that an assembly of bound pairs of electrons has some similarity to a Bose gas. If we define the operator that creates an electron in the state $\mathbf{k}\uparrow$ and one in the state $-\mathbf{k}\downarrow$ by

$$b_\mathbf{k}^\dagger = c_{\mathbf{k}\uparrow}^\dagger c_{-\mathbf{k}\downarrow}^\dagger$$

then we can show that

$$[b_\mathbf{k}, b_{\mathbf{k}'}^\dagger] = \delta_{\mathbf{k}\mathbf{k}'}(1 - n_{\mathbf{k}\uparrow} - n_{-\mathbf{k}\downarrow}).$$

Note that it is the *commutator*, and not the anticommutator, that vanishes when $\mathbf{k} \neq \mathbf{k}'$. When $\mathbf{k} = \mathbf{k}'$, the commutator is not the same as when the b are boson operators, and so in the case of superconductivity it is necessary to use these special commutation relations for electron pairs. An even better example in which it is suitable to approximate the operators for composite particles by boson operators is the case of liquid ^{4}He. The isotope of helium of atomic mass 4 is composed of an even number of fermions, has no net spin or magnetic moment, and – what is most important – is very tightly bound, so that the wavefunctions are well localized. This means that operators for atoms at different locations will commute. If it is valid to treat helium atoms as noninteracting bosons, then we should expect that at zero temperature all the atoms are in the state $\mathbf{k} = 0$, and we should have

$$|\Phi\rangle = |N, 0, 0, \ldots\rangle.$$

At finite temperatures we should expect to use Eq. (3.3.1) to predict the various properties of the system. There is, however, a difficulty involved in this in that we must choose a zero of energy for the single-particle states. That is, the Hamiltonian,

$$\mathcal{H} = \sum_{\mathbf{k}} \mathcal{E}_{\mathbf{k}} n_{\mathbf{k}} + \mathcal{H}_{\text{interactions}},$$

could equally well be written as

$$\hat{\mathcal{H}} = \sum_{\mathbf{k}} (\mathcal{E}_{\mathbf{k}} - \mu) n_{\mathbf{k}} + \mathcal{H}_{\text{interactions}} = \mathcal{H} - N\mu,$$

as we have no obvious way of deciding the absolute energy of a single-particle state. This was not a problem when N was not conserved, for then we knew exactly the energy $\mathcal{E}_{\mathbf{k}}$ required to create a phonon or a magnon, and we could take μ as being zero. The approach we take, which corresponds to the concept of the *grand canonical ensemble* in statistical mechanics, is to choose μ in such a way that Eq. (3.3.1) predicts the correct result, N_1, for the average value of the operator N when the trace includes a summation over all possible N. That is, we choose μ such that

$$N_1 = \frac{\operatorname{Tr} N e^{-\beta(\mathcal{H} - N\mu)}}{\operatorname{Tr} e^{-\beta(\mathcal{H} - N\mu)}}$$

$$= \frac{1}{\beta} \frac{\partial}{\partial \mu} \ln \operatorname{Tr} e^{-\beta(\mathcal{H} - N\mu)}.$$

The energy μ is known as the *chemical potential*.

We are now in a position to calculate explicitly various temperature-dependent properties of a system of independent bosons or fermions. For $\bar{n}_p$, the average number of bosons in the pth single-particle state, for example, we find

$$\bar{n}_p = \frac{\text{Tr } a_p^\dagger a_p e^{-\beta(\mathcal{H}-N\mu)}}{\text{Tr } e^{-\beta(\mathcal{H}-N\mu)}}$$

$$= \frac{\text{Tr } a_p e^{-\beta(\mathcal{H}-N\mu)} a_p^\dagger}{\text{Tr } e^{-\beta(\mathcal{H}-N\mu)}}.$$

We are allowed to permute cyclically the product of which we are taking the trace because the exponential makes the sum converge. Now

$$e^{-\beta(\mathcal{H}-N\mu)} a_p^\dagger = a_p^\dagger e^{-\beta(\mathcal{H}-N\mu)} e^{-\beta(\mathcal{E}_p-\mu)},$$

since $a_p^\dagger$ increases N by one, and alters the eigenvalues of $\mathcal{H}$ by an amount $\mathcal{E}_p$. Thus

$$\bar{n}_p = \frac{\text{Tr } a_p a_p^\dagger e^{-\beta(\mathcal{H}-N\mu)} e^{-\beta(\mathcal{E}_p-\mu)}}{\text{Tr } e^{-\beta(\mathcal{H}-N\mu)}}$$

$$= \frac{\text{Tr } (1 + a_p^\dagger a_p) e^{-\beta(\mathcal{H}-N\mu)} e^{-\beta(\mathcal{E}_p-\mu)}}{\text{Tr } e^{-\beta(\mathcal{H}-N\mu)}}$$

$$= e^{-\beta(\mathcal{E}_p-\mu)}(1 + \bar{n}_p),$$

from which

$$\bar{n}_p = \frac{1}{e^{\beta(\mathcal{E}_p-\mu)} - 1}. \qquad (3.3.2)$$

For fermions the anticommutator leads to a positive sign, giving

$$\bar{n}_{p(\text{fermions})} = \frac{1}{e^{\beta(\mathcal{E}_p+\mu)} + 1}. \qquad (3.3.3)$$

These functions $\bar{n}_p$ give the average value taken by the operator n_p.

The form of the boson distribution function, (3.3.2), has an interesting consequence for a system of independent particles in which the total number N is conserved, as in the case of ^{4}He. We have

$$N = \sum_p \bar{n}_p = \sum_p \frac{1}{e^{\beta(\mathcal{E}_p-\mu)} - 1}.$$

3.3 Quantum statistics at finite temperatures

We know that $\mu \leq 0$, because if it were not then some $\bar{n}_p$ would be negative, which would be nonsense. Thus

$$N \leq \sum_p \frac{1}{e^{\beta \mathcal{E}_p} - 1}. \tag{3.3.4}$$

Because we know (from Section 2.1) that the density of states in wave-number space is $\Omega/8\pi^3$ we can change the sum to an integral and write

$$N \leq \bar{n}_0 + \frac{\Omega}{8\pi^3} \int_{0+}^{\infty} \frac{4\pi k^2 \, dk}{\exp(\beta \hbar^2 k^2 / 2m) - 1}.$$

We consider $\bar{n}_k$ for $k = 0$ separately, since this term is not defined in (3.3.4). The integral is well behaved, and gives a number which we shall call $N_0(T)$. As T tends to zero, $N_0(T)$, which represents an upper bound to the number of particles in excited states, becomes indefinitely small, as illustrated in Fig. 3.3.1. When $N_0(T) < N$ it follows from the inequality that all the rest of the particles must be in the state for which $k = 0$. Thus there is a temperature T_c, defined by $N_0(T_c) = N$, below which the zero-energy state is occupied by a macroscopic number of particles. This phenomenon is known as the *Bose–Einstein condensation*, and is remarkable in being a phase transition that occurs in the absence of interparticle forces.

We might expect the introduction of forces between particles to destroy the transition to a condensed phase, but this is not the case. Bose–Einstein

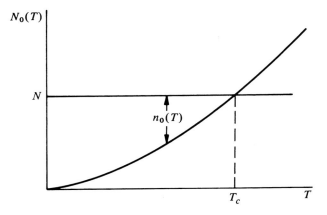

Figure 3.3.1. This curve represents the greatest possible number of particles that can be in excited states. When it falls below N, the actual number of particles present, we know that a macroscopic number of particles, $n_0(T)$, must be in the state for which $k = 0$.

condensation is observed in a wide variety of systems, including not only ^{4}He but also spin-polarized atomic hydrogen and gases of alkali atoms like ^{23}Na, which consist of an even number of fermions. We must now develop the formalism with which to attack this problem.

3.4 Bogoliubov's theory of helium

As early as 1946 Bogoliubov developed a theory of a system of interacting bosons of the number-conserving kind by making use of the fact that n_0 may be very large. He was able in this way to provide an insight into how a weak interaction may totally change the nature of the excitation spectrum of a system and also increased our understanding of the phenomenon of superfluidity.

We treat liquid ^{4}He as a system of interacting bosons. The Hamiltonian will look just like that for the spinless electron gas, except that we shall have to replace every c and $c^\dagger$ by an a or an $a^\dagger$, and, of course, the form of the interaction will be different. We have

$$\mathcal{H} = \sum_{\mathbf{k}} \mathcal{E}_{\mathbf{k}} a_{\mathbf{k}}^\dagger a_{\mathbf{k}} + \frac{1}{2} \sum_{\mathbf{k},\mathbf{k}',\mathbf{q}} V_{\mathbf{q}} a_{\mathbf{k}-\mathbf{q}}^\dagger a_{\mathbf{k}'+\mathbf{q}}^\dagger a_{\mathbf{k}'} a_{\mathbf{k}}.$$

The single-particle energies $\mathcal{E}_{\mathbf{k}}$ will be just

$$\mathcal{E}_{\mathbf{k}} = \frac{\hbar^2 \mathbf{k}^2}{2M},$$

with M being the mass of the helium atom, and $V_{\mathbf{q}}$ the Fourier transform of a short-range potential.

We can immediately arrive at an expression for the energy of this system at zero temperature by employing the same procedure that we used in deriving the Hartree–Fock approximation for the electron gas. That is, we write

$$\mathcal{E}_H = \langle \Phi | \mathcal{H} | \Phi \rangle,$$

where Φ is the wavefunction in the absence of interactions. Because all N particles are in the state having $k=0$ we simply find

$$\begin{aligned}\mathcal{E}_H &= N\mathcal{E}_0 + \tfrac{1}{2} V_0 \langle \Phi | a_0^\dagger a_0^\dagger a_0 a_0 | \Phi \rangle \\ &= N\mathcal{E}_0 + \tfrac{1}{2} N(N-1) V_0.\end{aligned}$$

3.4 Bogoliubov's theory of helium

We recall from the definition (2.3.14) that V_0 is inversely proportional to the volume Ω. If we restore this factor by writing $V_0 = V_0'/\Omega$ and approximate $N - 1$ by N we find

$$\mathcal{E}_H \simeq N\mathcal{E}_0 + \frac{N^2 V_0'}{2\Omega}.$$

Note that there are no exchange terms present in this approximation, as only one state is occupied.

We can now predict from the dependence of the energy on the volume that this system will support longitudinal sound waves of small wavenumber. For a classical fluid the velocity of sound is given by

$$v = \sqrt{\frac{R}{\rho}},$$

where R is the bulk modulus, $-\Omega(\partial P/\partial \Omega)_N$, and ρ is the mass density. If we interpret the pressure P as $-(\partial \mathcal{E}_H/\partial \Omega)_N$ we find that

$$v = \sqrt{\frac{\Omega \frac{\partial^2 \mathcal{E}_H}{\partial \Omega^2}}{\rho}} = \sqrt{\frac{NV_0'}{M\Omega}}.$$

Thus for a system of unit volume we expect there to be boson excitations for small $\mathbf{k}$ having energies $\hbar \omega_\mathbf{k}$ such that

$$\omega_\mathbf{k}^2 = \frac{NV_0 \mathbf{k}^2}{M}.$$

Bogoliubov's method shows how these excitations arise as a modification of the single-particle excitation spectrum.

In looking for the ground-state solution of this problem we invoke the fact that in the noninteracting system all the particles are in the state for which $k = 0$. We make the assumption that even in the interacting system *there is still a macroscopic number of particles in the zero-momentum state*. The number still in the zero-momentum state is the expectation value of $a_0^\dagger a_0$, which we write as N_0. Because we expect N_0 to be large we treat it as a number rather than an operator, and similarly take $a_0^\dagger a_0$ to be equal to N_0. In fact, the operator $a_0^\dagger a_0^\dagger$ operating on a wavefunction with N_0 particles in the $k = 0$ state would give $\sqrt{(N_0 + 1)(N_0 + 2)}$ times the state with $N_0 + 2$ particles, but since $2 \ll N_0$, we ignore this difference. We next rewrite the Hamiltonian dropping all terms that are of order less than N_0. Provided $V_\mathbf{q}$ is equal to $V_{-\mathbf{q}}$ this leaves us

$$\mathcal{H} \simeq \sum_\mathbf{k} \mathcal{E}_\mathbf{k} a_\mathbf{k}^\dagger a_\mathbf{k} + \tfrac{1}{2} N_0^2 V_0 + N_0 V_0 \sum_\mathbf{k} a_\mathbf{k}^\dagger a_\mathbf{k} + N_0 \sum_{\mathbf{k}'} V_{\mathbf{k}'} a_{\mathbf{k}'}^\dagger a_{\mathbf{k}'}$$
$$+ \tfrac{1}{2} N_0 \sum_\mathbf{q} V_\mathbf{q} (a_\mathbf{q} a_{-\mathbf{q}} + a_{-\mathbf{q}}^\dagger a_\mathbf{q}^\dagger),$$

where the sums exclude the zero term. We then put

$$N_0 + \sum_\mathbf{k} a_\mathbf{k}^\dagger a_\mathbf{k} = N; \quad N_0 V_\mathbf{k} = \eta_\mathbf{k}; \quad \mathcal{E}_\mathbf{k} + \eta_\mathbf{k} = \hbar\Omega_\mathbf{k},$$

and make the assumption that $N - N_0 \ll N_0$ (an assumption that is of dubious validity for real liquid helium). We then only make errors in terms of order $(N - N_0)/N_0$ if we write

$$\mathcal{H} = \tfrac{1}{2} N^2 V_0 + \sum_\mathbf{k} \hbar\Omega_\mathbf{k} a_\mathbf{k}^\dagger a_\mathbf{k} + \tfrac{1}{2} \sum_\mathbf{k} \eta_\mathbf{k} (a_\mathbf{k} a_{-\mathbf{k}} + a_\mathbf{k}^\dagger a_{-\mathbf{k}}^\dagger). \quad (3.4.1)$$

While the first term is a constant, and the second is an old friend, the third term is an awkward one. In perturbation theory it leads to divergences, the pictorial representations of which are aptly known as "dangerous diagrams." The major advance we have made, however, is to reduce our original Hamiltonian, which contained interactions represented by a product of four operators, to a *quadratic form*, in which only products of two operators are present. It is then in principle always possible to diagonalize the Hamiltonian.

The trick that Bogoliubov used to get rid of the off-diagonal terms $a_k a_{-k}$ and $a_k^\dagger a_{-k}^\dagger$ was to define a new set of operators. He wrote

$$\alpha_\mathbf{k} = (\cosh\theta_\mathbf{k}) a_\mathbf{k} - (\sinh\theta_\mathbf{k}) a_{-\mathbf{k}}^\dagger,$$

where the $\theta_\mathbf{k}$ are left arbitrary for the time being. One can show that the α obey the same commutation relations as the a,

$$[\alpha_\mathbf{k}, \alpha_{\mathbf{k}'}^\dagger] = \delta_{\mathbf{k}\mathbf{k}'}.$$

3.4 Bogoliubov's theory of helium

Now suppose we had a Hamiltonian

$$\mathcal{H} = \sum_{\mathbf{k}} \hbar \omega_{\mathbf{k}} \alpha_{\mathbf{k}}^{\dagger} \alpha_{\mathbf{k}}.$$

This would pose no difficulties; the energies would just be $\sum_{\mathbf{k}} n_{\mathbf{k}} \hbar \omega_{\mathbf{k}}$. Our approach now is to write out $\alpha_{\mathbf{k}}^{\dagger} \alpha_{\mathbf{k}}$ in terms of the a's and see if we can choose $\omega_{\mathbf{k}}$ and $\theta_{\mathbf{k}}$ in such a way as to make it equal to the **k**th component of our approximate Hamiltonian, (3.4.1). Substituting, we have

$$\alpha_{\mathbf{k}}^{\dagger} \alpha_{\mathbf{k}} = [(\cosh \theta_{\mathbf{k}}) a_{\mathbf{k}}^{\dagger} - (\sinh \theta_{\mathbf{k}}) a_{-\mathbf{k}}][(\cosh \theta_{\mathbf{k}}) a_{\mathbf{k}} - (\sinh \theta_{\mathbf{k}}) a_{-\mathbf{k}}^{\dagger}]$$

$$= (\cosh^2 \theta_{\mathbf{k}}) a_{\mathbf{k}}^{\dagger} a_{\mathbf{k}} + (\sinh^2 \theta_{\mathbf{k}}) a_{-\mathbf{k}} a_{-\mathbf{k}}^{\dagger} - (\cosh \theta_{\mathbf{k}} \sinh \theta_{\mathbf{k}})$$

$$\times (a_{\mathbf{k}}^{\dagger} a_{-\mathbf{k}}^{\dagger} + a_{-\mathbf{k}} a_{\mathbf{k}}).$$

Then, if $\omega_{\mathbf{k}} = \omega_{-\mathbf{k}}$ and $\theta_{\mathbf{k}} = \theta_{-\mathbf{k}}$,

$$\sum_{\mathbf{k}} \hbar \omega_{\mathbf{k}} \alpha_{\mathbf{k}}^{\dagger} \alpha_{\mathbf{k}} = \sum_{\mathbf{k}} \hbar \omega_{\mathbf{k}} (\cosh 2\theta_{\mathbf{k}}) a_{\mathbf{k}}^{\dagger} a_{\mathbf{k}} + \sum_{\mathbf{k}} \hbar \omega_{\mathbf{k}} \sinh^2 \theta_{\mathbf{k}}$$

$$- \tfrac{1}{2} \sum_{\mathbf{k}} \hbar \omega_{\mathbf{k}} (\sinh 2\theta_{\mathbf{k}})(a_{\mathbf{k}} a_{-\mathbf{k}} + a_{\mathbf{k}}^{\dagger} a_{-\mathbf{k}}^{\dagger}).$$

This is identical to (3.4.1) except for a constant if we choose ω and θ such that

$$\omega_{\mathbf{k}} \cosh 2\theta_{\mathbf{k}} = \Omega_{\mathbf{k}}; \quad \hbar \omega_{\mathbf{k}} \sinh 2\theta_{\mathbf{k}} = -\eta_{\mathbf{k}}.$$

Then

$$\hbar^2 \omega_{\mathbf{k}}^2 = \hbar^2 \Omega_{\mathbf{k}}^2 - \eta_{\mathbf{k}}^2$$

and

$$\hbar \omega_{\mathbf{k}} = [(\mathcal{E}_{\mathbf{k}} + N_0 V_{\mathbf{k}})^2 - (N_0 V_{\mathbf{k}})^2]^{1/2}.$$

Thus Bogoliubov's transformation from the a's to the α's has diagonalized the Hamiltonian. Within the approximation that N_0 is large compared with everything else in sight we can say that the excitations of the system above its ground state are equivalent to Bose particles of energy $\hbar \omega_{\mathbf{k}}$.

The interesting thing about these excitations is the way the energy varies with **k** for small **k**. We can write

$$\hbar \omega_{\mathbf{k}} = \sqrt{\mathcal{E}_{\mathbf{k}}^2 + 2\mathcal{E}_{\mathbf{k}} N_0 V_{\mathbf{k}}},$$

and since for small enough **k** we shall have that $\mathcal{E}_\mathbf{k}^2$, which varies as k^4, will be small compared with $\mathcal{E}_\mathbf{k} N_0 V_\mathbf{k}$, we shall find

$$\omega_\mathbf{k} \simeq \sqrt{\frac{NV_\mathbf{k}}{M}} k.$$

That is, the excitations will look more like phonons than like free particles, and will have the dispersion law predicted from the elementary arguments used at the beginning of this section. When **k** becomes large, so that $\mathcal{E}_\mathbf{k} \gg N_0 V_\mathbf{k}$, then the excitations will once again be like particles. The detailed shape of the graph of ω against **k** will depend upon the form of $V_\mathbf{k}$. If we choose a form of $V_\mathbf{k}$ like Fig. 3.4.1 then we should find that ω behaves as in Fig. 3.4.2, starting off with a finite gradient, but then dipping down again to a minimum at some value of **k**.

This is the form of the dispersion relation for liquid ^{4}He that is found experimentally, and is in accord with the superfluid properties of this substance at low temperatures. We consider a heavy particle of mass M_0 projected into a container of liquid helium at zero temperature, and investigate the mechanism by which the particle is slowed down. Since its energy $\mathcal{E}_M$ is $p^2/2M_0$, a heavy particle has a lot of momentum but not much energy, as shown in Fig. 3.4.3. If the particle is slowed down by the helium it will only give up a small amount of energy even though it loses a considerable amount of momentum. Now if the helium is in its ground state, then all the excitations available in Fig. 3.4.2 require a lot of energy for each bit of momentum they provide. The massive particle is not capable of providing this energy, and hence cannot cause an excitation and will experience no viscous force. It is only when

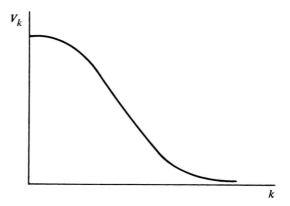

Figure 3.4.1. One possible form that the effective interaction between helium atoms might take.

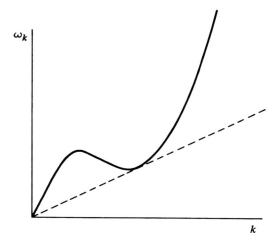

Figure 3.4.2. An interaction of the form shown in Fig. 3.4.1 would lead to a dispersion curve with a minimum as shown here.

Figure 3.4.3. For a given momentum p a heavy particle has very little energy.

the particle has such a large momentum, p_c, that its velocity is equal to the gradient of the dotted line in Fig. 3.4.2 that excitations will be caused.

In fact liquid ^{4}He at low temperatures is found to have superfluid properties for motions below a certain critical velocity, but the magnitude of this velocity is only about $1\,\mathrm{cm\,s^{-1}}$, rather than the $10^4\,\mathrm{cm\,s^{-1}}$ predicted by this theory. The discrepancy is accounted for by low-energy excitations in the form of vortex rings not included in the Bogoliubov theory.

3.5 Phonons in one dimension

In the case of the Bogoliubov theory of helium we started with a system containing a fixed number of Bose particles. It was the fact that the total number of particles had to be conserved that obliged the $\mathbf{k}=0$ state to contain a macroscopic number of particles, and which, in turn, gave the

94   Boson systems

system its remarkable properties. We now turn back to the situation that we encountered with the harmonic oscillator, where we start with a Hamiltonian and transform it in such a way that the excitations appear as the creation of an integral number of bosons. We return to a linear chain of interacting atoms as the first such system to consider.

Once again we let the displacements of the atoms from their equilibrium positions, l, be y_l, and abbreviate the notation $y_{l_1}, y_{l_2}, \ldots$ by writing $y_1, y_2, \ldots$. Then the Hamiltonian will be

$$\mathcal{H} = \sum_l \frac{p_l^2}{2m} + V(y_1, y_2, \ldots).$$

We expand V in a Maclaurin series to get

$$V(y_1, y_2, \ldots) = V(0, 0, \ldots) + \sum_l y_l \left[\frac{\partial}{\partial y_l} V(y_1, y_2, ,\ldots) \right]_{y_1 = y_2 = \cdots = 0}$$

$$+ \frac{1}{2!} \sum_{l,l'} y_l y_{l'} \left[\frac{\partial^2}{\partial y_l \partial y_{l'}} V(y_1, y_2, ,\ldots) \right]_{y_1 = y_2 = \cdots = 0}$$

$$+ \frac{1}{3!} \sum_{l,l',l''} y_l y_{l'} y_{l''} \left[\frac{\partial^3}{\partial y_l \partial y_{l'} \partial y_{l''}} V(y_1, y_2, ,\ldots) \right]_{y_1 = y_2 \cdots = 0}$$

$$+ \text{ higher terms.} \tag{3.5.1}$$

The first term on the right-hand side may be eliminated by suitable choice of the zero of energy, and all the terms in the summation forming the second term must be zero by virtue of the definition of $y = 0$ as the equilibrium positions of the atoms. Thus the first set of terms we need to consider are the set

$$\sum_{l,l'} y_l y_{l'} \frac{\partial^2 V}{\partial y_l \partial y_{l'}}.$$

We could write this double summation in matrix notation. If we abbreviate $\partial^2 V / \partial y_l \partial y_{l'}$ by $V_{ll'}$ then we can represent the double sum as

$$(y_1, y_2, \ldots) \begin{pmatrix} V_{11} & V_{12} & \cdots \\ V_{21} & & \\ \vdots & & \end{pmatrix} \begin{pmatrix} y_1 \\ y_2 \\ \vdots \end{pmatrix}.$$

3.5 Phonons in one dimension

Now we can always diagonalize a finite matrix like $V_{ll'}$. That is, we can find some matrix T such that TVT^{-1} is diagonal. If T has elements T_{ql} this means

$$TVT^{-1} = \sum_{l,l'} T_{ql} V_{ll'} (T^{-1})_{l'q'} = V_q \delta_{qq'}, \qquad (3.5.2)$$

where the V_q are a set of numbers defined by T and V. Then

$$\sum_{l,l'} y_l V_{ll'} y_{l'} = \sum_{\substack{l,l',l'',l''' \\ q,q'}} y_l (T^{-1})_{lq} T_{ql''} V_{l''l'''} (T^{-1})_{l'''q'} T_{q'l'} y_{l'}$$

$$= \sum_{l,l',q,q'} y_l (T^{-1})_{lq} V_q \delta_{qq'} T_{q'l'} y_{l'} \qquad (3.5.3)$$

$$= \sum_q y_q \tilde{y}_q V_q,$$

where

$$y_q = \sum_l y_l (T^{-1})_{lq}$$

and

$$\tilde{y}_q = \sum_l T_{ql} y_l.$$

Because y_l is a physical observable it must be its own conjugate, and $y_l = y_l^\dagger$. If we can choose T such that $(T^{-1})_{lq} = T_{ql}^*$ we should have that $\tilde{y}_q = y_q^\dagger$, and we could write the potential energy as $\frac{1}{2} \sum_q y_q y_q^\dagger V_q$.

What we have done here is really no more complicated than the elementary approach of Section 1.2 – we have changed from the particle coordinates y_l to the collective coordinates y_q. We can similarly define collective momenta, p_q, using the inverse transformation:

$$p_q = \sum_l T_{ql} p_l.$$

This follows from the fact that

$$p_l = -i\hbar \frac{\partial}{\partial y_l} = -i\hbar \sum_q \frac{\partial y_q}{\partial y_l} \frac{\partial}{\partial y_q} = -i\hbar \sum_q (T^{-1})_{lq} \frac{\partial}{\partial y_q}.$$

Thus on multiplication by T we have

$$-i\hbar \frac{\partial}{\partial y_q} = \sum_l T_{ql} p_l.$$

The kinetic energy remains diagonal in the new coordinates, since

$$\frac{1}{2M} \sum_l p_l^2 = \frac{1}{2M} \sum_{l,q,q'} p_q (T^{-1})_{lq} p_{q'}^\dagger (T^{-1})^*_{lq'}$$

$$= \frac{1}{2M} \sum_{l,q,q'} p_q T^*_{ql} (T^{-1})^*_{lq'} p_{q'}^\dagger$$

$$= \frac{1}{2M} \sum_q p_q p_q^\dagger.$$

Thus if we ignore all terms in the Hamiltonian that are of order y^3 or higher (this is known as the *harmonic approximation*) we can write

$$\mathcal{H} = \sum_q \left(\frac{1}{2M} p_q p_q^\dagger + \frac{1}{2} M \omega_q^2 y_q y_q^\dagger \right),$$

where $M\omega_q^2 = V_q$. Because $p_q = -i\hbar \partial / \partial y_q$ the commutation relations for the collective coordinates are similar to those for particles, and we have

$$[y_q, p_{q'}] = i\hbar \delta_{qq'}.$$

We then see that by defining operators

$$a_q = \sqrt{\frac{1}{2M\hbar\omega_q}} (M\omega_q y_q + i p_q^\dagger) \tag{3.5.4}$$

$$a_q^\dagger = \sqrt{\frac{1}{2M\hbar\omega_q}} (M\omega_q y_q^\dagger - i p_q), \tag{3.5.5}$$

which are a simple generalization of (3.2.1), we can write

$$\mathcal{H} = \sum_q \hbar\omega_q \left(a_q^\dagger a_q + \tfrac{1}{2} \right). \tag{3.5.6}$$

Thus to know all about the excitation spectrum of the linear chain we simply need to know the transformation matrix T_{ql}.

The matrix we need is, of course, the one that will make the y_q the collective coordinates for phonons. We thus need to have T_{ql} proportional to e^{iql} and so we write

$$y_q = N^{-1/2} \sum_l e^{-iql} y_l; \quad P_q = N^{-1/2} \sum_l e^{iql} p_l, \qquad (3.5.7)$$

where N is the total number of atoms in the chain. In order to avoid difficulties with the ends of the chain we adopt the device of introducing periodic boundary conditions, as was done in Section 2.1 for the electron wavefunctions. That is, we specify that

$$y_{l+Na} \equiv y_l,$$

with a once again the distance between atoms, so that the ends of the chain are effectively joined. This restricts the possible values of q, since from expression (3.5.7) we must have

$$e^{iql} = e^{iq(l+Na)}$$

if the y_q are to be uniquely defined. We then have

$$q = \frac{2\pi n}{Na},$$

where n is an integer. It then follows that

$$y_q \equiv y_{q+g}; \quad P_q \equiv P_{q+g}$$

where $g = 2\pi/a$, which shows that there are only N distinct collective coordinates. The inverse transformations are found to be

$$y_l = N^{-1/2} \sum_q e^{iql} y_q; \quad p_l = N^{-1/2} \sum_q e^{-iql} p_q, \qquad (3.5.8)$$

where the summations proceed over all N distinct values of q. We note that $y_q^\dagger = y_{-q}$ and $p_q^\dagger = p_{-q}$, so that since $\omega_q = \omega_{-q}$ we can, by making use of expressions (3.5.4) and (3.5.5), write

$$y_q = \sqrt{\frac{\hbar}{2M\omega_q}} (a^\dagger_{-q} + a_q) \qquad (3.5.9)$$

$$p_q = i\sqrt{\frac{M\hbar\omega_q}{2}} (a^\dagger_q - a_{-q}). \qquad (3.5.10)$$

These relations allow us to write any operator in terms of phonon annihilation and creation operators.

The frequencies ω_q that appear in the Hamiltonian (3.5.6) are given by

$$\omega_q = \sqrt{\frac{V_q}{M}}$$

and

$$V_q = \sum_{l,l'} T_{ql} V_{ll'} (T^{-1})_{l'q}$$

$$= N^{-1} \sum_{l,l'} e^{iql} V_{ll'} e^{-iql'}.$$

As $V_{ll'}$ is, by the translational invariance of the system, a function only of $(l - l')$, we have

$$V_q = \sum_L e^{iqL} V_L,$$

where we have written L for $l - l'$.

This result for the frequencies is identical to that which we obtained by classical methods in Section 1.2. For the particular case where there were interactions only between nearest neighbors we had

$$V = \sum_l \tfrac{1}{2} K(y_l - y_{l+a})^2 = \sum_l K(y_l^2 - y_l y_{l+a}),$$

so that

$$V_{ll'} = 2K \quad \text{if} \quad l = l'$$
$$= -K \quad \text{if} \quad l = l' \pm a$$
$$= 0 \quad \text{otherwise.}$$

Then
$$V_q = 2K - K(e^{iqa} + e^{-iqa})$$
$$= 4K \sin^2 \frac{qa}{2},$$

and
$$\omega_p = 2\sqrt{\frac{K}{M}} \sin \frac{qa}{2}$$

as before.

3.6 Phonons in three dimensions

The theory of phonons in three-dimensional crystals is not very much more difficult in principle than the one-dimensional theory. The basic results that we found merely become decorated with a wealth of subscripts and superscripts. We first consider the simplest type of crystal, known as a *Bravais lattice*, in which the vector distance l between any two atoms can always be written in the form

$$l = n_1 l_1 + n_2 l_2 + n_3 l_3.$$

Here the n are integers and the l_i are the *basis vectors* of the lattice. It is convenient to define a set of vectors g such that $e^{i g \cdot l} = 1$ for all l. These form the *reciprocal lattice*. We can calculate the useful property that sums of the form $\sum_l e^{i q \cdot l}$ vanish unless q is equal to some g, in which case the sum is equal to N, the total number of atoms. Thus we can define a function $\Delta(q)$ by the equation

$$\sum_l e^{i\mathbf{q} \cdot l} = N \sum_g \delta_{\mathbf{qg}} \equiv N\Delta(\mathbf{q}).$$

The Hamiltonian of a lattice of atoms interacting via simple potentials can be written in analogy with Eq. (3.5.1) as

$$\mathcal{H} = \sum_{l,i} \frac{1}{2M} (p_l^i)^2$$
$$+ \frac{1}{2!} \sum_{l,l',i,j} y_l^i y_{l'}^j V_{ll'}^{ij} + \frac{1}{3!} \sum_{l,l',l'',i,j,k} y_l^i y_{l'}^j y_{l''}^k V_{ll'l''}^{ijk} + \cdots, \qquad (3.6.1)$$

where p_l^i and y_l^i represent the ith Cartesian component of the momentum and displacement, respectively, of the atom whose equilibrium position is l. The tensor quantities $V_{ll'}^{ij}$, $V_{ll'l''}^{ijk}$, etc., are the derivatives of the potential energy with respect to the displacements as before. In the harmonic approximation only the first two terms are retained. The Hamiltonian can then be written in matrix notation as

$$\mathcal{H} = \sum_l \frac{1}{2M} (p_l^x, p_l^y, p_l^z) \begin{pmatrix} p_l^x \\ p_l^y \\ p_l^z \end{pmatrix} + \frac{1}{2} \sum_{l,l'} (y_l^x, y_l^y, y_l^z) \begin{pmatrix} V_{ll'}^{xx} & V_{ll'}^{xy} & V_{ll'}^{xz} \\ V_{ll'}^{yx} & V_{ll'}^{yy} & V_{ll'}^{yz} \\ V_{ll'}^{zx} & V_{ll'}^{zy} & V_{ll'}^{zz} \end{pmatrix} \begin{pmatrix} y_{l'}^x \\ y_{l'}^y \\ y_{l'}^z \end{pmatrix}.$$

(3.6.2)

Collective coordinates may be defined as in the one-dimensional problem. We put

$$y_\mathbf{q}^i = N^{-1/2} \sum_l e^{-i\mathbf{q}\cdot l} y_l^i; \quad p_\mathbf{q}^i = N^{-1/2} \sum_l e^{i\mathbf{q}\cdot l} p_l^i.$$

From these definitions one can see that

$$\mathbf{y}_{\mathbf{q}+\mathbf{g}} \equiv \mathbf{y}_\mathbf{q}; \quad \mathbf{p}_{\mathbf{q}+\mathbf{g}} \equiv \mathbf{p}_\mathbf{q},$$

for any reciprocal lattice vector $\mathbf{g}$, and so we only need to consider N non-equivalent values of $\mathbf{q}$. It is usually most convenient to consider those for which $|\mathbf{q}|$ is smallest, in which case we say that we take $\mathbf{q}$ as being in the *first Brillouin zone*. (We note also that since there are only $3N$ degrees of freedom in the problem it would be an embarrassment to have defined more than N coordinates $\mathbf{y}_\mathbf{q}$.) With this restriction on $\mathbf{q}$ the inverse transformations are

$$y_l^i = N^{-1/2} \sum_\mathbf{q} e^{i\mathbf{q}\cdot l} y_\mathbf{q}^i; \quad p_l^i = N^{-1/2} \sum_\mathbf{q} e^{-i\mathbf{q}\cdot l} p_\mathbf{q}^i,$$

and may be substituted into (3.6.2) to give

$$\mathcal{H} = \sum_\mathbf{q} \left\{ \frac{1}{2M} (p_\mathbf{q}^{x\dagger}, p_\mathbf{q}^{y\dagger}, p_\mathbf{q}^{z\dagger}) \begin{pmatrix} p_\mathbf{q}^x \\ p_\mathbf{q}^y \\ p_\mathbf{q}^z \end{pmatrix} \right. \\ \left. + \frac{1}{2} (y_\mathbf{q}^{x\dagger}, y_\mathbf{q}^{y\dagger}, y_\mathbf{q}^{z\dagger}) \begin{pmatrix} V_\mathbf{q}^{xx} & V_\mathbf{q}^{xy} & V_\mathbf{q}^{xz} \\ V_\mathbf{q}^{yx} & V_\mathbf{q}^{yy} & V_\mathbf{q}^{yz} \\ V_\mathbf{q}^{zx} & V_\mathbf{q}^{zy} & V_\mathbf{q}^{zz} \end{pmatrix} \begin{pmatrix} y_\mathbf{q}^x \\ y_\mathbf{q}^y \\ y_\mathbf{q}^z \end{pmatrix} \right\},$$

3.6 Phonons in three dimensions

where

$$V_{\mathbf{q}}^{ij} = \sum_{l'} e^{i\mathbf{q}\cdot(l-l')} V_{ll'}^{ij}.$$

The Hamiltonian has thus been separated into a sum of N independent terms governing the motions having different wavenumbers $\mathbf{q}$. To complete the solution we now just have to diagonalize the matrix $V_{\mathbf{q}}^{ij}$. This can be achieved merely by rotating the coordinate system. The matrix $V_{\mathbf{q}}^{ij}$ will have three mutually perpendicular eigenvectors which we can write as the unit vectors $\mathbf{s}_1$, $\mathbf{s}_2$, and $\mathbf{s}_3$, with eigenvalues $V_{\mathbf{q}}^1$, $V_{\mathbf{q}}^2$, and $V_{\mathbf{q}}^3$. Then in the coordinate system defined by the $\mathbf{s}$

$$\mathcal{H} = \sum_{\mathbf{q},s} \left\{ \frac{1}{2M} p_{\mathbf{q}}^{s\dagger} p_{\mathbf{q}}^s + \frac{1}{2} V_{\mathbf{q}}^s y_{\mathbf{q}}^{s\dagger} y_{\mathbf{q}}^s \right\}. \tag{3.6.3}$$

The three directions $\mathbf{s}$ that describe the eigenvectors of $V_{\mathbf{q}}^{ij}$ are the directions of *polarization* of the phonons, and are functions of $\mathbf{q}$. If it happens that one of the $\mathbf{s}$ is parallel to $\mathbf{q}$ we say that there can be *longitudinally polarized phonons* in the crystal. Since the $\mathbf{s}$ are mutually perpendicular it follows that there can also be *transversely polarized phonons* of the same wavenumber; for these $\mathbf{s} \cdot \mathbf{q} = 0$. It is usually only when $\mathbf{q}$ is directed along some symmetry direction of the lattice that this will occur. However, if $\mathbf{q}$ and $\mathbf{s}$ are approximately parallel it is still useful to retain the terminology of longitudinal and transverse polarizations.

The frequencies of the phonons described by expression (3.6.3) are given by

$$\omega_{\mathbf{q}s} = \sqrt{\frac{V_{\mathbf{q}}^s}{M}}.$$

We can write the Hamiltonian in the concise language of second quantization by defining annihilation and creation operators

$$\begin{aligned} a_{\mathbf{q}s} &= \frac{1}{\sqrt{2M\hbar\omega_{\mathbf{q}s}}} (M\omega_{\mathbf{q}s}\mathbf{y}_{\mathbf{q}} + i\mathbf{p}_{\mathbf{q}}^{\dagger}) \cdot \mathbf{s}_{\mathbf{q}} \\ a_{\mathbf{q}s}^{\dagger} &= \frac{1}{\sqrt{2M\hbar\omega_{\mathbf{q}s}}} (M\omega_{\mathbf{q}s}\mathbf{y}_{\mathbf{q}}^{\dagger} - i\mathbf{p}_{\mathbf{q}}) \cdot \mathbf{s}_{\mathbf{q}}. \end{aligned} \tag{3.6.4}$$

Then

$$\mathcal{H} = \sum_{\mathbf{q},s} \hbar\omega_{\mathbf{q}s}(a_{\mathbf{q}s}^{\dagger}a_{\mathbf{q}s} + \tfrac{1}{2}).$$

3.7 Acoustic and optical modes

In solving the dynamics of the Bravais lattice we diagonalized the Hamiltonian in two stages. First we transformed from the $\mathbf{y}_l$ to the $\mathbf{y}_\mathbf{q}$ and thereby reduced the double summation over l and l' to a single summation over $\mathbf{q}$. We then rotated the coordinate system for each $\mathbf{q}$ so as to eliminate terms off the diagonal of the matrix $V_\mathbf{q}^{ij}$. This completed the separation of the Hamiltonian into terms governing the motion in the $3N$ independent modes of vibration.

Not all lattices, however, are of the simple Bravais type, and this leads to a further stage that must be included in the task of diagonalization of the Hamiltonian. In a *lattice with a basis* the vectors l no longer define the equilibrium positions of atoms, but rather the positions of *identical groups of atoms*. The equilibrium position of an atom is then given by the vector $l + \mathbf{b}$, where l is a vector of the Bravais lattice, and $\mathbf{b}$ is a vector describing the position of the atom within the group (Fig. 3.7.1). There may be several different types of atom within the group, each having a different mass $M_\mathbf{b}$. The harmonic Hamiltonian then takes on the rather complicated form

$$\mathcal{H} = \sum_{l,\mathbf{b},i} \frac{1}{2M_\mathbf{b}} (p_{l,\mathbf{b}}^i)^2 + \sum_{l,\mathbf{b},l',\mathbf{b}',i,j} \frac{1}{2} y_{l\mathbf{b}}^i y_{l'\mathbf{b}'}^j V_{l\mathbf{b}l'\mathbf{b}'}^{ij}.$$

One can look upon a lattice with a basis as a set of interlocked Bravais lattices, and this suggests that we define collective coordinates for each

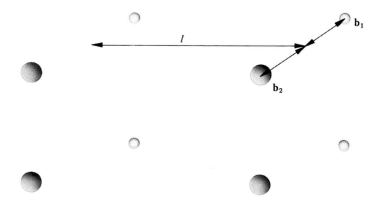

Figure 3.7.1. In a lattice with a basis the vectors l now define the position of some reference point of a group of atoms, while the vectors $\mathbf{b}$ define the positions of individual atoms of this group relative to the reference point.

3.7 Acoustic and optical modes

sublattice separately. We write for each of the n_b possible values of **b**

$$y_{\mathbf{qb}} = N^{-1/2} \sum_l y_{l\mathbf{b}} e^{-i\mathbf{q}\cdot l}$$

and

$$V^{ij}_{\mathbf{qbb'}} = \sum_{l'} e^{i\mathbf{q}\cdot(l-l')} V^{ij}_{l\mathbf{b}l'\mathbf{b'}},$$

which reduces the Hamiltonian to

$$\mathcal{H} = \sum_{\mathbf{q},\mathbf{b},i} \frac{1}{2M_\mathbf{b}} p^{i\dagger}_{\mathbf{qb}} p^i_{\mathbf{qb}} + \sum_{\mathbf{qbb'}ij} \frac{1}{2} y^i_{\mathbf{qb}} V^{ij}_{\mathbf{qbb'}} y^{j\dagger}_{\mathbf{qb'}}.$$

It is not enough now just to rotate the coordinate system to complete the diagonalization of $\mathcal{H}$; we also need to form some linear combination of the $y^i_{\mathbf{qb}}$ that will remove terms of the form $V_{\mathbf{qbb'}}$ when $\mathbf{b} \neq \mathbf{b'}$. We then find that for each **q** there are $3n_b$ distinct modes of vibration. The polarization directions of these modes are in general ill-defined since the n_b atoms that form the basis group may be moving in quite different directions. It is only the collective coordinate formed by the linear combination of the $y^i_{\mathbf{qb}}$ that has a specific direction in which it vibrates.

The $3n_b$ different modes that one finds in this way form the various *branches* of the phonon spectrum of the crystal. The lowest frequencies of vibration will be found in the three modes in which all the atoms within the basis move more or less in phase. For vanishingly small values of **q** these can be identified as the three modes of ordinary sound, for which ω is proportional to $|\mathbf{q}|$. For this reason these three are said to form the *acoustic branch* of the phonon spectrum. In the other modes the atoms within the basis move to some extent out of phase, and ω tends to a nonzero value as $|\mathbf{q}|$ tends to zero. (There is some parallel here with plasma oscillations, in which the ions and electrons also move out of phase.) Because the frequencies of these phonons may be high enough to be excited by infrared radiation, they are said to lie in the *optical branch* of the phonon spectrum (Fig. 3.7.2).

An understanding of the way in which the phonon spectrum splits into acoustic and optical branches is helped by considering the problem of the linear chain when alternate atoms have different masses. This is solved classically in many texts on solid state physics. An instructive variation of this system, to be solved quantum mechanically, is given in Problem 3.4.

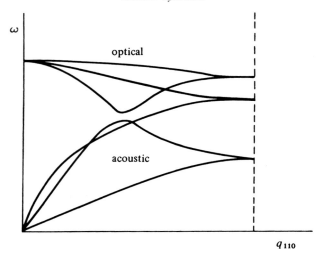

Figure 3.7.2. There are two atoms in the basis of the diamond lattice, and so this structure has a phonon dispersion curve with acoustic and optical branches.

3.8 Densities of states and the Debye model

We have found that in the harmonic approximation the lattice may be considered as a gas of independent phonons of energies $\hbar\omega_\mathbf{q}$, where now the subscript $\mathbf{q}$ is intended to specify the wavenumber and polarization of a phonon as well as the branch of the spectrum in which it lies in the case of a lattice with a basis. It is useful to define a function $D(\omega)$ to be the density of phonon states – that is, the number of states per unit frequency range near a given frequency. We write

$$D(\omega) = \sum_\mathbf{q} \delta(\omega - \omega_\mathbf{q}), \qquad (3.8.1)$$

from which it is seen that $\int_{\omega_1}^{\omega_2} D(\omega)\,d\omega$ is the number of phonon states with frequencies between ω_1 and ω_2.

This function is important in the interpretation of many experiments. There are, for instance, many processes that could occur in crystals but are forbidden because they do not conserve energy. Some of these nevertheless take place if it is possible to correct the energy imbalance by absorbing or emitting a phonon in the process. The probability of these phonon-assisted processes occurring will be proportional to $D(\omega)$ among other things. As another example we might consider the specific heat of the phonon gas, which we could calculate by finding the variation with temperature of the

3.8 Densities of states and the Debye model

average expectation value of the Hamiltonian. According to Section 3.3 we should have

$$\bar{\mathcal{E}} = \frac{\text{Tr}\,\mathcal{H}e^{-\beta\mathcal{H}}}{\text{Tr}\,e^{-\beta\mathcal{H}}} = \frac{\text{Tr}\sum_\mathbf{q} \hbar\omega_\mathbf{q}(n_\mathbf{q} + \tfrac{1}{2})e^{-\beta\mathcal{H}}}{\text{Tr}\,e^{-\beta\mathcal{H}}}$$

$$= \frac{\sum_\mathbf{q} \hbar\omega_\mathbf{q}\,\text{Tr}(n_\mathbf{q} + \tfrac{1}{2})e^{-\beta\mathcal{H}}}{\text{Tr}\,e^{-\beta\mathcal{H}}} = \sum_\mathbf{q} \hbar\omega_\mathbf{q}(\bar{n}_\mathbf{q} + \tfrac{1}{2}),$$

where

$$\bar{n}_\mathbf{q} = \frac{1}{\exp(\hbar\omega_\mathbf{q}/kT) - 1}.$$

Note that μ, the chemical potential, is zero in this case because the number of phonons is not conserved. Then

$$C_v = \frac{d\bar{\mathcal{E}}}{dT} = \frac{1}{kT^2}\sum_\mathbf{q}\left(\frac{(\hbar\omega_\mathbf{q})^2 \exp(\hbar\omega_\mathbf{q}/kT)}{[\exp(\hbar\omega_\mathbf{q}/kT) - 1]^2}\right)$$

$$= \frac{1}{kT^2}\int_0^\infty \frac{(\hbar\omega)^2 \exp(\hbar\omega/kT)}{[\exp(\hbar\omega/kT) - 1]^2} D(\omega)\,d\omega. \quad (3.8.2)$$

Thus the function $D(\omega)$ is all that we require to calculate the specific heat of a harmonic crystal.

Unfortunately, it is a tedious job to calculate $D(\omega)$ for even the simplest crystal structure and set of force constants. One would like, however, to have some model for $D(\omega)$ in order to interpret experiments. A popular and convenient model is the one first proposed by Debye in 1912, in which $D(\omega)$ is proportional to ω^2 below a certain cutoff frequency, ω_D, above which it is zero (Fig. 3.8.1). The foundation for this model comes from consideration of the form of $\omega_\mathbf{q}$ when q^{-1} is much greater than the lattice spacing. Then ω is proportional to $|\mathbf{q}|$ in the acoustic branch of the spectrum, so that the density of states in frequency is proportional to the density of states as a function of $|\mathbf{q}|$. By arguments similar to those we used in considering electron states (Section 2.1), one can show that the density of states is uniform in q-space. Thus one knows that the exact $D(\omega)$ certainly varies as ω^2 in the limit of small ω. The Debye model is an extrapolation of this behavior to all ω up to ω_D.

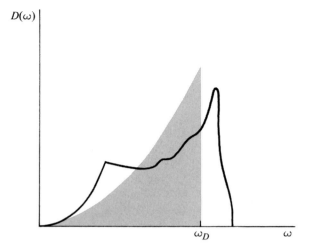

Figure 3.8.1. In the Debye model the phonon density of states $D(\omega)$, which may be a very intricate shape, is approximated by part of a parabola.

It is convenient to express the cutoff parameter in temperature units rather than frequency units. This is achieved by defining

$$\hbar\omega_D = k\Theta,$$

where Θ is known as the Debye temperature. The cutoff frequency is expected to correspond to a wavelength of the order of the lattice spacing, a, and so one has the useful approximate relation for the Debye model

$$\frac{\hbar\omega_\mathbf{q}}{kT} \simeq \frac{qa\Theta}{T}.$$

The constant of proportionality of $D(\omega)$ to ω^2 is fixed by stipulating that the total number of modes must be equal to $3N$, where N is the number of atoms in the crystal. Thus if $D(\omega) = A\omega^2$ one has

$$A\int_0^{\omega_D} \omega^2\, d\omega = 3N,$$

so that

$$D(\omega) = 9N\left(\frac{\hbar}{k\Theta}\right)^3 \omega^2 \quad (\omega \leq \omega_D). \tag{3.8.3}$$

Substitution of Eq. (3.8.3) into the specific heat formula (3.8.2) gives the well known Debye result

$$C_v = 9Nk\left(\frac{T}{\Theta}\right)^3 \int_0^{\Theta/t} \frac{x^4 e^x}{(e^x - 1)^2}\, dx,$$

from which C_v is found to vary as T^3 at very low temperatures.

In some physical problems in which the phonon spectrum only enters in a minor way, it is occasionally desirable to have an even simpler approximation for $D(\omega)$. In these cases one may use the *Einstein model*, in which it is assumed that a displaced atom experiences a restoring force caused equally by every other atom in the crystal, rather than by the near neighbors alone. Then all vibrations have the same frequency, and

$$D(\omega) = 3N\,\delta(\omega - \omega_E). \tag{3.8.4}$$

Because this model neglects all the vibrational modes of low frequency, its use is appropriate only for describing the optical modes of vibration.

3.9 Phonon interactions

While the picture of a lattice as a gas of independent phonons may be an excellent approximation with which to calculate the specific heat, there are many physical properties that it completely fails to explain. We know, for instance, that sound waves are attenuated in passing through a crystal, which shows that phonons have a finite lifetime. We also know that if we heat a substance then its elastic constants will change, or it may even undergo a martensitic transformation and change its crystal structure. The fact that the elastic constants change implies that the frequencies of the long-wavelength phonons also change. This means that $\omega_\mathbf{q}$ must be a function not only of $\mathbf{q}$, but also of all the occupation numbers of the other phonon states. To explain these phenomena we must return to the lattice Hamiltonian (3.6.1), and rescue the higher-order terms that we previously neglected.

The term of third order in the displacements was

$$\mathcal{H}_3 = \frac{1}{3!} \sum_{l,l',l'',i,j,k} y_l^i y_{l'}^j y_{l''}^k V_{l\,l'\,l''}^{ijk},$$

where for simplicity we consider a Bravais lattice, so that there is no summation over **b**. We can substitute for the y^i_l with the $y^i_\mathbf{q}$ and write

$$\sum_{l',l''} e^{i\mathbf{q}' \cdot (l'-l)} e^{i\mathbf{q}'' \cdot (l''-l)} V^{ijk}_{ll'l''} = V^{ijk}_{\mathbf{q}'\mathbf{q}''}$$

to obtain

$$\mathcal{H}_3 = \frac{1}{3!N^{3/2}} \sum_{l,\mathbf{q},\mathbf{q}',\mathbf{q}'',i,j,k} e^{i(\mathbf{q}+\mathbf{q}'+\mathbf{q}'') \cdot l} y^i_\mathbf{q} y^j_{\mathbf{q}'} y^k_{\mathbf{q}''} V^{ijk}_{\mathbf{q}'\mathbf{q}''}$$

$$= \frac{1}{3!N^{1/2}} \sum_{\substack{\mathbf{q},\mathbf{q}',\mathbf{q}'' \\ i,j,k}} y^i_\mathbf{q} y^j_{\mathbf{q}'} y^k_{\mathbf{q}''} V^{ijk}_{\mathbf{q}'\mathbf{q}''} \Delta(\mathbf{q}+\mathbf{q}'+\mathbf{q}'').$$

From (3.6.4)

$$y^i_\mathbf{q} = \sum_s \sqrt{\frac{\hbar}{2M\omega_{\mathbf{q}s}}} (a^\dagger_{-\mathbf{q}s} + a_{\mathbf{q}s}) s^i,$$

where s^i is the ith Cartesian component of the unit polarization vector **s**, and so

$$\mathcal{H}_3 = \frac{1}{3!N^{1/2}} \left(\frac{\hbar}{2M}\right)^{3/2} \sum_{\substack{\mathbf{q},\mathbf{q}',\mathbf{q}'',i,j,k \\ s,s',s''}} (\omega_{\mathbf{q}s}\omega_{\mathbf{q}'s'}\omega_{\mathbf{q}''s''})^{-1/2} s^i s'^j s''^k$$

$$\times V^{ijk}_{\mathbf{q}'\mathbf{q}''} \Delta(\mathbf{q}+\mathbf{q}'+\mathbf{q}'')(a^\dagger_{-\mathbf{q}s} + a_{\mathbf{q}s})(a^\dagger_{-\mathbf{q}'s'} + a_{\mathbf{q}'s'})(a^\dagger_{-\mathbf{q}''s''} + a_{\mathbf{q}''s''}). \quad (3.9.1)$$

The third-order term in the Hamiltonian thus appears as a sum of products of three annihilation or creation operators, and can be interpreted as representing interactions between phonons. As in the case of electron–electron interactions we can draw diagrams to represent the various components of (3.9.1), although the form of these will be different in that the number of phonons is not conserved. In the case of electron interactions the diagrams always depicted the mutual scattering of two electrons, as there were always an equal number of annihilation and creation operators in each term in the Hamiltonian. The interactions represented by expression (3.9.1), however, are of the four types shown in Fig. 3.9.1. Some terms will be products of three creation operators, and will be represented by Fig. 3.9.1(a). It is, of course, impossible to conserve energy in processes such as these, and so

3.9 Phonon interactions

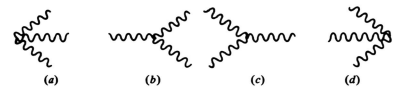

(a) (b) (c) (d)

Figure 3.9.1. The anharmonic term in the Hamiltonian that is of third order in the atomic displacements gives rise to processes involving three phonons. These are the four possible types of three-phonon interactions.

the three phonons created in this way would have to be very short-lived. They might be quickly annihilated by a process such as that shown in Fig. 3.9.1(d), which represents a product of three annihilation operators. The processes of Figs. 3.9.1(b) and 3.9.1(c) are more like scattering events, except that one of the phonons is created or destroyed in the process. Such interactions may conserve energy if the wavenumbers and polarizations are appropriate, and would then represent real transitions.

The fact that the term $\Delta(\mathbf{q} + \mathbf{q}' + \mathbf{q}'')$ appears in the expression for $\mathcal{H}_3$ implies a condition that is equivalent to the conservation of momentum in particle interactions. Because this function vanishes unless the vector $\mathbf{q} + \mathbf{q}' + \mathbf{q}''$ is zero or a reciprocal lattice vector, $\mathbf{g}$, the total wavenumber must be conserved, modulo $\mathbf{g}$. Thus in Fig. 3.9.1(a) the sum of the wavenumbers of the three created phonons must either vanish, in which case we call the interaction a normal process, or N-process, or else the total wavenumber is equal to a nonzero reciprocal lattice vector, in which case we call the interaction an Umklapp process, or U-process.

The distinction between N-processes and U-processes is to some extent artificial, in that whether a scattering is designated as N or U depends on the definition of the range of allowed values of $\mathbf{q}$. It remains a useful concept, however, in discussing phonon interactions by virtue of the fact that there *is* a well defined distinction between N- and U-processes within the framework of the Debye model. This is of importance in the theory of thermal conductivity as a consequence of a theorem first proved by Peierls. He pointed out that the heat current density is calculated from the group velocity $\partial \omega / \partial \mathbf{q}$ of the phonons as

$$\mathbf{J} = \Omega^{-1} \sum_{\mathbf{q},s} \hbar \omega_{\mathbf{q}s} \left(\frac{\partial \omega_{\mathbf{q}s}}{\partial \mathbf{q}} \right) n_{\mathbf{q}s}.$$

In the Debye model $\omega_{\mathbf{q}s} = v|\mathbf{q}|$, where the velocity of sound, v, is independent of **q** or **s**, so that

$$\mathbf{J} = \Omega^{-1} \sum_{\mathbf{q},s} \hbar v^2 \mathbf{q} n_{\mathbf{q}s}.$$

This quantity is conserved when $\mathcal{H}_3$ contains only terms describing N-processes, and so the energy current should remain constant in time. This indicates that thermal resistivity – the ability of a solid to support a steady temperature gradient – must be due to U-processes or impurities in this model.

Now that we have expressed the third-order anharmonic part, $\mathcal{H}_3$, of the Hamiltonian in terms of the $a_\mathbf{q}$ and $a_\mathbf{q}^\dagger$, it is straightforward in principle to use perturbation theory to find the change in energy of the system caused by phonon interactions. If the unperturbed lattice is in the eigenstate $|\{n_i\}\rangle$, then first-order perturbation theory gives an energy shift of

$$\langle\{n_i\}|\mathcal{H}_3|\{n_i\}\rangle,$$

which clearly vanishes because of the fact that each term in $\mathcal{H}_3$ is a product of an odd number of annihilation or creation operators. Just as in the anharmonic oscillator of Section 3.2, the perturbation cannot recreate the same state $|\{n_i\}\rangle$ that it operates upon. We must then go to second order in perturbation theory, allowing the possibility of $\mathcal{H}_3$ causing transitions into virtual intermediate states $|\{n_j\}\rangle$. The qualitative result of including phonon interactions in the Hamiltonian is to give the energy a set of terms that will not be linear in the occupation numbers, $n_\mathbf{q}$. As in the case of the interacting electron system, it is meaningless to talk about the energy of one particular phonon in an interacting system. But we can ask how the energy of the whole system changes when we remove one phonon from the unperturbed state, and to evaluate this we need to form $\partial\mathcal{E}/\partial n_{\mathbf{q}s}$. The result we find will contain a term $\hbar\omega_{\mathbf{q}s}$ arising from the differentiation of the unperturbed energy, and also a set of terms arising from differentiation of products like $n_{\mathbf{q}s}n_{\mathbf{q}'s'}$. The energy required to introduce an extra phonon into the **q**th mode is thus a function of the occupation numbers of the other modes. For a crystal in equilibrium these occupation numbers are functions of the temperature, as dictated by the Bose–Einstein distribution formula for their average value $\bar{n}_{\mathbf{q}s}$. In particular the energy required to introduce phonons of long wavelength, as in a measurement of the elastic constants of the material, will depend on the temperature. The inclusion of phonon interactions is thus necessary for the calculation of all properties at temperatures near the Debye temperature, and in particular for the thermal expansion and thermal conductivity.

3.10 Magnetic moments and spin

The classical idea of a magnetic substance is that of an assembly of atoms containing circulating electrons. By using the laws of electromagnetism one may show that the average magnetic field **h** due to a single circulating electron of mass m and charge e is of the form associated with a magnetic dipole, i.e.,

$$\mathbf{h} = \frac{3(\boldsymbol{\mu} \cdot \mathbf{r})\mathbf{r} - r^2 \boldsymbol{\mu}}{r^5}$$

at large distances **r** from the atom. Here the magnetic dipole moment, **μ**, is given by

$$\boldsymbol{\mu} = \frac{e}{2c} \mathbf{r} \times \mathbf{v}, \qquad (3.10.1)$$

averaged over a period of the particle's orbital motion. The magnetization **M** of a macroscopic sample of unit volume is then given by

$$\mathbf{M} = \sum_i \boldsymbol{\mu}_i,$$

where the sum proceeds over all contributing electrons. While the definition (3.10.1) is quite adequate for the calculation of magnetic moments of classical systems, it is not sufficiently general to be useful in the framework of quantum mechanics. We can, however, derive an expression for **μ** in terms of the Hamiltonian of the electron which may then be interpreted as defining the magnetic moment operator of a quantum-mechanical system.

To achieve this we consider the motion of the electron from the point of view of formal classical mechanics. In the presence of an externally applied magnetic field **H** an electron experiences the *Lorentz force*,

$$\mathbf{F} = \frac{e}{c} \mathbf{v} \times \mathbf{H},$$

so that in a potential $V(\mathbf{r})$ the equation of motion is

$$m\dot{\mathbf{v}} = -\nabla V + \frac{e}{c} \mathbf{v} \times \mathbf{H}. \qquad (3.10.2)$$

(Note that we are considering effects on a microscopic scale here, and do not make any distinction between the magnetic induction **B** and the magnetic field **H**. If the atom we are considering is located within a sample of magnetic material we should say that **H** is the sum of an applied field $\mathbf{H}_0$ and the dipole fields $\mathbf{h}_i$ of the other atoms. It is only when one is considering the average field in a macroscopic body that it is useful to make the distinction between **B** and **H**.) Now Lagrange's equation states that

$$\frac{d}{dt}\left(\frac{\partial \mathcal{L}}{\partial \mathbf{v}}\right) = \frac{\partial \mathcal{L}}{\partial \mathbf{r}},$$

and in order for this to be equivalent to Eq. (3.10.2) it is sufficient to write the Lagrangian

$$\mathcal{L} = \frac{1}{2}m v^2 - V + \frac{e}{c}\mathbf{v}\cdot\mathbf{A},$$

where **A** is a vector potential defined by $\mathbf{H} = \nabla \times \mathbf{A}$. The momentum **p** is then defined by

$$\mathbf{p} = \frac{\partial \mathcal{L}}{\partial \mathbf{v}} = m\mathbf{v} + \frac{e}{c}\mathbf{A}, \qquad (3.10.3)$$

and the classical Hamiltonian is

$$\mathcal{H} = \mathbf{v}\cdot\frac{\partial \mathcal{L}}{\partial \mathbf{v}} - \mathcal{L} = \frac{1}{2}m v^2 + V. \qquad (3.10.4)$$

If one then differentiates the Hamiltonian with respect to the applied magnetic field, keeping **p** and **r** constant, one finds

$$\frac{\partial \mathcal{H}}{\partial \mathbf{H}} = \frac{\partial}{\partial \mathbf{H}}\left[\frac{1}{2}m\sum_i v_i^2 + V(\mathbf{r})\right]$$

$$= m\sum_i v_i\left(\frac{\partial v_i}{\partial \mathbf{H}}\right)_{\mathbf{p},\mathbf{r}}$$

$$= -\frac{e}{c}\sum_i v_i\left(\frac{\partial A_i}{\partial \mathbf{H}}\right)_{\mathbf{r}}.$$

For a uniform field, **H**, it is convenient to write

$$\mathbf{A} = \tfrac{1}{2}\mathbf{H}\times\mathbf{r}, \qquad (3.10.5)$$

which is consistent with the definition of **A**. Then

$$\frac{\partial \mathcal{H}}{\partial \mathbf{H}} = -\frac{e}{2c}\sum_i v_i\frac{\partial(\mathbf{H}\times\mathbf{r})_i}{\partial \mathbf{H}} = -\frac{e}{2c}\mathbf{r}\times\mathbf{v}.$$

3.10 Magnetic moments and spin

By comparison with (3.10.1) we then have

$$\boldsymbol{\mu} = -\frac{\partial \mathcal{H}}{\partial \mathbf{H}}. \tag{3.10.6}$$

It is this expression that is taken as a definition of the quantum-mechanical operator that represents the magnetic moment of a system. If in particular a system is in an eigenstate of energy $\mathcal{E}_i$ then its magnetic moment is $-\partial \mathcal{E}_i / \partial \mathbf{H}$. If it is a member of an ensemble of systems at temperature T then by Eq. (3.3.1) its average magnetic moment is

$$\bar{\boldsymbol{\mu}} = -\frac{\text{Tr}[(\partial \mathcal{H}/\partial \mathbf{H})e^{-\beta \mathcal{H}}]}{\text{Tr}[e^{-\beta \mathcal{H}}]}$$

$$= -\left(\frac{\partial \mathcal{F}}{\partial \mathbf{H}}\right)_\beta, \tag{3.10.7}$$

where the Helmholtz energy, $\mathcal{F}$, is given by

$$\mathcal{F} = -\beta^{-1} \ln[\text{Tr}(e^{-\beta \mathcal{H}})].$$

To illustrate this we might consider the magnetic moment due to a single spinless electron. In the absence of a magnetic field the Hamiltonian is

$$\mathcal{H}_0 = \frac{1}{2m} \mathbf{p}^2 + V(\mathbf{r}).$$

As is seen from substituting for $\mathbf{v}$ from Eq. (3.10.3) in (3.10.4), the presence of a magnetic field modifies the Hamiltonian to

$$\mathcal{H} = \frac{1}{2m} \left(\mathbf{p} - \frac{e}{c} \mathbf{A}\right)^2 + V(\mathbf{r}), \tag{3.10.8}$$

which is equivalent to adding to $\mathcal{H}_0$ a perturbation

$$\mathcal{H}_1 = \frac{e}{2mc} \left(\frac{e}{c} \mathbf{A}^2 - \mathbf{p} \cdot \mathbf{A} - \mathbf{A} \cdot \mathbf{p}\right).$$

Use of the relation (3.10.5) then gives

$$\mathcal{H}_1 = -\frac{e}{2mc} \mathbf{r} \times \mathbf{p} \cdot \mathbf{H} + \frac{e^2}{8mc^2} (\mathbf{H} \times \mathbf{r})^2.$$

We then find that when **H** is taken in the z-direction

$$\mu_z = -\frac{\partial \mathcal{H}}{\partial H} = \frac{e}{2mc}(\mathbf{r} \times \mathbf{p})_z - \frac{e^2 H}{4mc^2}(x^2 + y^2). \tag{3.10.9}$$

This shows that in the limit of small applied fields the magnetic moment due to a spinless nonrelativistic electron is proportional to its orbital angular momentum, $\mathbf{L} = \mathbf{r} \times \mathbf{p}$, with a constant of proportionality equal to $e/2mc$.

In solids the form of the potential $V(\mathbf{r})$ that acts on an electron bound to a particular atom is frequently so lacking in symmetry that the eigenstates have no net orbital angular momentum. That is to say, the states of the electron are formed out of mixtures of equal amounts of the two degenerate wavefunctions corresponding to orbital angular momenta $\mathbf{L}$ and $-\mathbf{L}$. The only magnetic moment observed is then due to second-order effects such as the second term in expression (3.10.9), which, being intrinsically negative, leads to diamagnetic effects in which the induced moment is in the opposite direction to **H**. One says that the strong magnetic moment one would expect from the first term in expression (3.10.9) is *quenched*.

Having thus considered and disposed of the orbital angular momentum as an important source of magnetic effects in solids we now restore to the electron its spin, and ask how this property modifies our picture. As spin has been shown to be a consequence of the relativistic nature of the electron we might take as our starting point the *Dirac equation*, which describes the relativistic motion of an electron or positron by means of a wavefunction having four components. In the nonrelativistic limit the electron and positron parts of this equation may be separated by means of the *Foldy–Wouthuysen transformation* to give an equation for the two-component wavefunction describing the electron alone. The fact that the electron wavefunction does have two components is consistent with the electron possessing a degree of freedom corresponding to a spin angular momentum **s** of $\frac{1}{2}\hbar$ that can point either up or down.

The most important terms from our point of view that are contained in this reduction of the Dirac equation are given by the Hamiltonian

$$\begin{aligned}
\mathcal{H} = {} & \frac{1}{2m}\left(\mathbf{p} - \frac{e}{c}\mathbf{A}\right)^2 - \frac{1}{8m^3c^2}\mathbf{p}^4 + V(\mathbf{r}) - \frac{e}{mc}\mathbf{s}\cdot\nabla\times\mathbf{A} \\
& + \frac{1}{2m^2c^2}\mathbf{s}\cdot\left[\nabla V(\mathbf{r})\times\left(\mathbf{p} - \frac{e}{c}\mathbf{A}\right)\right] + \frac{\hbar^2}{8m^2c^2}\nabla^2 V.
\end{aligned} \tag{3.10.10}$$

3.10 Magnetic moments and spin

This differs from expression (3.10.8) in having a term in $\mathbf{p}^4$, which is a relativistic correction to the kinetic energy, and in having two terms involving the spin angular momentum $\mathbf{s}$. In this section we consider only the first of the terms containing $\mathbf{s}$. Since $\nabla \times \mathbf{A} = \mathbf{H}$ the presence of this term in the Hamiltonian shows the electron to have a magnetic moment of $e\mathbf{s}/mc$ due to its spin. The ratio of the magnetic moment of a substance to its angular momentum is a quantity that can be determined by experiment, and is found to be close to a value of e/mc in many ferromagnetic substances. This indicates that it is the spin of the electron rather than its orbital motion, which from (3.10.9) would have led to a value of $e/2mc$ for this ratio, that is principally responsible for magnetic properties in these materials.

The operator $\mathbf{s}$, being an angular momentum, has the same commutation properties as the orbital angular momentum $\mathbf{L}$. Because of the definition

$$\mathbf{L} = \mathbf{r} \times \mathbf{p}$$

and the relation

$$[\mathbf{r}, \mathbf{p}] = i\hbar$$

it follows that

$$[L_x, L_y] = i\hbar L_z; \quad [L_y, L_z] = i\hbar L_x; \quad [L_z, L_x] = i\hbar L_y,$$

or, more concisely,

$$\mathbf{L} \times \mathbf{L} = i\hbar \mathbf{L}.$$

We then also have the relation

$$\mathbf{s} \times \mathbf{s} = i\hbar \mathbf{s}. \tag{3.10.11}$$

It is useful to define two new operators, s^+ and s^-, known as *spin raising and lowering operators*, by writing

$$s^+ = s_x + is_y; \quad s^- = s_x - is_y. \tag{3.10.12}$$

We then find that

$$[s_z, s^+] = \hbar s^+; \quad [s_z, s^-] = -\hbar s^-. \tag{3.10.13}$$

It then follows that if the state $|\uparrow\rangle$ is an eigenfunction of s_z having eigenvalue $\frac{1}{2}\hbar$, then $s^-|\uparrow\rangle$ is the eigenfunction $|\downarrow\rangle$ of s_z having eigenvalue $-\frac{1}{2}\hbar$.

$$s_z(s^-|\uparrow\rangle) = (s^- s_z - \hbar s^-)|\uparrow\rangle$$
$$= (\tfrac{1}{2}\hbar s^- - \hbar s^-)|\uparrow\rangle$$
$$= -\tfrac{1}{2}\hbar(s^-|\uparrow\rangle).$$

The naming of s^+ may be similarly justified by showing that it transforms $|\downarrow\rangle$ into $|\uparrow\rangle$. As these are the only two possible states for a spin-$\frac{1}{2}$ particle we have

$$s^+ s^+ = s^- s^- = 0.$$

The spin raising and lowering operators remind us rather strongly of boson creation and annihilation operators. We recall that the number operator for a boson state can have an eigenvalue equal to any one of the infinite spectrum of natural numbers (Fig. 3.10.1(b)). The operator s_z, which has eigenvalues $\pm\frac{1}{2}\hbar$ (Fig. 3.10.1(a)) could be considered as operating within the "ladder" of the boson system if we could somehow disconnect the bottom two rungs from the rest of the spectrum. The procedure that enables one to make the correspondence between the spin system and the boson system is known as the *Holstein–Primakoff transformation*. We define boson operators $a^\dagger$ and a as usual so that

$$[a, a^\dagger] = 1; \quad a^\dagger a = n,$$

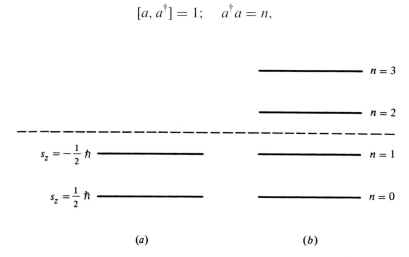

Figure 3.10.1. In the Holstein–Primakoff transformation a direct correspondence is achieved between the two possible states of a particle of spin $\frac{1}{2}\hbar$ and the $n=0$ and $n=1$ states of a harmonic oscillator.

and make the identification

$$s^+ = \hbar(1-n)^{1/2}a; \quad s^- = \hbar a^\dagger(1-n)^{1/2}. \tag{3.10.14}$$

Substitution of these expressions in (3.10.11), (3.10.12), and (3.10.13) shows that these relations satisfy the commutation relations for spin operators, and that

$$s_z = \hbar(\tfrac{1}{2} - n). \tag{3.10.15}$$

This form for s_z seems to suggest that it can have all the eigenvalues $\tfrac{1}{2}\hbar$, $-\tfrac{1}{2}\hbar$, $-\tfrac{3}{2}\hbar$, etc. While this is so one can also see that the operator s^-, which transforms the state with $s_z = +\tfrac{1}{2}\hbar$ to that with $s_z = -\tfrac{1}{2}\hbar$, is not capable of further lowering the spin, as the factor $(1-n)^{1/2}$ then gives zero. There is in effect a barrier separating the lowest two levels of the boson system from the other states. This correspondence paves the way for the description of a ferromagnet in terms of a gas of interacting bosons. We shall in particular consider a model of a ferromagnetic insulator. This is distinguished from the ferromagnetic conductor considered in Section 2.8 by the fact that the spins are considered as bound to a particular lattice site in the manner of the classical model of Section 1.4.

3.11 Magnons

In a ferromagnet an atom carrying a magnetic moment is not free to orient itself at random, but is influenced by the moments carried by other atoms in the crystal. The simplest model of such a situation is due to Weiss, who assumed that there was an effective magnetic field $\mathbf{H}_m$ acting on each atom proportional to the macroscopic magnetization of the whole crystal. This model is very similar in concept to the Einstein model of lattice dynamics introduced at the end of Section 3.8, where it was assumed that the restoring force on a displaced atom was due equally to every other atom in the crystal. The term *mean field model* is now used generically to refer to theories such as the Weiss or Einstein models in which a sum of different forces is approximated by an overall average. Because a mean field theory ignores the dominance of interactions between neighboring atoms, there are no low-frequency phonons in the Einstein model of lattice dynamics. As a consequence the lattice specific heat is incorrectly predicted to vary exponentially at low temperatures. In a similar way the Weiss model of ferromagnetism does not support the existence of magnons of low frequency, and the magnetization of a ferromagnet of spinning electrons is also incorrectly predicted to vary exponentially at low temperatures. What is needed to rectify this situation is a theory based on a model in which the interaction between near neighbors is emphasized.

The simplest such model of a ferromagnet is one in which neighboring spins interact only through the z-component of their magnetic moments. A lattice of N fixed electrons (fixed so that we can neglect kinetic and potential energies) would then have the Hamiltonian

$$\mathcal{H} = -\sum_l s_z(l)\left[|\omega_0| + \sum_{l'} J_{ll'} s_z(l')\right],$$

where $\omega_0 = e\mathbf{H}_z/mc$, the l and l' are lattice sites, and the $J_{ll'}$ are functions only of $l - l'$. This is known as the *Ising model*, and is of great interest to those who study the statistical mechanics of phase transitions. It is of less interest to us, however, as it is no more able to support magnons than was the Weiss model. We could say in classical terms that because the x- and y-components of magnetic moment are ignored, the tilting of one moment does not induce its neighbor to change its orientation. It is thus necessary to introduce interactions between the x- and y-components of spin, leading us to the *Heisenberg model*, for which

$$\mathcal{H} = -\sum_l \mathbf{s}(l) \cdot \left[\omega_0 + \sum_{l'} J_{ll'} \mathbf{s}(l')\right]. \tag{3.11.1}$$

We rewrite this in terms of boson operators by making use of the Holstein–Primakoff transformation. From Eqs. (3.10.12), (3.10.14), and (3.10.15) we find

$$\begin{aligned}\mathbf{s}(l) \cdot \mathbf{s}(l') &= s_x(l)s_x(l') + s_y(l)s_y(l') + s_z(l)s_z(l') \\ &= \tfrac{1}{2}[s^+(l)s^-(l') + s^-(l)s^+(l')] + s_z(l)s_z(l') \\ &= \tfrac{1}{2}\hbar^2[(1-n_l)^{1/2}a_l a_{l'}^\dagger(1-n_{l'})^{1/2} + a_l^\dagger(1-n_l)^{1/2}(1-n_{l'})^{1/2}a_{l'}] \\ &\quad + \hbar^2(\tfrac{1}{2}-n_l)(\tfrac{1}{2}-n_{l'}). \end{aligned} \tag{3.11.2}$$

At very low temperatures the magnetization of the specimen, which will be directed in the z-direction, will be close to its saturation value of $(Ne/mc)\tfrac{1}{2}\hbar$.

3.11 Magnons

That is, the total z-component of spin will be close to $\frac{1}{2}N\hbar$. In terms of the boson number operators, n_l, we have from (3.10.15) that

$$\sum_l s_z(l) = \tfrac{1}{2}N\hbar - \sum_l n_l \hbar,$$

showing that the expectation value of the n_l will all be small compared with unity at low temperatures. We take this as justification for neglecting terms of order n^2 in expression (3.11.2). Replacing $(1-n_l)^{1/2}$ by unity in this way we find that

$$\mathbf{s}(l)\cdot\mathbf{s}(l') \simeq \tfrac{1}{2}\hbar^2 \left(a_l a_{l'}^\dagger + a_l^\dagger a_{l'} + \tfrac{1}{2} - n_l - n_{l'} \right).$$

As we are looking for spin waves, now is clearly the time to transform from local to collective coordinates. The magnon creation and annihilation operators are defined by

$$a_\mathbf{q}^\dagger = N^{-1/2}\sum_l e^{i\mathbf{q}\cdot l}a_l^\dagger; \quad a_\mathbf{q} = N^{-1/2}\sum_l e^{-i\mathbf{q}\cdot l}a_l,$$

from which

$$a_l^\dagger = N^{-1/2}\sum_\mathbf{q} e^{-i\mathbf{q}\cdot l}a_\mathbf{q}^\dagger; \quad a_l = N^{-1/2}\sum_\mathbf{q} e^{i\mathbf{q}\cdot l}a_\mathbf{q}.$$

The sum over $\mathbf{q}$ is restricted as for the case of phonons to N distinct allowed values, such as are contained within the first Brillouin zone defined in Section 3.6. We then find that

$$\mathbf{s}(l)\cdot\mathbf{s}(l') \simeq \frac{\hbar^2}{2N}\sum_{\mathbf{q},\mathbf{q}'}[a_\mathbf{q}a_{\mathbf{q}'}^\dagger e^{i(\mathbf{q}\cdot l - \mathbf{q}'\cdot l')} + a_\mathbf{q}^\dagger a_{\mathbf{q}'} e^{-i(\mathbf{q}\cdot l - \mathbf{q}'\cdot l')}$$
$$- a_\mathbf{q}^\dagger a_{\mathbf{q}'}(e^{-i(\mathbf{q}-\mathbf{q}')\cdot l} + e^{-i(\mathbf{q}-\mathbf{q}')\cdot l'})] + \tfrac{1}{4}\hbar^2.$$

If we define

$$J_\mathbf{q} = \sum_{l'} e^{i\mathbf{q}\cdot(l'-l)}J_{ll'}$$

we have for the Heisenberg Hamiltonian (3.11.1)

$$\mathcal{H} = \hbar\omega_0 \sum_\mathbf{q}\left(a_\mathbf{q}^\dagger a_\mathbf{q} - \tfrac{1}{2}\right) + \tfrac{1}{2}\hbar^2 \sum_\mathbf{q}\left[(J_0-J_\mathbf{q})a_\mathbf{q}^\dagger a_\mathbf{q} + (J_0-J_{-\mathbf{q}})a_\mathbf{q}a_\mathbf{q}^\dagger - \tfrac{3}{2}J_0\right].$$

For a Bravais lattice it will be true that $J_\mathbf{q} = J_{-\mathbf{q}}$, giving us

$$\mathcal{H} = \mathcal{E}_0 + \sum_\mathbf{q} \hbar\omega_\mathbf{q} a_\mathbf{q}^\dagger a_\mathbf{q},$$

where $\mathcal{E}_0$ is a constant and

$$\omega_\mathbf{q} = \omega_0 + \hbar(J_0 - J_\mathbf{q}).$$

Within the approximations we have made we can thus consider the magnet as a system of independent bosons. Because

$$J_0 - J_\mathbf{q} = 2\sum_{l'} J_{ll'} \sin^2\left[\tfrac{1}{2}\mathbf{q}\cdot(\mathbf{l} - \mathbf{l}')\right]$$

the magnon frequency ω will always increase as q^2 for small values of q, in agreement with the classical approach of Section 1.4.

This simple theory is adequate to account for a number of the low-temperature properties of ferromagnets, when only a few magnons are excited. The total magnetization, M_z, for instance, is given by

$$M_z = \frac{e}{mc}\sum_l s_z(\mathbf{l}) = \frac{e}{mc}\left(\frac{1}{2}N\hbar - \sum_l n_l \hbar\right) = \frac{e\hbar}{mc}\left(\frac{1}{2}N - \sum_\mathbf{q} n_\mathbf{q}\right), \quad (3.11.3)$$

showing that each magnon carries a magnetic moment of $(e\hbar/mc)$. The Bohr magneton μ_B is defined as $e\hbar/2mc$, and so we can write the deviation from saturation of the magnetization as

$$M_0 - M_z = 2\mu_B \sum_\mathbf{q} n_\mathbf{q}.$$

Because the magnons behave as bosons the average number present in any mode will be given by (3.3.2), from which

$$\sum_\mathbf{q} \bar{n}_\mathbf{q} = \sum_\mathbf{q} \frac{1}{\exp(\hbar\omega_\mathbf{q}/kT) - 1}. \quad (3.11.4)$$

At low temperatures only magnons of low energy will be present, and so in a cubic crystal in the absence of an applied field $\omega_\mathbf{q}$ may be approximated by αq^2, where α is a constant. Writing the summation over $\mathbf{q}$ as an integral in

3.11 Magnons

q-space we find

$$\sum_q \bar{n}_q \propto \int \frac{q^2 \, dq}{\exp(\alpha \hbar q^2 / kT) - 1}.$$

Changing the variable of integration from q to $x = \alpha \hbar q^2 / kT$ gives the low-temperature relation

$$M_0 - M_z \propto T^{3/2}, \tag{3.11.5}$$

which is a result well verified experimentally. Mean field theories predict either a linear or exponential variation, according to whether a classical or quantized picture of the magnetic moment is used.

As the temperature is increased and the magnetization begins to deviate from its saturation value, the approximation of replacing $(1 - n_l)^{1/2}$ by 1 will become less valid. If we expand this expression binomially, writing

$$(1 - n_l)^{1/2} = 1 - \tfrac{1}{2} n_l - \tfrac{1}{8} n_l^2 - \cdots$$

$$= 1 - \frac{1}{2N} \sum_{\mathbf{q},\mathbf{q}'} a_\mathbf{q}^\dagger a_{\mathbf{q}'} e^{-i(\mathbf{q}-\mathbf{q}')\cdot \mathbf{l}} - \cdots$$

we can interpret the exact Hamiltonian as describing a magnon system *with interactions*. We could then use perturbation theory to calculate a better estimate of how the magnetization should vary with temperature at low T. However, we know that there exists a *Curie temperature*, T_C, at which the magnetization vanishes. It will thus be a fruitless task to pursue the perturbation approach too far in this direction, as convergence will become very slow as soon as T becomes comparable to T_C. There are also complications that arise from the upper limit $\omega_{\max}$ to the frequencies $\omega_\mathbf{q}$ over which one sums in Eq. (3.11.4). This introduces terms of the form $e^{-\hbar\omega_{\max}/kT}$, which are not expressible as any sort of power series in T, and makes comparison with experiment very difficult.

As a final note on magnons it should be mentioned that more complicated magnetic structures than the ferromagnet also have elementary excitations in the form of spin waves. Simple, helical, and canted antiferromagnetism and ferrimagnetism are examples of phenomena that arise when various interactions occur between localized spins in various crystal structures. All these exhibit magnon excitations of one form or another, and show a variety of forms of $\omega(\mathbf{q})$.

Problems

3.1 Verify that for the Bogoliubov operators defined in Section 3.4

$$[\alpha_{\mathbf{k}}, \alpha_{\mathbf{k}'}^{\dagger}] = \delta_{\mathbf{kk}'}.$$

3.2 Calculate the excitation spectrum of a gas of charged bosons interacting through the Coulomb potential.

3.3 An alternative approach to the Bogoliubov theory of interacting bosons first expresses the Hamiltonian (3.4.1) in terms of α and $\alpha^{\dagger}$. One then argues that the ground-state energy is found by evaluating $\langle\phi|\mathcal{H}|\phi\rangle$, where $|\phi\rangle$ is the vacuum state such that $\alpha_{\mathbf{k}}|\phi\rangle = 0$ for all $\mathbf{k}$. Show that minimization of this ground-state energy with respect to the $\theta_{\mathbf{k}}$ leads to the same results as the approach given in the text.

3.4 *Optical and Acoustic Modes* The problem of the chain of masses and springs is modified by the introduction of extra springs connecting every second particle. Then, with $l = na$,

$$\mathcal{H} = \sum_{\text{all } n}\left[\frac{1}{2M}p_l^2 + \frac{1}{2}K_1(y_l - y_{l+a})^2\right]$$

$$+ \sum_{\text{even } n}\frac{1}{2}K_2(y_l - y_{l+2a})^2.$$

Find the phonon frequencies for this system. [Hint: First make the transformations:

$$y_q^{(1)} = \sum_{n \text{ even}} e^{-iql}y_l; \quad y_q^{(2)} = \sum_{n \text{ odd}} e^{iql}y_l.]$$

3.5 A particle is bound in a one-dimensional potential, $V(x)$, which can be approximated for small x by

$$V = \tfrac{1}{2}m\omega^2 x^2 - \alpha x^3.$$

Show how the mean position of the particle, $\int \psi^* x \psi\, dx$, changes with the energy of the eigenstates when α is small. [Hint: Use perturbation theory on the harmonic oscillator states by writing x^3 and x in terms of $a^{\dagger}$ and a.] This illustrates the fact that the thermal expansion of a crystal is due to anharmonic terms in the potential energy.

3.6 *Magnon–phonon Interactions* If we allow the spins in the Heisenberg Hamiltonian (3.11.1) to be displaced by the presence of phonons in the lattice, then we must allow the constants $J_{ll'}$ to be functions of the displacements $\mathbf{y}_l$ and $\mathbf{y}_{l'}$. At low temperatures these displacements will be small and one can put

$$J_{ll'}(\mathbf{y}_l, \mathbf{y}_{l'}) = J_{ll'}(0, 0) + \mathbf{y}_l \cdot \mathbf{K}_{ll'} - \mathbf{y}_{l'} \cdot \mathbf{K}_{ll'}.$$

Rewrite the total Hamiltonian in terms of magnon and phonon annihilation and creation operators, $a_{\mathbf{q}}^{(\text{magnon})}$, $a_{\mathbf{q}}^{\dagger(\text{magnon})}$, $a_{\mathbf{q}}^{(\text{phonon})}$, and $a_{\mathbf{q}}^{\dagger(\text{phonon})}$.

3.7 Show that the Hamiltonian for magnon–phonon interactions derived in Problem 3.6 exhibits conservation of the total number of magnons, in that

$$\left[\mathcal{H}, \sum_{\mathbf{q}} a_{\mathbf{q}}^{\dagger} a_{\mathbf{q}} \right] = 0$$

when $a_{\mathbf{q}}^{\dagger}$ and $a_{\mathbf{q}}$ are magnon operators.

3.8 The result of Problem 3.7 is no more than an expression of the conservation of total angular momentum in the z-direction. Nonconservation of total magnon number can occur when there is interaction between the electron spin and the spin of the nucleus at a particular site. Express in terms of magnon operators the Hamiltonian of a Heisenberg ferromagnet interacting with a nuclear spin of $\frac{1}{2}\hbar$ at one particular site.

3.9 Evaluate the expectation value of $n_\mathbf{k} = a_\mathbf{k}^{\dagger} a_\mathbf{k}$ in the ground state $|\phi\rangle$ of the Bogoliubov picture of helium. [Hint: Express $n_\mathbf{k}$ in terms of the α-operators and then make use of the fact that $\alpha_\mathbf{k} |\phi\rangle = 0$ for all $\mathbf{k}$.]

3.10 Express the ground-state $|\phi\rangle$ of the Bogoliubov picture of helium in terms of the operators $a_\mathbf{k}^{\dagger}$ and the vacuum state $|0\rangle$.

3.11 Prove the statement preceding Eq. (3.6.3) which says that the matrix $V_\mathbf{q}^{ij}$ has three mutually perpendicular eigenvectors.

3.12 Consider a large number N of spinless interacting bosons of mass m in a large one-dimensional box of length L. There are periodic boundary conditions. The particles interact via a delta-function potential, and so the Hamiltonian is

$$\mathcal{H} = \sum_{k} Ak^2 a_k^\dagger a_k + (V/2L) \sum_{k,k',q} a_{k-q}^\dagger a_{k'+q}^\dagger a_{k'} a_k$$

with A and V constants. The sums proceed over all permitted values of k, k', and q. That is, the terms with $q = 0$ are not excluded from the sum.

(a) Calculate the energy of the ground state of the noninteracting system.
(b) Calculate the energy of the ground state of the interacting system in the Hartree approximation.
(c) Estimate the speed of low-frequency sound waves in this system.

Chapter 4
One-electron theory

4.1 Bloch electrons

The only model of a metal that we have considered so far has been the gas of interacting electrons. A real metal, of course, contains ions as well as electrons, and we should really include the ionic potentials in the Hamiltonian, rather than the uniform background of positive charge that we used to approximate them. The difficulty of the many-electron problem is such, however, that the loss of translational invariance caused by adding an ionic potential $V(\mathbf{r})$ to the Hamiltonian proves disastrous. Even in the Hartree–Fock approximation, for example, it becomes impossible to write the energy in a closed form. When there was no ionic potential we could write the wavefunction of the noninteracting system as

$$|\Phi\rangle = \prod_{|\mathbf{k}|<k_F} c_{\mathbf{k}}^{\dagger}|0\rangle, \qquad (4.1.1)$$

where the operators $c_{\mathbf{k}}^{\dagger}$ created electrons in plane-wave states. But if there is an additional potential applied to the system we might find the energy to be lower in the state Φ if we replaced the $c_{\mathbf{k}}^{\dagger}$ by operators $c_{\alpha}^{\dagger}$ that create electrons in states that are not plane waves. The definition of the Hartree–Fock approximation in this general case is then taken to be that Φ must be a Slater determinant and must make the expectation value of the Hamiltonian a minimum. It becomes very laborious to work out what states the $c_{\alpha}^{\dagger}$ should create to fulfill this condition.

We are saved from what seems to be an impossible task by three fortunate features of the problem. The first of these is the fact that we can initially ignore the thermal motion of the lattice, and study the motion of the electrons in the potential of a stationary array of ions. This is known as the *adiabatic* or *Born–Oppenheimer* approximation. Its justification appears when we apply the motion of the ions as a perturbation, and calculate the effects of the gas of phonons interacting with the electrons. This is discussed in detail in Chapter 6. There we find the corrections to the electron energies to be comparatively small, and of importance only in special circumstances like superconductivity.

The second useful fact is that the electrons in the core of an atom are so tightly bound that they are not significantly perturbed by the motion of electrons at the Fermi energy. The effects of the core electrons on the properties of a solid are limited to their repulsion of other electrons through electrostatic forces, and to the effective repulsion that arises from the demands of the Exclusion Principle that no two electrons of the same spin occupy the same location. The consequences of the requirement that the wavefunctions of the higher-energy electrons be orthogonal to those of the core states will be discussed in Section 4.4.

The third happy feature of the problem is the most important. It is the fact that the properties of a system of interacting electrons in a static lattice potential can be found by solving a set of related one-electron Schrödinger equations. The formalism by means of which this equivalence can be proved is known as *density functional theory*, and is the topic of Chapter 5. For now we assume this result, and turn our attention to the solution of the Schrödinger equation for a single electron moving in the potential $V(\mathbf{r})$ due to a periodic crystal lattice. That is, we need to solve the equation

$$-\frac{\hbar^2}{2m}\nabla^2\psi(\mathbf{r}) + V(\mathbf{r})\psi(\mathbf{r}) = \mathcal{E}\psi(\mathbf{r}), \qquad (4.1.2)$$

where the lattice potential has the property that, for all lattice vectors $\mathbf{l}$, $V(\mathbf{r}) = V(\mathbf{r}+\mathbf{l})$.

If the potential were zero, the wavefunctions would be of the form discussed in Section 2.1,

$$\phi_\mathbf{k} = \Omega^{-1/2} e^{i\mathbf{k}\cdot\mathbf{r}},$$

with energies given by

$$\mathcal{E}_\mathbf{k} = \frac{\hbar^2 \mathbf{k}^2}{2m}.$$

If we now slowly switch on the lattice potential the wavefunctions are perturbed to a new form

$$\psi_\mathbf{k} = \Omega^{-1/2} u_\mathbf{k}(\mathbf{r}) e^{i\mathbf{k}\cdot\mathbf{r}}. \qquad (4.1.3)$$

4.1 Bloch electrons

Because the lattice potential is periodic, this modification of the wavefunction is also periodic, and so

$$u_{\mathbf{k}}(\mathbf{r}) = u_{\mathbf{k}}(\mathbf{r} + \mathbf{l}).$$

The fact that the wavefunctions can be written in the form (4.1.3) is known as *Bloch's theorem*, and can be proved as follows. Let us consider any solution ψ of the Schrödinger equation

$$\mathcal{H}(\mathbf{r})\psi(\mathbf{r}) = \mathcal{E}\psi(\mathbf{r}).$$

On relocating the origin of $\mathbf{r}$ we find

$$\mathcal{H}(\mathbf{r} + \mathbf{l})\psi(\mathbf{r} + \mathbf{l}) = \mathcal{E}\psi(\mathbf{r} + \mathbf{l}).$$

But since $V(\mathbf{r} + \mathbf{l}) = V(\mathbf{r})$ it follows that

$$\mathcal{H}(\mathbf{r})\psi(\mathbf{r} + \mathbf{l}) = \mathcal{E}\psi(\mathbf{r} + \mathbf{l}).$$

Thus any linear combination of the $\psi(\mathbf{r} + \mathbf{l})$ for different $\mathbf{l}$ gives a valid eigenstate of energy $\mathcal{E}$. Let us in particular choose the combination

$$\psi_{\mathbf{k}}(\mathbf{r}) = \sum_{\mathbf{l}} \psi(\mathbf{r} + \mathbf{l}) e^{-i\mathbf{k}\cdot\mathbf{l}}$$

$$= e^{i\mathbf{k}\cdot\mathbf{r}} \sum_{\mathbf{l}} \psi(\mathbf{r} + \mathbf{l}) e^{-i\mathbf{k}\cdot(\mathbf{r}+\mathbf{l})}.$$

Because the sum is over all $\mathbf{l}$ it must be a periodic function of $\mathbf{r}$ with the period of the lattice, and can be identified with the function $u_{\mathbf{k}}(\mathbf{r})$ of Eq. (4.1.3). If we now impose the cyclic boundary conditions that we used in Section 2.1 for free electrons, and demand that

$$\psi_{\mathbf{k}}(\mathbf{r}) = \psi_{\mathbf{k}}(\mathbf{r} + \mathbf{L})$$

for three different large lattice vectors $\mathbf{L}$, we shall clearly have the condition that all the components of $\mathbf{k}$ must be real.

An electron having a wavefunction of the form (4.1.3), where $u_{\mathbf{k}}(\mathbf{r})$ is periodic in the lattice, is known as a *Bloch electron*. In terms of $u_{\mathbf{k}}(\mathbf{r})$ the Schrödinger equation (4.1.2) becomes

$$\left[-\frac{\hbar^2}{2m}(\nabla + i\mathbf{k})^2 + V(\mathbf{r})\right] u_{\mathbf{k}}(\mathbf{r}) = \mathcal{E}_{\mathbf{k}} u_{\mathbf{k}}(\mathbf{r}), \qquad (4.1.4)$$

which we write as

$$\mathcal{H}_\mathbf{k} u_\mathbf{k}(\mathbf{r}) = \mathcal{E}_\mathbf{k} u_\mathbf{k}(\mathbf{r}).$$

The reciprocal lattice vectors, **g**, defined in Section 3.6, have the property that $e^{i\mathbf{g}\cdot\mathbf{l}} = 1$ for all **g** and **l**. It follows that any function that may be written in the form $\sum_\mathbf{g} a_\mathbf{g} e^{i\mathbf{g}\cdot\mathbf{r}}$ is periodic with the periodicity of the lattice. The converse may also be shown to be true for any well behaved periodic function, which allows us to expand $V(\mathbf{r})$ and $u_\mathbf{k}(\mathbf{r})$ in Fourier series of the plane waves $e^{i\mathbf{g}\cdot\mathbf{r}}$. We can thus write

$$V(\mathbf{r}) = \sum_\mathbf{g} V_\mathbf{g} e^{i\mathbf{g}\cdot\mathbf{r}}; \quad u_\mathbf{k}(\mathbf{r}) = \sum_\mathbf{g} u_\mathbf{g}(\mathbf{k}) e^{i\mathbf{g}\cdot\mathbf{r}},$$

where

$$V_\mathbf{g} = \Omega^{-1} \int e^{-i\mathbf{g}\cdot\mathbf{r}} V(\mathbf{r})\, d\mathbf{r}; \quad u_\mathbf{g}(\mathbf{k}) = \Omega^{-1} \int e^{-i\mathbf{g}\cdot\mathbf{r}} u_\mathbf{k}(\mathbf{r})\, d\mathbf{r}.$$

If we substitute these expressions into (4.1.4) and equate the various Fourier components we find the infinite set of equations

$$\frac{\hbar^2}{2m}(\mathbf{g}+\mathbf{k})^2 u_\mathbf{g}(\mathbf{k}) + \sum_{\mathbf{g}'} V_{\mathbf{g}'} u_{\mathbf{g}-\mathbf{g}'}(\mathbf{k}) = \mathcal{E}_\mathbf{k} u_\mathbf{g}(\mathbf{k}),$$

which can in principle be solved for $\mathcal{E}_\mathbf{k}$ and $u_\mathbf{g}(\mathbf{k})$. The graph of $\mathcal{E}_\mathbf{k}$ against **k** is known as the *band structure*, for reasons that will soon be apparent. It is interesting to note that it is only certain Fourier components of the atomic potentials that contribute to these equations which determine the $\mathcal{E}_\mathbf{k}$. If the lattice potential is supposed to be due to a superposition of atomic potentials $V_a(\mathbf{r})$, so that

$$V(\mathbf{r}) = \sum_l V_a(\mathbf{r} - \mathbf{l}),$$

one can write the Fourier transform of $V(\mathbf{r})$ as

$$V(\mathbf{q}) = \Omega^{-1} \int e^{-i\mathbf{q}\cdot\mathbf{r}} V(\mathbf{r})\, d\mathbf{r}$$

$$= \Omega^{-1} \left(\sum_l e^{-i\mathbf{q}\cdot\mathbf{l}}\right) \int V_a(\mathbf{r}') e^{-i\mathbf{q}\cdot\mathbf{r}'}\, d\mathbf{r}'.$$

4.1 Bloch electrons

Thus the Fourier transform of the lattice potential is expressed in terms of the Fourier transform, $V_a(\mathbf{q})$, of the atomic potential $V_a(\mathbf{r})$. However, because the sum over l vanishes unless $\mathbf{q}$ is equal to a reciprocal lattice vector, $\mathbf{g}$, the energies of the Bloch electrons depend only on these particular terms in $V(\mathbf{q})$. This is illustrated in Fig. 4.1.1.

If one were to take a free electron and slowly switch on the lattice potential, the wavefunction would be gradually transformed from a plane wave to a Bloch wave of the form (4.1.3). In general the value of $\mathbf{k}$ is then well defined, since it does not change from its original value, and the plane wave merely becomes modulated by the function $u_\mathbf{k}(\mathbf{r})$. It is, however, possible to write the

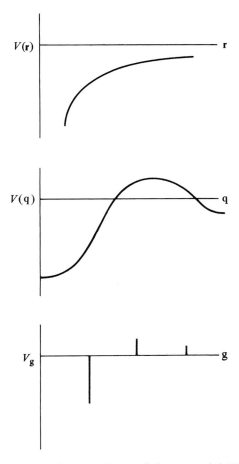

Figure 4.1.1. $V(\mathbf{q})$ is the Fourier transform of the potential $V(\mathbf{r})$ due to a single atom. When one sums the contributions of all the atoms in the lattice the only parts of $V(\mathbf{q})$ that do not vanish occur when $\mathbf{q}$ is equal to a reciprocal lattice vector $\mathbf{g}$.

Bloch wavefunction in the form

$$\psi_{\mathbf{k}}(\mathbf{r}) = \Omega^{-1/2}[u_{\mathbf{k}}(\mathbf{r})e^{i\mathbf{g}\cdot\mathbf{r}}]e^{i(\mathbf{k}-\mathbf{g})\cdot\mathbf{r}}.$$

Because both $u_{\mathbf{k}}(\mathbf{r})$ and $e^{i\mathbf{g}\cdot\mathbf{r}}$ are periodic, so is their product. This means that the wavefunction can be considered as obeying the Bloch condition (4.1.3) not just for one value of $\mathbf{k}$, but for any value such as $\mathbf{k} - \mathbf{g}$. This is not the same thing as the equivalence that was found in Section 3.6 of the coordinate $y_{\mathbf{q}}$ to $y_{\mathbf{q+g}}$, for here we are dealing with a set of wavefunctions that are all different although they happen to obey the Bloch equation with the same $\mathbf{k}$. The function $u_{\mathbf{k}}(\mathbf{r})e^{i\mathbf{g}\cdot\mathbf{r}}$ is not the same as $u_{\mathbf{k+g}}(\mathbf{r})$. The difference arises from the fact that the phonon problem has only a limited number of degrees of freedom, while the Schrödinger equation has an infinity of solutions. The Bloch theorem allows these solutions to be classified either according to the original value of $\mathbf{k}$ before the potential was switched on, or else according to the value of $\mathbf{k}$ after some reciprocal lattice vector has been added to it. The first Brillouin zone was defined in Section 3.6 as the volume in k-space containing all those points for which $|\mathbf{k}| \leq |\mathbf{k}+\mathbf{g}|$ for all $\mathbf{g}$. It is sometimes convenient to classify Bloch states by specifying the value of $\mathbf{k}$ reduced to lie in the first Brillouin zone. To define the state completely it is then also necessary to define a *band index* which is related to the value of $\mathbf{g}$ necessary to achieve this reduction. In general the band index is defined as the number of different values of $\mathbf{g}$ (including zero) for which $|\mathbf{k}| \geq |\mathbf{k}+\mathbf{g}|$. This is illustrated in Fig. 4.1.2 for a hexagonal lattice in two dimensions. The point a cannot be

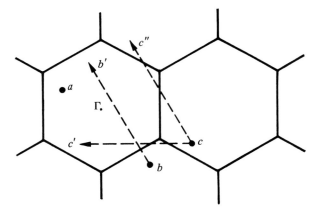

Figure 4.1.2. In this two-dimensional hexagonal lattice the point a lies in the first Brillouin zone. The point b lies in the second Brillouin zone while the point c lies in the third zone.

brought closer to the origin Γ by addition of any nonzero reciprocal lattice vector. It thus has band index 1, and is said to be in the first Brillouin zone. The point b can be brought closer to Γ only by addition of the vector $\mathbf{g}$ that takes it to b'. It thus has band index 2 and is in the second Brillouin zone. Point c is in the third Brillouin zone, since it may be reduced either to c' or c'', and so on. One thus has two alternative schemes for depicting the band structure $\mathcal{E}_\mathbf{k}$. The first scheme, in which $\mathbf{k}$ is allowed to take on any value consistent with the boundary conditions, is known as the *extended zone scheme*. In the second, the band structure is written in the form $\mathcal{E}_\mathbf{k}^{(n)}$, where n is the band index, and now $\mathbf{k}$ is the reduced wavenumber, restricted to lie within the first Brillouin zone. This is known as the *reduced zone scheme*. The usefulness of the reduced zone scheme is a consequence of the most important property of Bloch electrons, namely that surfaces of discontinuity in general exist in $\mathcal{E}_\mathbf{k}$ in the extended zone scheme at all the boundaries between Brillouin zones. In the reduced zone scheme, however, $\mathcal{E}_\mathbf{k}^{(n)}$ is always a continuous function of $\mathbf{k}$.

This may be made plausible by the following argument. When $\mathbf{k}$ does not lie on a Brillouin zone boundary, so that there is no nonzero $\mathbf{g}$ for which $|\mathbf{k}| = |\mathbf{k} + \mathbf{g}|$, then $\psi_\mathbf{k}$ is certainly different from its complex conjugate $\psi_\mathbf{k}^*$. Now a complex wavefunction always carries a current, since then $\psi_\mathbf{k}^* \nabla \psi_\mathbf{k} - \psi_\mathbf{k} \nabla \psi_\mathbf{k}^*$ cannot vanish, and so the application of an electric field to the system will change the energy of the electron even if the field is vanishingly small. This energy change comes about by the mixing of $\psi_\mathbf{k}$ with states of neighboring wavenumber whose energies are arbitrarily close to $\mathcal{E}_\mathbf{k}$. If $\mathbf{k}$ lies on a zone boundary, however, $\psi_\mathbf{k}$ may be real and still satisfy the Bloch condition. There is then no current carried by these states, and so there are not necessarily states whose energies are arbitrarily close to $\mathcal{E}_\mathbf{k}$ both from above and below. One may then find ranges of energy over which there are no states. One may define a *density of states* in energy $D(\mathcal{E})$ by the relation

$$D(\mathcal{E}) = \sum_\mathbf{k} \delta(\mathcal{E} - \mathcal{E}_\mathbf{k}),$$

where the sum proceeds over all values of $\mathbf{k}$ in the extended zone scheme. For an infinite crystal this spectrum of delta-functions becomes a continuum, and can be plotted to give a curve that might, for example, be of the form shown in Fig. 4.1.3. The states for which $\mathbf{k}$ is in the first Brillouin zone contribute to the shaded part of $D(\mathcal{E})$, while states in the second and higher Brillouin zones give the rest. The fact that there are bands of energy for which $D(\mathcal{E})$ vanishes justifies the naming of $\mathcal{E}(\mathbf{k})$ as the band structure of the crystal.

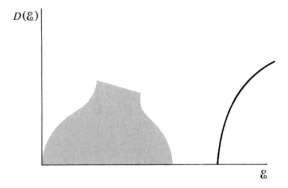

Figure 4.1.3. In this model all the states in the valence band (shown shaded) are occupied, while all those in the conduction band are empty. This system would be an insulator, since a finite amount of energy would be required to create a current-carrying elementary excitation.

4.2 Metals, insulators, and semiconductors

If a crystal is composed of N atoms then each Brillouin zone contains N allowed values of $\mathbf{k}$. As an electron has two spin eigenstates each zone may then contain $2N$ electrons, which is equivalent to saying that there are two states per zone per atom in a Bravais lattice. In some elements it happens that the number of electrons that each atom possesses is just sufficient to populate all the states below one of the gaps in $D(\mathcal{E})$. This has a very important effect on a number of the properties of these elements, and especially on the electrical properties. At zero temperature all the states below one of the band gaps are filled, while all the states above it are empty. One thus is required to provide an energy equal to the width, 2Δ, of the band gap if one is to excite an electron from the lower band (known as the *valence band*) to the upper one (known as the *conduction band*). Thus the crystal will not absorb electromagnetic radiation of frequency ω if $\hbar\omega < 2\Delta$. In particular the crystal will not absorb energy from a weak electric field of zero frequency – that is, it is an *insulator*. In a metal, on the other hand (in which term we include such good conductors as ReO_3 or RuO_2), the number of electrons is such that at zero temperature there are occupied and unoccupied states differing in energy by an arbitrarily small amount. The energy of the most energetic electron (the *Fermi energy*, $\mathcal{E}_F$, which is the chemical potential at zero temperature) does not coincide with a gap in the density of states. The crystal can then absorb radiation of low frequency, which makes it a conductor.

The Fermi surface of a metal of Bloch electrons is defined as the locus of values of **k** for which $\mathcal{E}(\mathbf{k}) = \mathcal{E}_F$. Whereas for free electrons this surface was a sphere of radius $\hbar^{-1}(2m\mathcal{E}_F)^{1/2}$, for Bloch electrons it is distorted to a greater or lesser extent from spherical shape, particularly by the effects of the discontinuities in $\mathcal{E}(\mathbf{k})$. Because the discontinuities of $\mathcal{E}(\mathbf{k})$ occur only at the zone boundaries, the Fermi surface remains continuous in the reduced zone scheme. If, however, there are states with different band indices contributing to the Fermi surface then the surface will consist of two or more sheets. In experiments on metals in magnetic fields one can sometimes separately distinguish the effects of the various parts of the Fermi surface.

We started our discussion of Bloch electrons by considering the effects of the lattice potential as a perturbation of the free-electron wavefunction. We might, however, have approached the problem from the opposite extreme, and looked at Bloch states as perturbations of atomic orbitals. A useful picture of this viewpoint is given in Fig. 4.2.1(a), which shows the degenerate energy levels of a crystal of widely spaced atoms broadening into bands as the lattice spacing a is reduced from some initial large value. In the case of a simple monovalent metal like sodium or potassium, with only one electron in its outer shell, the lower band shown in Fig. 4.2.1(b) would always be only half-filled, since there are two possible spin states for each value of **k**. The

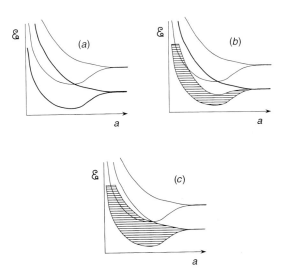

Figure 4.2.1. Atomic energy levels found when the interatomic spacing a is large broaden into bands when a is reduced (a). Monovalent atoms have half-filled bands (b). Divalent atoms have filled valence bands at large a but become conductors when a is small and the bands overlap (c).

electrical conductivity would then be high, provided a were sufficiently small to allow passage from one atom to another.

For a crystal of divalent atoms like magnesium, the lower band will be filled, as in Fig. 4.2.1(c). For large a the crystal would be an insulator, as by symmetry the occupied states would carry zero net total current. That magnesium is a metal must be a consequence of a being small enough that two bands overlap, allowing each to be partially filled. Magnesium does indeed crystallize in the hexagonal close-packed structure, and is a good metal. Carbon, on the other hand, can crystallize into the much more open diamond structure. Being tetravalent it then has a filled valence band, and is an excellent electrical insulator, with a band gap of over 5 eV.

Silicon and germanium also crystallize in the diamond structure, but have much smaller band gaps, of the order of 1.1 eV for Si and 0.67 eV for Ge. This makes them *semiconductors*, as it is possible at room temperature to excite a few electrons from the valence band into the conduction band. Thermal energies at room temperature are only about 0.03 eV, giving a probability of thermal excitation of only about e^{-22}, which is 10^{-10}, in germanium. Intrinsic semiconductors, like pure Ge and Si, thus have quite low conductivities.

A much larger conductivity, and hence more technological usefulness, can be obtained by adding impurities to produce an *extrinsic* semiconductor. Adding small amounts of a pentavalent element such as arsenic to silicon increases the number of electrons available for conduction. This doping with *donor* atoms moves the chemical potential up from its previous position in the middle of the energy gap, and into the conduction band. The material is now known as n-type silicon. The converse process of doping with trivalent *acceptor* atoms produces p-type silicon, whose chemical potential lies in the valence band.

The union of p-type and n-type material to form a p–n junction enables a wealth of useful phenomena to occur. In a photovoltaic cell, electrons fed into the valence band of p-type material may be elevated into the conduction band by a photon of sunlight. They then emerge from the n-type material at a higher potential and can do useful work before being returned to the lower potential of the p-type material. In a light-emitting diode, or LED, the reverse process occurs. Electrons at a high enough potential to be fed into the conduction band of the n-type material can move to the p-type side and then fall into the valence band, emitting a photon of light as they do so. From more complex arrangements of p-type and n-type semiconductors and metals, that ubiquitous foundation of our technological society – the transistor – can be constructed.

4.3 Nearly free electrons

There are now powerful computer codes that can produce credible forms for $\mathcal{E}(\mathbf{k})$ for any given periodic potential. In materials in which there are several atoms in each unit cell these band structures may be exceedingly complex, and so it is useful to start by looking at some much simpler situations. In this way we can appreciate some of the concepts that play important roles in determining material properties. The simplest approximation one can make is to neglect the lattice potential altogether, except in as much as to allow the existence of infinitesimal discontinuities in $\mathcal{E}(\mathbf{k})$ at the zone boundaries. The Fermi surface one obtains then consists of portions of the free-electron sphere reduced to lie in the first Brillouin zone. This is known as the *remapped free-electron model*. An example of this construction in two dimensions is shown in Fig. 4.3.1 for a hexagonal reciprocal lattice. The circle of the extended zone scheme is reduced to a central portion, A, derived from the second Brillouin zone and a group of small regions, B, derived from the third Brillouin zone. The occupied electron states are always on the concave side of the boundaries, and so the surface A is seen to contain unoccupied states. It is consequently known as a "hole surface." The portions B, on the other hand, contain electrons. We note that the use here of the term "hole" is quite

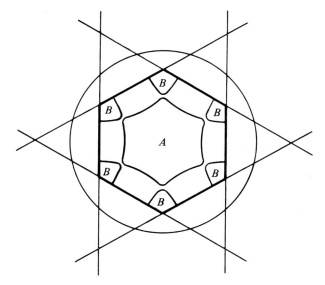

Figure 4.3.1. The circle represents the free-electron Fermi surface in the extended zone scheme. In the remapped representation of the reduced zone scheme it forms a central "hole" surface and a group of small electron surfaces.

distinct from that of Chapter 2, where a hole was simply the absence of an electron from a state below the Fermi energy. In the context of band structures, a hole is also an unoccupied state, but one that has an additional property: the energy of the state *decreases* as one moves away from the interior of the constant-energy surface, as is the case for the surface A in Fig. 4.3.1. A hole state in band-structure parlance may have an energy either below or above the Fermi energy. We shall see some examples of hole states in later sections of this chapter.

The way in which the remapped free-electron Fermi surface is derived from the Fermi surface in the extended zone scheme is seen most easily when the first Brillouin zone of Fig. 4.3.1 is repeated periodically to form the scheme shown in Fig. 4.3.2. In this *repeated zone scheme* one sees that the various parts of the Fermi surface are formed in the first Brillouin zone when a free-electron sphere is drawn around each point in the reciprocal lattice.

The simplest approximation that can be made that includes the effect of the lattice potential is known as the model of *nearly free electrons*. Here one assumes that only a certain small number of different plane waves combine to form the Bloch wave $\psi_\mathbf{k}$. The relative coefficients of these plane waves are then varied to minimize $\int \psi_\mathbf{k}^* \mathcal{H} \psi_\mathbf{k} \, \mathbf{dr}$, and this gives an approximation for $\mathcal{E}(\mathbf{k})$ and the wavefunction in the first Brillouin zone. The form of $\mathcal{E}(\mathbf{k})$ in the second zone is found by minimizing the integral by varying wavefunctions restricted to be orthogonal to those in the first zone, and so on. In practice

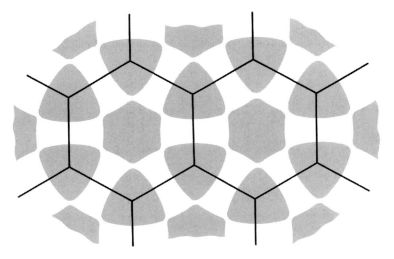

Figure 4.3.2. This picture shows the model of Fig. 4.3.1 in the repeated zone scheme, and is formed by periodically repeating the first Brillouin zone of the reduced zone scheme.

4.3 Nearly free electrons

the labor involved can be greatly reduced if one chooses to include in $\psi_{\mathbf{k}}$ only those plane waves that one thinks will enter with large coefficients. Thus if a point $\mathbf{k}$ in the extended zone scheme is much nearer to one zone boundary than any other, then one might approximate $\psi_{\mathbf{k}}$ by a mixture of $e^{i\mathbf{k}\cdot\mathbf{r}}$ and $e^{i(\mathbf{k}+\mathbf{g}_1)\cdot\mathbf{r}}$ where $\mathbf{g}_1$ is that reciprocal lattice vector that makes $|\mathbf{k}|$ close to $|\mathbf{k}+\mathbf{g}|$. Physically this is equivalent to saying that the plane wave $e^{i\mathbf{k}\cdot\mathbf{r}}$ will only have mixed with it other plane waves whose energies are close to its own. Thus, if we write

$$\psi_{\mathbf{k}} = \Omega^{-1/2}[u_0(\mathbf{k})e^{i\mathbf{k}\cdot\mathbf{r}} + u_1(\mathbf{k})e^{i(\mathbf{k}-\mathbf{g}_1)\cdot\mathbf{r}}] \tag{4.3.1}$$

and

$$V(\mathbf{r}) = \sum_{\mathbf{g}} V_{\mathbf{g}} e^{i\mathbf{g}\cdot\mathbf{r}}$$

with the zero of potential energy defined to make V_0 vanish, then

$$\int \psi^* \mathcal{H} \psi \, d\mathbf{r} = \frac{\hbar^2}{2m}[u_0^* u_0 \mathbf{k}^2 + u_1^* u_1 (\mathbf{k}-\mathbf{g}_1)^2] + u_0^* u_1 V_{\mathbf{g}_1} + u_1^* u_0 V_{-\mathbf{g}_1}.$$

The normalization condition is

$$\int \psi^* \psi \, d\mathbf{r} = u_0^* u_0 + u_1^* u_1 = 1.$$

We now minimize the energy by varying the wavefunction. According to Lagrange's method of undetermined multipliers we can take account of the normalization condition by writing

$$\delta\left[\int \psi^* \mathcal{H} \psi \, d\mathbf{r} - \lambda \int \psi^* \psi \, d\mathbf{r}\right] = 0.$$

But now we can immediately identify the multiplier λ with the energy $\mathcal{E}$, for this is the only way we can ensure that just multiplying ψ by a constant will leave the term in brackets unchanged. On differentiating partially with respect to u_0^* and u_1^* and putting $\lambda = \mathcal{E}$ we find

$$\frac{\hbar^2 \mathbf{k}^2}{2m} u_0 + V_{\mathbf{g}_1} u_1 = \mathcal{E} u_0$$

$$\frac{\hbar^2 (\mathbf{k}-\mathbf{g}_1)^2}{2m} u_1 + V_{-\mathbf{g}_1} u_0 = \mathcal{E} u_1.$$

For these equations to be consistent the determinant of the coefficients must vanish, and so

$$\begin{vmatrix} \dfrac{\hbar^2 k^2}{2m} - \mathcal{E} & V_{g_1} \\ V_{-g_1} & \dfrac{\hbar^2 (k - g_1)^2}{2m} - \mathcal{E} \end{vmatrix} = 0$$

and thus

$$\mathcal{E} = \frac{1}{2} \left\{ \frac{\hbar^2 k^2}{2m} + \frac{\hbar^2 (k - g_1)^2}{2m} \right. \\ \left. \pm \sqrt{\left(\frac{\hbar^2 (k - g_1)^2}{2m} - \frac{\hbar^2 k^2}{2m}\right)^2 + 4 V_{g_1} V_{-g_1}} \right\}. \quad (4.3.2)$$

The two possible signs of the square root correspond to the nonuniqueness of the wave vector $\mathbf{k}$ that characterizes the Bloch state. That is to say, this expression tells us the energy of the electron in the first and second energy bands. If $\mathbf{k}$ is chosen to lie in the first Brillouin zone then the negative square root will give the energy of the state that is formed from the wavefunction $e^{i \mathbf{k} \cdot \mathbf{r}}$ when the lattice potential is turned on slowly. The positive square root will refer to the state formed from $e^{i(\mathbf{k} - \mathbf{g}_1) \cdot \mathbf{r}}$, and which was originally in the second Brillouin zone.

It is interesting to note that this formula (4.3.2) for the energy of a Bloch state is identical to the one we should obtain from the use to second order of the Brillouin–Wigner perturbation theory described in Section 2.5. This result is peculiar to the two-plane-wave assumption of expression (4.3.1), and should not be looked upon as indicating that an approach using perturbation theory is necessarily equivalent to a variational approach.

The general expression of the model of nearly free electrons is found when any finite sum of plane waves is chosen as the trial wavefunction. Then if

$$\psi_\mathbf{k} = \Omega^{-1/2} \sum_\mathbf{g} u_\mathbf{g}(\mathbf{k}) e^{i(\mathbf{k}+\mathbf{g}) \cdot \mathbf{r}},$$

minimization of the energy leads to the series of equations

$$\sum_\mathbf{g} \left\{ \left[\mathcal{E} - \frac{\hbar^2}{2m}(\mathbf{k}+\mathbf{g})^2 \right] \delta_{\mathbf{g}\mathbf{g}'} - V_{\mathbf{g}'-\mathbf{g}} \right\} u_\mathbf{g}(\mathbf{k}) = 0.$$

4.3 Nearly free electrons

For these to be consistent the determinant of the coefficients of the $u_g(\mathbf{k})$ must vanish, and so

$$\left\|\left[\mathcal{E} - \frac{\hbar^2}{2m}(\mathbf{k}+\mathbf{g})^2\right]\delta_{\mathbf{gg}'} - V_{\mathbf{g}'-\mathbf{g}}\right\| = 0. \tag{4.3.3}$$

This polynomial in $\mathcal{E}$ has as many solutions as there are plane waves in the expansion of $\psi_\mathbf{k}$, and reduces to expression (4.3.2) when that number is only two.

We can observe some of the effects of the lattice potential in the simplest three-dimensional model, which is known as "sandwichium." Here the lattice potential is just $2V\cos gx$, and so the loci of the points $|\mathbf{k}| = |\mathbf{k}+\mathbf{g}|$ are just the planes defined by $k_x = \pm(n/2)g$. The two-plane-wave version of the nearly-free-electron model then gives expression (4.3.2) for the energy, which in the neighborhood of $k_x = \frac{1}{2}g$ becomes

$$\mathcal{E} = \frac{\hbar^2}{2m}(k_y^2 + k_z^2) + \frac{1}{2}\left\{\frac{\hbar^2 k_x^2}{2m} + \frac{\hbar^2(k_x-g)^2}{2m}\right.$$

$$\left.\pm\sqrt{\left[\frac{\hbar^2(k_x-g)^2}{2m} - \frac{\hbar^2 k_x^2}{2m}\right]^2 + 4V^2}\right\}, \tag{4.3.4}$$

which we write as

$$\mathcal{E} = \frac{\hbar^2}{2m}(k_y^2 + k_z^2) + \mathcal{E}_x(k_x).$$

The form of $\mathcal{E}_x(k_x)$ is shown in Fig. 4.3.3 for two different values of V. The sign of the square root has been chosen so that $\mathcal{E}_x \to \hbar^2 k_x^2/2m$ as $V \to 0$, which means that we are using the extended zone scheme. One sees that it is only when k_x is in the vicinity of $\frac{1}{2}g$ that $\mathcal{E}_x$ deviates appreciably from its free-electron value, and that a discontinuity in $\mathcal{E}_x$ does indeed occur when $k_x = \frac{1}{2}g$. The magnitude of this discontinuity is $2V$.

The shapes of the surfaces of constant energy are shown in Fig. 4.3.4, where their intersections with the plane $k_z = 0$ are plotted. For low energies the surfaces are close to spherical; then as k_x approaches the zone boundary $\mathcal{E}_x(k_x)$ starts to fall below the free-electron value, and the magnitude of k_x becomes correspondingly greater for a given energy. One says that the constant-energy surfaces are "pulled out" towards the zone boundary. The

140   One-electron theory

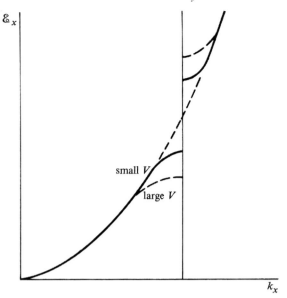

Figure 4.3.3. The variation of energy with k_x in the sandwichium model is shown by the solid line for the case where V is small compared with $\hbar^2 g^2/8m$. As V is increased, the discontinuity at the zone boundary becomes larger, as illustrated by the dashed curve.

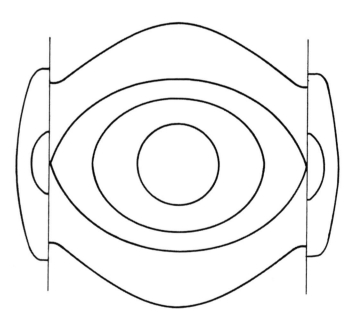

Figure 4.3.4. The variation of energy with the x–y component of wavenumber in the sandwichium model is shown in this picture, in which the lines of constant energy are drawn in the plane in which $k_z = 0$.

lowest-energy surface to meet the zone boundary is, in fact, pulled out to a conical point at the place where it does so. We can verify this by writing

$$\mathbf{k} = (\tfrac{1}{2}g - \kappa_x, k_y, k_z)$$

and expanding the square root in (4.3.4). For the negative root we find

$$\mathcal{E} \simeq \frac{\hbar^2}{2m}\left(\frac{1}{2}g\right)^2 - V + \frac{\hbar^2}{2m}(k_y^2 + k_z^2) - \frac{\hbar^2}{2m}\kappa_x^2\left[\frac{\hbar^2 g^2}{4mV} - 1\right]$$

$$= \mathcal{E}_g - V + \frac{\hbar^2}{2m}(k_y^2 + k_z^2) - \frac{\hbar^2 \kappa_x^2}{2m}\left(\frac{2\mathcal{E}_g}{V} - 1\right) \quad (4.3.5)$$

where we have used the abbreviation

$$\frac{\hbar^2}{2m}\left(\frac{1}{2}g\right)^2 = \mathcal{E}_g.$$

Thus when

$$\mathcal{E} = \mathcal{E}_g - V$$

the energy surfaces are given by

$$k_y^2 + k_z^2 \simeq \left(\frac{2\mathcal{E}_g}{V} - 1\right)\kappa_x^2.$$

This equation defines a cone whose axis is in the x-direction.

When the energy is greater than $\mathcal{E}_g + V$ one also finds energy states in the second band. Then the positive square root is chosen in Eq. (4.3.4) and

$$\mathcal{E} \simeq \mathcal{E}_g + V + \frac{\hbar^2}{2m}(k_y^2 + k_z^2) + \frac{\hbar^2 \kappa_x^2}{2m}\left(\frac{2\mathcal{E}_g}{V} + 1\right).$$

The constant-energy surfaces are thus approximately spheroidal in this region of k-space. If it were not for the factor of $(2\mathcal{E}_g/V) + 1$ that multiplies the term in κ_x^2 the surfaces would be spherical, and the free-electron band structure that one finds near $k = 0$ would merely be repeated at the bottom of the second band. It is possible to exploit this similarity by considering the electron energy to be given by the free-electron relationship, with the exception that the inverse of the electron mass must now be considered a *tensor*.

Thus we can write

$$\mathcal{E} \simeq \mathcal{E}_g + V + \tfrac{1}{2}\hbar^2 \mathbf{k} \cdot \mathbf{M}^{-1} \cdot \mathbf{k},$$

with the understanding that the origin of $\mathbf{k}$ is taken to be the point $(\tfrac{1}{2}g, 0, 0)$ and that

$$\mathbf{M}^{-1} \equiv \begin{pmatrix} \dfrac{1}{m}\left(\dfrac{2\mathcal{E}_g}{V}+1\right) & 0 & 0 \\ 0 & \dfrac{1}{m} & 0 \\ 0 & 0 & \dfrac{1}{m} \end{pmatrix}.$$

The inverse-effective-mass tensor in this problem is thus *anisotropic* in that the energy increases more rapidly as a function of k_x than of k_y or k_z. If the lattice potential is weak enough, then $(M^{-1})_{xx}$ may be many times larger than $(M^{-1})_{yy}$. In this case one says that it is a light electron for motion in the x-direction.

It is also possible to interpret the band structure in the first band in these terms by writing Eq. (4.3.5) in the form

$$\mathcal{E} \simeq \mathcal{E}_g - V + \tfrac{1}{2}\hbar^2 \mathbf{k} \cdot \mathbf{M}^{-1} \cdot \mathbf{k}.$$

In this case $(M^{-1})_{xx}$ is negative while $(M^{-1})_{yy}$ is still positive. The electron is said to exhibit *hole-like* behavior for motion in the x-direction. If there were also a periodic potential $2V\cos gy$ then there would also be the possibility of $(M^{-1})_{yy}$ being negative for some points in k-space [Problem 4.5], and for a three-dimensional crystal the states at the corners of the first Brillouin zone will be completely hole-like.

The particular constant-energy surface that represents the boundary between filled and empty states in a metal is again known as the Fermi surface. The shape of the Fermi surface depends on the crystal structure, the lattice potential, and the electron density, and is different for every metal. For some, such as sodium or potassium, the lattice potential is weak and the Fermi surface deviates little from a sphere. For others, and in particular the polyvalent metals, the Fermi surface is far from spherical, and may be formed from regions in several different Brillouin zones.

4.4 Core states and the pseudopotential

In using the nearly-free-electron approximation we have confined our interest for the most part to the first few Brillouin zones. We have tacitly assumed that the conduction-electron states can be found by solving the Schrödinger equation for electrons moving in a weak potential that is composed of the Coulomb attraction of the nuclei screened by the presence of the electrons in the filled atomic shells.

This picture is not justifiable on two counts. Firstly, we must remember that the Exclusion Principle demands that the wavefunctions of the conduction electrons be orthogonal to those of the electrons in the filled atomic shells, or *core states* as we shall call them. Secondly, we can calculate that for real solids the lattice potential is too strong for the nearly-free-electron approach to be valid when only a few plane waves are used. That is to say, the lowest Fourier components $V_\mathbf{g}$ of the lattice potential are not small compared with $\hbar^2 \mathbf{g}^2/8m$ for the smallest reciprocal lattice vectors. However, while either one of these considerations alone would prevent us from using the nearly-free-electron approximation, it happens that taken together they present a tractable situation. Because we are now going to take account of the Bloch states of the core electrons, the determinant (4.3.3) must now be much larger than the 2×2 form that we have just been using. If we wish to apply this method to potassium, for instance, we must calculate that with an atomic number of 19 this metal has enough electrons to fill $9\frac{1}{2}$ Brillouin zones. This means that a large number of Fourier components of the lattice potential must be included if we are to find energy discontinuities at all the relevant zone boundaries. But if the conduction states and the Fermi surface are to be located in higher Brillouin zones than the first few, then their k-vectors in the extended zone scheme must be very large. This means that if they are to be scattered by the lattice to a state of approximately equal energy, then it will be mostly large reciprocal lattice vectors that will describe the difference in wavenumbers of the two states. Consequently it will be the Fourier components $V_\mathbf{g}$ of the lattice potential corresponding to large $\mathbf{g}$ that will describe the energy discontinuities at the zone boundaries. Because these components are much smaller than those corresponding to small $\mathbf{g}$, the validity of the nearly-free-electron approximation is restored as a means of calculating the band structure of the conduction bands.

The use of plane-wave expansions of all the electron states in solids has, however, one big disadvantage. We have not so far made use of the fact that the core states are very highly localized around the nuclei of the atoms. If the core level in the free atom is tightly bound, then its kinetic energy starts to become negative at a short distance from the nucleus, and the wavefunction decays rapidly outside this distance. When many of these atoms are brought together to form a solid, the wavefunctions of the core states of different atoms do not overlap appreciably, and the tendency of the degenerate core states to broaden into a band is very small. This suggests that it would be more appropriate to expand the $\psi_\mathbf{k}$, not in terms of plane waves, but in terms of the atomic wavefunctions, $\phi_i(\mathbf{r})$. Because we know that Bloch's theorem must still be obeyed, we first form linear combinations of normalized atomic wavefunctions centered on different atoms by writing

$$\phi_i^\mathbf{k}(\mathbf{r}) = N^{-1/2} \sum_l e^{i\mathbf{k}\cdot l}\phi_i(\mathbf{r}-l). \tag{4.4.1}$$

We then expand $\psi_\mathbf{k}$ in terms of these, and write

$$\psi_\mathbf{k}(\mathbf{r}) = \sum_i u_i(\mathbf{k})\phi_i^\mathbf{k}(\mathbf{r}).$$

This formalism is known as the Linear Combination of Atomic Orbitals, or LCAO, method.

For an exact solution we should include not only the bound atomic states, but also states of positive energy, so that we have a complete set in which to expand $\psi_\mathbf{k}$. In practice, however, this method is still useful when only a few atomic states are assumed to contribute. We take matrix elements of the Hamiltonian between the states $\phi_i^\mathbf{k}(\mathbf{r})$ and write a secular equation analogous to Eq. (4.3.3) of the form

$$|D| = 0$$

where

$$D_{ij} = \int \phi_i^{\mathbf{k}*}(\mathbf{r})(\mathcal{H} - \mathcal{E})\phi_j^\mathbf{k}(\mathbf{r})\,\mathbf{dr}.$$

We note that the nonorthogonality of the $\phi_i^\mathbf{k}$ must be taken into account.

The LCAO method is widely used for practical computations, as are variations of it in which $\psi_\mathbf{k}$ is expanded in eigenstates of other spherically symmetric potentials. Our goal in this section, however, is not to provide detailed instructions for performing these calculations. It is rather to point out the physical significance of the presence of the core states in reducing the effective lattice potential. To this end we make the most drastic simplification possible, which is known as the method of *tight binding*, and assume that only the diagonal elements contribute to this determinant. The energy is then given simply by the expectation value of $\mathcal{H}$ in the state $\phi_i^{\mathbf{k}}(\mathbf{r})$. Now

$$\mathcal{H}\phi_i^{\mathbf{k}}(\mathbf{r}) = N^{-1/2} \sum_l e^{i\mathbf{k}\cdot l} \mathcal{H}\phi_i(\mathbf{r} - l),$$

and it is convenient to consider the lattice potential $V(\mathbf{r})$ that acts on $\phi_i(\mathbf{r} - l)$ as the sum of two separate terms – the potential due to an atom located at l and that due to all the other atoms. We thus write

$$\mathcal{H} = -\frac{\hbar^2}{2m} \nabla^2 + V_a(\mathbf{r} - l) + W(\mathbf{r} - l)$$

where $V_a(\mathbf{r})$ is the atomic potential and $W(\mathbf{r})$ is the difference between the lattice potential and the atomic potential (Fig. 4.4.1). We expect $W(\mathbf{r})$ to be

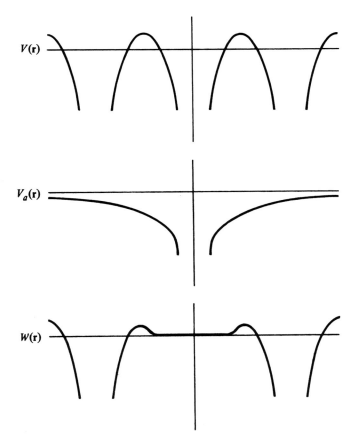

Figure 4.4.1. In the method of tight binding the periodic lattice potential $V(\mathbf{r})$ is considered to be the sum of an atomic potential $V_a(\mathbf{r})$ and a correction $W(\mathbf{r})$ that is small in the neighborhood of the origin. The potential $W(\mathbf{r})$ is then treated as a perturbation acting on the known atomic wavefunctions.

small when $|\mathbf{r}|$ is less than half the interatomic distance. If we define the energy of the atomic state $\phi_i(\mathbf{r})$ as $\mathcal{E}_i$, then

$$\mathcal{H}\phi_i^\mathbf{k}(\mathbf{r}) = N^{-1/2} \sum_l e^{i\mathbf{k}\cdot l}[\mathcal{E}_i + W(\mathbf{r}-l)]\phi_i(\mathbf{r}-l)$$

$$= \mathcal{E}_i\phi_i^\mathbf{k}(\mathbf{r}) + N^{-1/2} \sum_l e^{i\mathbf{k}\cdot l} W(\mathbf{r}-l)\phi_i(\mathbf{r}-l),$$

so that

$$\int \phi_i^{\mathbf{k}*} \mathcal{H}\phi_i^\mathbf{k} \, \mathbf{dr} = \mathcal{E}_i \int \phi_i^{\mathbf{k}*}\phi_i^\mathbf{k} \, \mathbf{dr}$$

$$+ N^{-1} \sum_{l,l'} e^{i\mathbf{k}\cdot(l-l')} \int \phi_i^*(\mathbf{r}-l') W(\mathbf{r}-l)\phi_i(\mathbf{r}-l) \, \mathbf{dr}.$$

If we assume that the $\phi_i(\mathbf{r}-l)$ overlap appreciably only when they are centered on adjacent atoms, the double summation reduces to a sum over pairs of neighboring atoms. With the further approximation that this overlap is small we find the $\phi_i^\mathbf{k}$ to be normalized, and so

$$\mathcal{E} = \mathcal{E}_i + \int \phi_i^*(\mathbf{r})W(\mathbf{r})\phi_i(\mathbf{r}) \, \mathbf{dr}$$

$$+ \sum_\mathbf{L} e^{-i\mathbf{k}\cdot\mathbf{L}} \int \phi_i^*(\mathbf{r}-\mathbf{L})W(\mathbf{r})\phi_i(\mathbf{r}) \, \mathbf{dr},$$

where $\mathbf{L}$ are the different lattice vectors connecting nearest neighbors. As the integrals are just constants, one finds for a Bravais lattice a result of the form

$$\mathcal{E} = \mathcal{E}_0 + W \sum_\mathbf{L} \cos \mathbf{k}\cdot\mathbf{L}. \qquad (4.4.2)$$

The tight-binding method is suitable only when the overlap between atomic wavefunctions is small, and this is appropriate only for states whose energies are well below the Fermi energy. Metals whose energy bands are composed of low-lying core states, which may be approximated by tight-binding wavefunctions, well separated from conduction states, which may be described in the nearly-free-electron approach, are known as *simple metals*. For these it is possible to reformulate the nearly-free-electron description of the conduction states by using as basis functions plane waves that have been modified so as to be automatically orthogonal to the tight-binding states of the core electrons. This is known as the method of orthogonalized plane waves (the

4.4 Core states and the pseudopotential

OPW method). One first defines a set of OPW functions which are formed from plane waves by subtracting their projections on the Bloch waves of the occupied core states. In the tight-binding approximation for the core states

$$\phi_{\text{OPW}}^{\mathbf{k}}(\mathbf{r}) = \Omega^{-1/2}\left[e^{i\mathbf{k}\cdot\mathbf{r}} - \sum_c \phi_c^{\mathbf{k}}(\mathbf{r}) \int \phi_c^{\mathbf{k}*}(\mathbf{r})e^{i\mathbf{k}\cdot\mathbf{r}}\,d\mathbf{r}\right],$$

or, in a briefer notation,

$$|\phi_{\text{OPW}}^{\mathbf{k}}\rangle = |\mathbf{k}\rangle - \sum_c |\phi_c^{\mathbf{k}}\rangle\langle\phi_c^{\mathbf{k}}|\mathbf{k}\rangle.$$

We note that the $\phi_c^{\mathbf{k}}$ were defined in the repeated zone scheme, since by (4.4.1)

$$\phi_c^{\mathbf{k}} = \phi_c^{\mathbf{k}+\mathbf{g}},$$

while the wavenumber $\mathbf{k}$ in the term $e^{i\mathbf{k}\cdot\mathbf{r}}$, is allowed to take on all values, and is thus considered to be in the extended zone scheme. One then again takes matrix elements of $\mathcal{H} - \mathcal{E}$ between the various OPW functions and sets

$$|D| = 0$$

where

$$D_{\mathbf{gg}'} = \langle\phi_{\text{OPW}}^{\mathbf{k}+\mathbf{g}}|\mathcal{H} - \mathcal{E}|\phi_{\text{OPW}}^{\mathbf{k}+\mathbf{g}'}\rangle.$$

The most noticeable difference between this equation and Eq. (4.3.3), in which matrix elements were taken between pure plane waves, is in the off-diagonal elements. Because the OPWs are not mutually orthogonal, we now find terms involving the energy as well as terms involving the tight-binding Bloch energies, $\mathcal{E}_c^{\mathbf{k}}$. When $\mathbf{g} \neq \mathbf{g}'$

$$D_{\mathbf{gg}'} = V_{\mathbf{gg}'} + \sum_c (\mathcal{E} - \mathcal{E}_c^{\mathbf{k}})\langle\mathbf{k}+\mathbf{g}|\phi_c^{\mathbf{k}}\rangle\langle\phi_c^{\mathbf{k}}|\mathbf{k}+\mathbf{g}'\rangle.$$

Because $\mathcal{E} > \mathcal{E}_c^{\mathbf{k}}$, the summation over core states has a tendency to be positive, while $V_{\mathbf{gg}'}$, which is just the Fourier transform of the lattice potential, tends to be negative. The off-diagonal elements of D, and hence the energy discontinuities at the zone boundaries, are thus smaller than we should expect from using the model based on plane waves.

We can see this another way by explicitly separating the core functions from the sum of OPWs that form the complete wavefunction. Let us first abbreviate the operator that projects out the core Bloch states by the symbol $P_\mathbf{k}$. Thus

$$P_\mathbf{k} \equiv \sum_c |\phi_c^\mathbf{k}\rangle \langle \phi_c^\mathbf{k}|$$

and

$$|\phi_{\text{OPW}}^{\mathbf{k+g}}\rangle = (1 - P_\mathbf{k})|\mathbf{k+g}\rangle.$$

The exact wavefunction is a sum of OPWs, and so

$$|\psi_\mathbf{k}\rangle = \sum_\mathbf{g} u_\mathbf{g}^{\text{OPW}}(\mathbf{k})(1 - P_\mathbf{k})|\mathbf{k+g}\rangle,$$

which we write as

$$|\psi_\mathbf{k}\rangle = (1 - P_\mathbf{k})|\chi_\mathbf{k}\rangle$$

where

$$|\chi_\mathbf{k}\rangle = \sum_\mathbf{g} u_\mathbf{g}^{\text{OPW}}(\mathbf{k})|\mathbf{k+g}\rangle.$$

Then

$$(\mathcal{H} - \mathcal{E})|\psi_\mathbf{k}\rangle = (\mathcal{H} - \mathcal{E})(1 - P_\mathbf{k})|\chi_\mathbf{k}\rangle = 0 \tag{4.4.3}$$

and we may look upon the problem not as one of finding the states $|\psi_\mathbf{k}\rangle$ that are eigenfunctions of $\mathcal{H} - \mathcal{E}$, but as one of finding the states $|\chi_\mathbf{k}\rangle$ that are eigenfunctions of $(\mathcal{H} - \mathcal{E})(1 - P_\mathbf{k})$. Now $\mathcal{H}$ is composed of kinetic energy T and potential V, so that $\chi_\mathbf{k}$ must be an eigenfunction of

$$(T + V - \mathcal{E})(1 - P_\mathbf{k}) = T - \mathcal{E} + V(1 - P_\mathbf{k}) - (T - \mathcal{E})P_\mathbf{k}.$$

Thus (4.4.3) can be written

$$(T + U_\mathbf{k})|\chi_\mathbf{k}\rangle = \mathcal{E}|\chi_\mathbf{k}\rangle,$$

where the operator

$$U_\mathbf{k} = V(1 - P_\mathbf{k}) - (T - \mathcal{E})P_\mathbf{k} \tag{4.4.4}$$

is known as a *pseudopotential* operator. We can argue that we expect it to have only a small effect on the *pseudo-wavefunction* $\chi_\mathbf{k}$ by noting two points.

4.4 Core states and the pseudopotential

Firstly we expect $V(1 - P_k)$ to have small matrix elements, since it is just what is left of V after all the core states have been projected out of it. The strongest part of V will be found in the regions near the atomic nuclei, and so the core states, which are concentrated in the same regions, will be suitable functions in which to expand V. The combination $T - \mathcal{E}$, on the other hand, is not so drastically affected by the operation $1 - P_k$. On the contrary, it becomes reasonable to assume that $(T - \mathcal{E})P_k$ has only a small effect, for there will be little overlap between χ_k and the core states if χ_k is indeed just a combination of a few plane waves of small wavenumber. It thus is self-consistent to assume that the pseudopotential is weak and that χ_k is a smoothly varying function. One should remember, however, that although U_k may be small it remains an operator rather than a simple potential, and has a dependence on energy that must sometimes be treated carefully.

Yet another way of looking at the pseudopotential is obtained by defining a new Hamiltonian $\mathcal{H}'$ formed by adding $(\mathcal{E} - \mathcal{H})P_k$ to the original Hamiltonian. Then

$$\mathcal{H}' = \mathcal{H} + (\mathcal{E} - \mathcal{H})P_k$$
$$= \mathcal{H} + \sum_c (\mathcal{E} - \mathcal{E}_c^k)|\phi_c^k\rangle\langle\phi_c^k|.$$

The extra terms added to $\mathcal{H}$ have artificially raised the energies of the core states to be equal to $\mathcal{E}$, as can be seen by letting $\mathcal{H}'$ act on the ϕ_c^k. Now since the lowest energy levels of $\mathcal{H}'$ are degenerate, we can state that any linear combination of ψ_k and the ϕ_c^k are eigenstates of $\mathcal{H}'$, and we are at liberty to choose that combination χ_k that is most smoothly varying, and hence which can be best approximated by the fewest plane waves. This expresses the fact that the pseudo-wavefunction χ_k is not uniquely defined by (4.4.3), which only says that the part of χ_k that is orthogonal to the core states must be equal to ψ_k.

When pseudopotentials are used in numerical calculations, their character as operators makes itself felt. One must then deal with a nonlocal form of the pseudopotential in which the interaction between an electron and a nucleus depends on their coordinates separately, and not only on their relative coordinates. Fortunately, the pseudopotential can usually be split into factors, each of which depends on only one separate coordinate. This greatly reduces the memory requirements for computer calculations. Pseudopotentials have been developed in which the normalization of the pseudo-wavefunction $|\chi_k\rangle$ has been relaxed in favor of making the pseudopotential as soft as possible. While this leads to a slight complication in calculating the electron charge density, the advantage of these so-called *ultrasoft pseudopotentials* is that many fewer plane waves are required in expansions of the electron valence states.

In summary, then, pseudopotential theory serves to show that the band structure of simple metals may be much closer to that of the remapped free-electron model than one would be led to believe by considering the strength of the lattice potential alone.

4.5 Exact calculations, relativistic effects, and the structure factor

Although pseudopotential theory provides a useful short cut for the calculation of band structures and Fermi surfaces of simple metals, there remain many cases for which it is difficult to implement. In *transition metals*, for example, the electron states of interest are formed from atomic s-states and d-states, and thus mix core-like and free-electron-like behavior. To account correctly for the magnetic properties of transition metals, care has to be taken to include adequately the interactions between bands formed from 3d and 4s states and deeper-lying bands formed from atomic 3s and 3p states. For these cases a variety of ways of solving the Schrödinger equation have been derived, and these are discussed in great detail in the many books now available that are devoted solely to band structure calculations. Here we shall outline just one such method which follows fairly naturally from Eq. (2.5.6), the starting point of Brillouin–Wigner perturbation theory.

Equation (2.5.6) may be written in the form

$$|\psi\rangle = a|\phi\rangle + (\mathcal{E} - \mathcal{H}_0)^{-1} V |\psi\rangle,$$

where a is a constant whose value is determined by the condition that the presence of the term $a|\phi\rangle$ ensure that $|\psi\rangle$ reduces to the unperturbed state $|\phi\rangle$ as V tends to zero. For the present problem we take $\mathcal{H}_0$ to be the kinetic energy of a single electron and V the lattice potential, so that for the Bloch state $|\psi_\mathbf{k}\rangle$

$$|\psi_\mathbf{k}\rangle = a|\mathbf{k}\rangle + (\mathcal{E} - \mathcal{H}_0)^{-1} V |\psi_\mathbf{k}\rangle$$

$$= a|\mathbf{k}\rangle + \sum_{\mathbf{k}'} \left(\mathcal{E} - \frac{\hbar^2 \mathbf{k}'^2}{2m}\right)^{-1} |\mathbf{k}'\rangle \langle \mathbf{k}'| V |\psi_\mathbf{k}\rangle.$$

But since $\psi_\mathbf{k}$ is a Bloch state and $V(\mathbf{r})$ is a periodic function, the matrix element $\langle \mathbf{k}'|V|\psi_\mathbf{k}\rangle$ must vanish unless, for some reciprocal lattice vector $\mathbf{g}$,

4.5 Exact calculations, relativistic effects, and the structure factor

we find that $\mathbf{k}' = \mathbf{k} + \mathbf{g}$. This follows from the direct substitution

$$\langle \mathbf{k}'|V|\psi_\mathbf{k}\rangle \propto \int e^{-i\mathbf{k}'\cdot\mathbf{r}} V(\mathbf{r}) e^{i\mathbf{k}\cdot\mathbf{r}} u_\mathbf{k}(\mathbf{r})\, d\mathbf{r}$$

$$= \int e^{i(\mathbf{k}-\mathbf{k}')\cdot(\mathbf{r}-\mathbf{l})} V(\mathbf{r}-\mathbf{l}) u_\mathbf{k}(\mathbf{r}-\mathbf{l})\, d\mathbf{r}$$

$$= \frac{1}{N} \sum_l e^{-i(\mathbf{k}-\mathbf{k}')\cdot\mathbf{l}} \int e^{i(\mathbf{k}-\mathbf{k}')\cdot\mathbf{r}} V(\mathbf{r}) u_\mathbf{k}(\mathbf{r})\, d\mathbf{r}$$

$$= 0 \quad \text{unless } \mathbf{k} - \mathbf{k}' = \mathbf{g}.$$

Thus

$$|\psi_\mathbf{k}\rangle = a|\mathbf{k}\rangle + \sum_\mathbf{g} \left[\mathcal{E} - \frac{\hbar^2(\mathbf{k}+\mathbf{g})^2}{2m}\right]^{-1} |\mathbf{k}+\mathbf{g}\rangle\langle\mathbf{k}+\mathbf{g}|V|\psi_\mathbf{k}\rangle.$$

Let us define an operator $G_\mathbf{k}(\mathcal{E})$ by writing

$$G_\mathbf{k}(\mathcal{E}) \equiv \sum_\mathbf{g} \left[\mathcal{E} - \frac{\hbar^2(\mathbf{k}+\mathbf{g})^2}{2m}\right]^{-1} |\mathbf{k}+\mathbf{g}\rangle\langle\mathbf{k}+\mathbf{g}|.$$

That is, $G_\mathbf{k}(\mathcal{E})$ is just the operator $(\mathcal{E} - \mathcal{H}_0)^{-1}$ restricted to act only on states that are of the Bloch form with wavenumber $\mathbf{k}$. Then

$$|\psi_\mathbf{k}\rangle = a|\mathbf{k}\rangle + G_\mathbf{k}(\mathcal{E})V|\psi_k\rangle.$$

We can verify by making use of the normalization condition $\langle\mathbf{k}|\psi_\mathbf{k}\rangle = 1$ that in this case the constant a can be put equal to zero (Problem 4.15), so that

$$|\psi_\mathbf{k}\rangle = G_\mathbf{k}(\mathcal{E})V|\psi_k\rangle. \tag{4.5.1}$$

It then follows that if one defines a quantity Λ by

$$\Lambda = \langle\psi_\mathbf{k}|V|\psi_\mathbf{k}\rangle - \langle\psi_\mathbf{k}|VG_\mathbf{k}(\mathcal{E})V|\psi_\mathbf{k}\rangle, \tag{4.5.2}$$

then from (4.5.1) we find that Λ vanishes when ψ_k is a solution of the Schrödinger equation. That is, since

$$V|\psi_\mathbf{k}\rangle = (\mathcal{E} - \mathcal{H}_0)|\psi_\mathbf{k}\rangle$$

we are in effect writing

$$\Lambda = \langle \psi_\mathbf{k} | V | \psi_\mathbf{k} \rangle - \langle \psi_\mathbf{k} | (\mathcal{E} - \mathcal{H}_0) \left(\frac{1}{\mathcal{E} - \mathcal{H}_0} \right) (\mathcal{E} - \mathcal{H}_0) | \psi_\mathbf{k} \rangle$$

$$= \langle \psi_\mathbf{k} | (\mathcal{E} - \mathcal{H}) | \psi_\mathbf{k} \rangle.$$

But now we not only know that Λ vanishes, but also that we may determine $\psi_\mathbf{k}$ by a variational approach that minimizes Λ.

In terms of integrals in r-space, expression (4.5.2) may be written

$$\Lambda = \int \psi_\mathbf{k}^*(\mathbf{r}) V \mathbf{r}) \psi_\mathbf{k}(\mathbf{r}) \, d\mathbf{r}$$

$$- \iint \psi_\mathbf{k}^*(\mathbf{r}) V(\mathbf{r}) G_\mathbf{k}(\mathbf{r} - \mathbf{r}') V(\mathbf{r}') \psi_\mathbf{k}(\mathbf{r}') \, d\mathbf{r} \, d\mathbf{r}' \quad (4.5.3)$$

where

$$G_\mathbf{k}(\mathbf{r} - \mathbf{r}') = \Omega^{-1} \sum_\mathbf{g} \left[\mathcal{E} - \frac{\hbar^2 (\mathbf{k} + \mathbf{g})^2}{2m} \right]^{-1} e^{i(\mathbf{k}+\mathbf{g}) \cdot (\mathbf{r}-\mathbf{r}')}.$$

An advantage claimed for this method is that $G_\mathbf{k}(\mathbf{r} - \mathbf{r}')$ depends only $\mathcal{E}$, $\mathbf{k}$, $\mathbf{r} - \mathbf{r}'$ and the positions of the reciprocal lattice sites $\mathbf{g}$, and may thus be computed once and for all for any particular crystal structure. One may then use this function in conjunction with whatever lattice potential $V(\mathbf{r})$ is appropriate to the material under consideration. This approach is known as the Korringa–Kohn–Rostoker (KKR) or Green's-function method.

One effective approach of this type is known as the Linear Muffin-Tin Orbital method, or LMTO method. If we look back to Eq. (4.5.1) we may be reminded of equations used to describe the scattering of a particle by a single spherically-symmetric scatterer. Particle-scattering theory describes the perturbed wavefunction ψ in terms of *phase shifts* for the various angular-momentum components of the scattering. This suggests that the operator $G_\mathbf{k}$ ($\mathcal{E}$) could be recast in a basis of spherical-harmonic components that are solutions of the Schrödinger equation for a single spherically-symmetric potential. It is convenient to choose a potential that approximates the actual effective potential of the atom, but which vanishes outside a certain radius (hence the term "muffin-tin"!). This reduces the computational effort needed, and makes possible the calculation of band structures in crystals having bases of hundreds of atoms per unit cell.

Before leaving the topic of the calculation of band structures, we should glance briefly at the question of when it is valid to ignore *relativistic effects*. The Fermi energies of metals, by which we generally mean the energy difference between the lowest and highest filled conduction state, are of the order of a few electron volts. As this figure is smaller than the rest-mass energy by a factor of about 10^{-5}, it might at first be thought that we could always neglect such effects. However, one must remember that the potential wells near the nuclei of heavy atoms are very deep, and that only a small change in energy may sometimes cause qualitative differences in band structure in semiconductors in which the band gaps are small.

Accordingly, we turn to the Dirac equation, which describes the motion of a relativistic electron in terms of a four-component wavefunction. Because the Dirac Hamiltonian, like the Schrödinger Hamiltonian, has the periodicity of the lattice, each component of the wavefunction obeys Bloch's theorem, and one may associate a wavenumber $\mathbf{k}$ with each state. It is thus formally possible to recast the OPW method in terms of the orthogonalization of four-component plane waves to the four-component tight-binding core states, a procedure that was first carried through for thallium. However, because the relativistic effects usually contribute only a small amount to the total energy of a Bloch electron, it is often possible to treat them as a perturbation of the nonrelativistic band structure. In Section 3.10 we noted that the Dirac equation could be reduced by means of the Foldy–Wouthuysen transformation to an equation of the form

$$\mathcal{H}\psi = \mathcal{E}\psi$$

where now ψ is a two-component wavefunction describing an electron with spin $\frac{1}{2}$, and $\mathcal{H}$, in the absence of a magnetic field, is given by

$$\mathcal{H} = \frac{\mathbf{p}^2}{2m} + V(\mathbf{r}) - \frac{\mathbf{p}^4}{8m^3c^2}$$
$$+ \frac{1}{2m^2c^2}\mathbf{s}\cdot[\nabla V(\mathbf{r})\times\mathbf{p}] + \frac{\hbar^2}{8m^2c^2}\nabla^2 V + \cdots.$$

The third term in this expression simply reflects the relativistic increase in mass of the electron, and is known as the mass–velocity term. The next term contains the spin angular momentum $\mathbf{s}$ of the electron, and is the spin–orbit coupling term. It may be qualitatively understood as the energy of alignment of the intrinsic magnetic moment of the electron in the magnetic field caused by its own orbital motion. The fifth term is known as the Darwin term, and can be thought of as a correction due to the finite radius of the electron.

If these three correction terms are to be treated as perturbations to the Schrödinger Hamiltonian, then it is necessary to calculate not only the energies of the Bloch states in the various bands, but also their wavefunctions, which involves considerably more labor. Some calculations have consequently been made using the tight-binding approximation to describe the wavefunctions. Because the relativistic terms are important only in the vicinity of the atomic nuclei, the tight-binding model provides a wavefunction whose shape is a very good approximation to that of the true Bloch state in the region that is important. However, the amplitude of the wavefunction and consequently the size and k-dependence of these effects may be less accurately predicted.

Since the Darwin, mass–velocity, and spin–orbit terms give energy shifts of comparable magnitude, they must all be considered in semiconductors such as PbTe in which such small perturbations may qualitatively change the band structure. Because their effect is strongest close to the atomic nuclei, the Darwin and mass–velocity terms tend to lower the energies of s-states relative to p- and d-states. Some of the differences in properties between copper, silver, and gold arise in this way. In the hexagonal metals it is the spin–orbit term which, because of its lack of symmetry, most often causes observable effects. The detailed study of the effect of the spin–orbit term on band structure is a difficult topic which requires some knowledge of group theory, but the nature of the effects can be seen from the following simple examples.

Let us consider first the zone boundary at $k_x = g$ in the sandwichium model used in Section 4.3 and in Problems 4.1 and 4.2. If we were to use the nearly-free-electron approximation with only the two plane waves $e^{ik_x x}$ and $e^{i(k_x - 2g)x}$ we should find no discontinuity in the energy at $k_x = g$. This would be a consequence of the vanishing of the matrix element

$$V_{2g} \propto \int e^{-ik_x x} 2V \cos gx \, e^{i(k_x - 2g)x} \, \mathbf{dr}.$$

If, however, we were to use the three plane waves $e^{ik_x x}$, $e^{i(k_x - g)x}$, and $e^{i(k_x - 2g)x}$, then we should find a discontinuity in the energy at $k_x = g$ (Problem 4.9). In physical terms we could say that the electron is scattered by the lattice first from k_x to $k_x - g$, and then from $k_x - g$ to $k_x - 2g$. Accordingly the discontinuity in energy is proportional to V^2, rather than to V as was the case at the zone boundary at $k_x = \frac{1}{2}g$.

4.5 Exact calculations, relativistic effects, and the structure factor

In contrast to this we now consider a square lattice of side a in which there are two identical atoms per unit cell, one at $(\frac{1}{4}a, \frac{1}{4}a)$ and one at $(-\frac{1}{4}a, -\frac{1}{4}a)$, as shown in Fig. 4.5.1(a). The first Brillouin zone is then a square of side $2\pi/a$, while the second Brillouin zone is contained by a square of side $2\sqrt{2}\pi/a$, as shown in Fig. 4.5.1(b). Once again we find that certain Fourier components of the potential vanish, so that, for example, $V_{\mathbf{g}} = 0$ when $\mathbf{g} = (0, 2\pi/a)$ or $(2\pi/a, 0)$. More generally one may suppose the lattice potential to be composed of atomic potentials V_a centered on the various sites, so that

$$V(\mathbf{r}) = \sum_{l,b} V_a(\mathbf{r} - \mathbf{l} - \mathbf{b}), \tag{4.5.4}$$

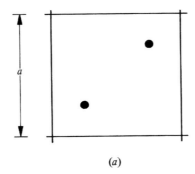

(a)

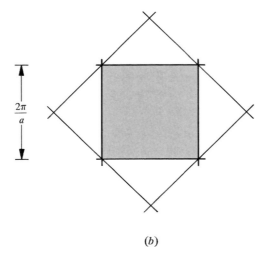

(b)

Figure 4.5.1. In this two-dimensional model the unit cell in r-space (a) is a square of side a containing two atoms at the points $\pm(\frac{1}{4}a, \frac{1}{4}a)$. The first Brillouin zone (b) is then a square of side $2\pi/a$.

where the *l* describe the positions of the centers of the unit cells and the **b** describe the positions of the atoms within the cell, so that in this case

$$\mathbf{b}_1 = (\tfrac{1}{4}a, \tfrac{1}{4}a); \quad \mathbf{b}_2 = (-\tfrac{1}{4}a, -\tfrac{1}{4}a).$$

Then

$$V_\mathbf{g} = \frac{1}{\Omega} \int e^{-i\mathbf{g}\cdot\mathbf{r}} V(\mathbf{r})\, d\mathbf{r}$$

$$= \frac{N}{\Omega} \sum_\mathbf{b} e^{-i\mathbf{g}\cdot\mathbf{b}} \int e^{-i\mathbf{g}\cdot\mathbf{r}} V_a(\mathbf{r})\, d\mathbf{r}.$$

The summation

$$S_\mathbf{g} = \sum_\mathbf{b} e^{-i\mathbf{g}\cdot\mathbf{b}}$$

is known as the *structure factor*, and in this case vanishes when

$$g_x + g_y = (2n+1)\frac{2\pi}{a}$$

for all integral n.

The vanishing of the structure factor, and hence of $V_\mathbf{g}$, for $\mathbf{g} = (0, 2\pi/a)$ and $(2\pi/a, 0)$ means that there is no discontinuity in energy to first order in the lattice potential at the boundaries of the first Brillouin zone. But this is not all. In this model we should find that *to all orders in the lattice potential*, there is no discontinuity at these zone boundaries. This fact becomes obvious if we merely tilt our heads on one side and notice that, in fact, we are really just considering a square Bravais lattice of side $a/\sqrt{2}$ (Fig. 4.5.2) whose first Brillouin zone is bounded by the same square of side $2\sqrt{2}\pi/a$ that was the boundary of the second Brillouin zone in our first way of looking at the model. We note the distinction between the vanishing of the structure factor, which is a property only of the crystal structure, and the vanishing of $V_{2\mathbf{g}}$ in sandwichium, which was an accident of our choice of potential.

This possibility that the energy discontinuity may vanish identically at some Brillouin zone boundaries is not confined to such artificial models as the present one. In such common structures as hexagonal close-packed, in which more than a dozen elements crystallize, and in the diamond and graphite structures, this very phenomenon occurs. This makes it reasonable to define a new set of zones that are separated by planes on which energy discontinuities *do* occur. These are known as *Jones zones*. The construction by which one defines which Jones zone a particular state is in is the following.

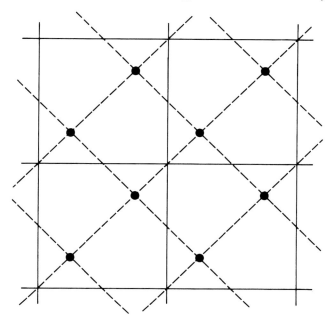

Figure 4.5.2. The model of which a unit cell was shown in Fig. 4.5.1(a) is here seen to be merely a square lattice of side $2^{-1/2}a$. This explains why no energy discontinuities were found at the boundaries of the first Brillouin zone shown in Fig. 4.5.1(b).

A straight line is drawn from the origin of k-space to the point **k** in the extended zone scheme. If this line passes through n discontinuities in energy, then **k** is in the $(n+1)$th Jones zone.

The relevance of spin–orbit coupling to these considerations lies in the fact that the lack of symmetry in this term in the Hamiltonian can cause the reappearance of energy discontinuities within the Jones zones of some crystal structures. The hexagonal close-packed structure is a particularly important example of a structure in which such effects have been observed. Although this particular lattice is rather complicated to investigate here, we can understand the way in which the energy gaps are restored by the spin–orbit interaction by considering a modification of the square lattice shown in Fig. 4.5.1. We retain the square cell of side a, but this time we place the two identical atoms at $(\frac{1}{6}a, \frac{1}{4}a)$ and $(-\frac{1}{6}a, -\frac{1}{4}a)$, as shown in Fig. 4.5.3. The structure factor will now be

$$S_{\mathbf{g}} = 2\cos\left(\frac{ag_x}{6} + \frac{ag_y}{4}\right),$$

158   One-electron theory

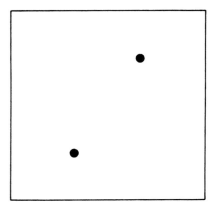

Figure 4.5.3. In this modification of the model shown in Fig. 4.5.1(a) the atoms are now placed at the points $\pm(\frac{1}{6}a, \frac{1}{4}a)$.

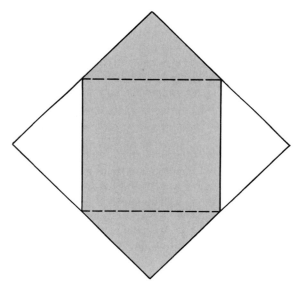

Figure 4.5.4. Although the Brillouin zone for the lattice shown in Fig. 4.5.3 is a square, no discontinuities in energy occur at the dashed lines when a nonrelativistic Hamiltonian is used.

which will still vanish for $\mathbf{g} = (0, 2\pi/a)$ but no longer for $\mathbf{g} = (2\pi/a, 0)$. In first order the energy discontinuities will then occur at the solid lines of Fig. 4.5.4. Let us now suppose that we use the nearly-free-electron approximation to find the wavefunctions that result from considering the Fourier component $V_\mathbf{g}$ of the lattice potential for $\mathbf{g} = (2\pi/a, 0)$. We then could write

4.5 Exact calculations, relativistic effects, and the structure factor

the wavefunctions in the form

$$\psi_{\mathbf{k}} = \Omega^{-1/2}[u_0 e^{i\mathbf{k}\cdot\mathbf{r}} + u_1 e^{i(\mathbf{k}\cdot\mathbf{r} - 2\pi x/a)}]$$

with u_0 and u_1 a pair of real coefficients which we could determine explicitly in terms of $\mathbf{k}$, if we so wished. We might now look for a second-order discontinuity along the lines $k_y = \pi/a$ (the dashed lines in Fig. 4.5.4) by seeing whether the lattice potential can mix the states of wavenumbers $\mathbf{k}$ and $\mathbf{k} + \mathbf{g}$ with $\mathbf{g} = (0, -2\pi/a)$. We thus form

$$W = \int \psi_{\mathbf{k}+\mathbf{g}}^* V(\mathbf{r}) \psi_{\mathbf{k}} \, d\mathbf{r}$$

$$= \Omega^{-1} \int (u_0 + u_1 e^{2\pi i x/a}) e^{2\pi i y/a} V(\mathbf{r})(u_0 + u_1 e^{-2\pi i x/a}) \, d\mathbf{r}.$$

A substitution of the form (4.5.4) then serves to show that W vanishes because of the form of the structure factor; there are terms in u_0^2 and in u_1^2 which vanish because $S_{\mathbf{g}} = 0$ for $\mathbf{g} = (0, 2\pi/a)$ and two terms in $u_0 u_1$ which cancel because the value of $S_{\mathbf{g}}$ when $\mathbf{g} = (2\pi/a, -2\pi/a)$ is the negative of its value when $\mathbf{g} = (-2\pi/a, -2\pi/a)$. If, however, we add to $V(\mathbf{r})$ the spin–orbit term we shall find a different result. Then

$$W = \frac{1}{2m^2 c^2} \mathbf{s} \cdot \int \psi_{\mathbf{k}+\mathbf{g}}^* [\nabla V(\mathbf{r}) \times \mathbf{p}] \psi_{\mathbf{k}} \, d\mathbf{r}.$$

The terms in u_0^2 and u_1^2 still vanish, but the cross term leaves a contribution from the different values $\mathbf{p}$ takes when acting on the two plane-wave components of $\psi_{\mathbf{k}}$. One finds

$$W = -\frac{\hbar u_0 u_1}{2m^2 c^2 \Omega} \mathbf{s} \cdot \int e^{2\pi i (y-x)/a} \nabla V \times (2\pi/a, 0) \, d\mathbf{r}.$$

The integral is proportional to $S_{\mathbf{g}}$ for $\mathbf{g} = (2\pi/a, -2\pi/a)$, which does not vanish. The degenerate states $\psi_{\mathbf{k}}$ and $\psi_{\mathbf{k}+\mathbf{g}}$ are thus mixed by the spin–orbit interaction, and energy discontinuities reappear at the Brillouin zone boundaries. Although these splittings are usually small they are still sufficient to alter the topology of the Fermi surface, and thus cause effects which are readily observable.

4.6 Dynamics of Bloch electrons

In what we have considered so far, the wavenumber **k** has been little more than a label for the Bloch states. Experiments, however, are concerned with such measurable properties as the electric current carried by a system of electrons in the presence of applied fields. We accordingly now turn to a consideration of the *velocity* of Bloch electrons and the modification of this quantity by applied electric and magnetic fields.

The velocity of an electron in the absence of a magnetic field is proportional to the expectation value of its momentum

$$\mathbf{v} = \frac{1}{m}\int \psi^* \mathbf{p}\psi \, d\mathbf{r}$$

$$= -\frac{i\hbar}{m}\int \psi^* \nabla\psi \, d\mathbf{r}. \qquad (4.6.1)$$

We can relate this to the band structure by returning to the Schrödinger equation written in the form (4.1.4)

$$\mathcal{H}_\mathbf{k} u_\mathbf{k}(\mathbf{r}) = \mathcal{E}_\mathbf{k} u_\mathbf{k}(\mathbf{r}) \qquad (4.6.2)$$

where

$$\mathcal{H}_\mathbf{k} \equiv -\frac{\hbar^2}{2m}(\nabla + i\mathbf{k})^2 + V(\mathbf{r}).$$

We differentiate (4.6.2) with respect to **k** (that is, we take the gradient in k-space) to find

$$(\mathcal{H}_\mathbf{k} - \mathcal{E}_\mathbf{k})\frac{\partial}{\partial \mathbf{k}} u_\mathbf{k}(\mathbf{r}) = -\left[\frac{\partial}{\partial \mathbf{k}}(\mathcal{H}_\mathbf{k} - \mathcal{E}_\mathbf{k})\right] u_\mathbf{k}(\mathbf{r})$$

$$= \left[\frac{i\hbar}{m}(\nabla + i\mathbf{k}) + \frac{\partial \mathcal{E}_\mathbf{k}}{\partial \mathbf{k}}\right] u_\mathbf{k}(\mathbf{r}).$$

But since

$$\psi_\mathbf{k} = e^{i\mathbf{k}\cdot\mathbf{r}} u_\mathbf{k}(\mathbf{r}),$$

then from (4.6.1)

$$\hbar \mathbf{v}_\mathbf{k} = \frac{\partial \mathcal{E}_\mathbf{k}}{\partial \mathbf{k}} - \int u_\mathbf{k}^*(\mathbf{r})(\mathcal{H}_k - \mathcal{E}_k)\frac{\partial}{\partial \mathbf{k}} u_\mathbf{k}(\mathbf{r}) \, d\mathbf{r}.$$

4.6 Dynamics of Bloch electrons

The integral vanishes because of the Hermitian nature of $\mathcal{H}_\mathbf{k}$, as can be seen by integrating by parts, leaving the result

$$\mathbf{v}_\mathbf{k} = \frac{1}{\hbar} \frac{\partial \mathcal{E}_\mathbf{k}}{\partial \mathbf{k}}. \tag{4.6.3}$$

This result appears more familiar if we define a frequency $\omega_\mathbf{k}$ by writing $\mathcal{E}_\mathbf{k} = \hbar \omega_\mathbf{k}$. Then $\mathbf{v}_\mathbf{k} = \partial \omega_\mathbf{k} / \partial \mathbf{k}$, which is the usual result for the group velocity of a wave of angular frequency $\omega_\mathbf{k}$ in a dispersive medium.

We now know the total electric current carried by the conduction electrons if we know which k-states are occupied. The current density due to a single electron in the state $\mathbf{k}$ will be $e\mathbf{v}_\mathbf{k}/\Omega$, so that the total current density is

$$\mathbf{j} = \Omega^{-1} \sum_\mathbf{k} n_\mathbf{k} e \mathbf{v}_\mathbf{k}. \tag{4.6.4}$$

In this independent-particle model the occupation number $n_\mathbf{k}$ takes on only the values 0 or 1.

In equilibrium $\mathbf{j}$, which is of course a macroscopic quantity, vanishes, and if we are to set up a current flow we must first apply an electric field by, for example, adding to the Hamiltonian a potential $-e\mathbf{E} \cdot \mathbf{r}$. There are now two paths open to us in investigating the effect of the electric field – the time-dependent approach and the time-independent approach. At first it seems that one should treat the applied field as a perturbation and look for the eigenstates of the perturbed system. Because the Hamiltonian is constant in time there appears no reason to use time-dependent methods. Unfortunately, however, this approach is a very difficult one, the chief difficulty arising from the fact that no matter how small $\mathbf{E}$ is, the potential $-e\mathbf{E} \cdot \mathbf{r}$ cannot be treated as a perturbation in an infinite system because $\mathbf{r}$ then becomes indefinitely large. A similar difficulty arises when one applies a magnetic field, the vector potential then becoming large at large distances. We shall consequently leave the question of the eigenstates of Bloch electrons in applied fields and turn to the time-dependent approach.

A wave packet traveling with velocity $\mathbf{v}_\mathbf{k}$ in a uniform force field $e\mathbf{E}$ might be expected to increase its energy at the rate $e\mathbf{E} \cdot \mathbf{v}_\mathbf{k}$. On the other hand, if this change in energy reflects a change in the wavenumber of the Bloch states forming the wave packet, we could write

$$\frac{d\mathcal{E}_\mathbf{k}}{dt} = \frac{\partial \mathcal{E}_\mathbf{k}}{\partial \mathbf{k}} \cdot \frac{d\mathbf{k}}{dt} = \hbar \mathbf{v}_\mathbf{k} \cdot \frac{d\mathbf{k}}{dt}$$

from (4.6.3). For these two pictures to be equivalent we must have

$$\hbar \frac{d\mathbf{k}}{dt} = e\mathbf{E}. \qquad (4.6.5)$$

This result is not quite correct, as it is really only the kinetic energy that we should expect to increase at the rate $e\mathbf{E} \cdot \mathbf{v_k}$, and the potential energy of the Bloch state will also be changing if $\mathbf{k}$ is changing. To see this more clearly we can consider the time-dependent Schrödinger equation for an electron initially in a Bloch state of wavenumber $\mathbf{k}$. Then

$$\Psi_\mathbf{k}(\mathbf{r}, t) = \psi_\mathbf{k}(\mathbf{r})e^{-i\mathcal{E}t/\hbar}$$
$$= e^{i\mathbf{k}\cdot\mathbf{r}} u_\mathbf{k}(\mathbf{r}) e^{-i\mathcal{E}t/\hbar}$$

satisfies the Schrödinger equation in the absence of the applied field. If we now add the potential $-e\mathbf{E}\cdot\mathbf{r}$ to the Bloch Hamiltonian $\mathcal{H}_0$, then at $t=0$

$$i\hbar \frac{\partial \Psi_\mathbf{k}}{\partial t} = (\mathcal{H}_0 - e\mathbf{E}\cdot\mathbf{r})\Psi_\mathbf{k}$$
$$= (\mathcal{E}_\mathbf{k} - e\mathbf{E}\cdot\mathbf{r})\Psi_\mathbf{k}. \qquad (4.6.6)$$

But if the only change in $\Psi_\mathbf{k}$ is to be a change in $\mathbf{k}$ at the rate given by (4.6.5) we should find at $t=0$

$$i\hbar \frac{\partial \Psi_\mathbf{k}}{\partial t} = i\hbar \left[\left(\frac{\partial \Psi_k}{\partial t}\right)_\mathbf{k} + \left(\frac{\partial \Psi_\mathbf{k}}{\partial \mathbf{k}}\right)_t \cdot \frac{d\mathbf{k}}{dt} \right]$$
$$= \mathcal{E}_\mathbf{k}\Psi_\mathbf{k} + ie\mathbf{E}\cdot e^{i\mathbf{k}\cdot\mathbf{r}}\left(i\mathbf{r} + \frac{\partial}{\partial \mathbf{k}}\right)u_\mathbf{k}(\mathbf{r})$$
$$= \left(\mathcal{E}_\mathbf{k} - e\mathbf{E}\cdot\mathbf{r} + ie\mathbf{E}\cdot\frac{\partial \ln u_\mathbf{k}(\mathbf{r})}{\partial \mathbf{k}}\right)\Psi_\mathbf{k}. \qquad (4.6.7)$$

When the third term in (4.6.7) is neglected, this expression becomes identical to (4.6.6) and one may say that the wavenumber of a Bloch electron is changed by the field at just the same rate as that of a free electron. We note, however, that the rate of change of the *velocity* of the electron bears no similarity to that of the free particle, in that as $\mathbf{k}$ approaches a zone boundary the velocity may fall to zero. This would be the case in sandwichium for $\mathbf{k} = (k_x, 0, 0)$, as shown in Fig. 4.6.1. The discontinuity in slope of $\mathbf{v_k}$ at the zone boundary draws attention to the fact that we do not expect a weak steady field to be able to provide the energy to enable the electron to

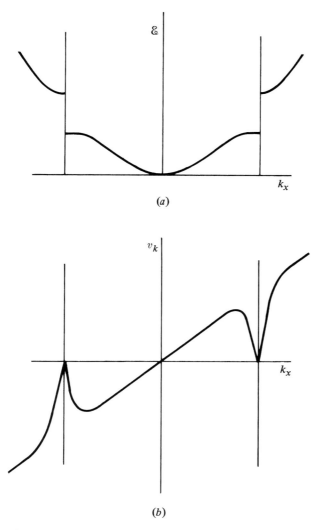

Figure 4.6.1. When the energy (a) varies in the k_x direction in sandwichium in the usual way, the velocity (b) in this direction falls to zero at the zone boundaries.

move from the first to the second Brillouin zone. That is, we cannot interpret (4.6.5) in the extended zone scheme, but must look more closely at the Bloch states for which **k** lies directly on the zone boundary.

In the two-plane-wave approximation for sandwichium, for example,

$$u_{\mathbf{k}}(\mathbf{r}) = u_0(\mathbf{k}) + u_1(\mathbf{k})e^{i\mathbf{g}\cdot\mathbf{r}},$$

and as **k** approaches $(\tfrac{1}{2}g, 0, 0)$ we choose for **g** the reciprocal lattice vector

$(-g, 0, 0)$. It may be verified by solving the equations of Section 4.3 for u_0 and u_1 that when **k** lies on the zone boundary, then $u_0 = -u_1$ for the solution of lowest energy when $V > 0$, so that

$$u_{g/2}(\mathbf{r}) = 2^{-1/2}(1 - e^{-igx}).$$

When **k** approaches $(-\tfrac{1}{2}g, 0, 0)$, on the other hand, we choose $\mathbf{g} = (+g, 0, 0)$. Again we find that also on this zone boundary $u_0 = -u_1$, and

$$u_{-g/2}(\mathbf{r}) = 2^{-1/2}(1 - e^{igx})$$
$$= -e^{igx} u_{g/2}(\mathbf{r}).$$

We now note that these two wavefunctions are identical, in that when we multiply by $e^{i\mathbf{k}\cdot\mathbf{r}}$ (with the appropriate **k**) we find just the same wavefunction $\psi_\mathbf{k}$, apart from an unimportant constant factor. Thus the action of the electric field is to cause the wavenumber of the electron to change at a constant rate until the zone boundary is reached, at which point the wavenumber is ambiguous. The electron may then be considered to have wavenumber $\mathbf{k} - \mathbf{g}$, and so the whole process may be repeated, with **k** increasing until the same zone boundary is again reached. Alternatively we may use the repeated zone scheme, and say that **k** is changing steadily with time, although the electron always remains in the first band. This is illustrated in Fig. 4.6.2(a) which shows the variation of the various components of $u_\mathbf{k}(\mathbf{r})$ with **k**. In Fig. 4.6.2(b) the electron velocity $v_\mathbf{k}$ in the first band is plotted in the repeated zone scheme. The fact that it is a periodic function of **k** shows that the electron would exhibit oscillatory motion in a crystal so perfect that no scattering occurred. In the region near the zone boundary $v_\mathbf{k}$ becomes more negative with increasing **k** as a consequence of the hole-like behavior characterized by the negative curvature of the function $\mathcal{E}_x(k_x)$.

The term in (4.6.7) that we neglected was of the form

$$i e \mathbf{E} \cdot \frac{\partial}{\partial \mathbf{k}} \ln u_\mathbf{k}(\mathbf{r}).$$

It is only when this term is small that the approximation (4.6.5) is valid, and this will only be the case when $u_\mathbf{k}(\mathbf{r})$ is a slowly varying function of **k**. Now if the lattice potential is very weak then the Bloch wave is very similar to a plane wave over most of the Brillouin zone. At the zone boundary, however, u_0 and u_1 will always be of equal magnitude irrespective of the strength of the

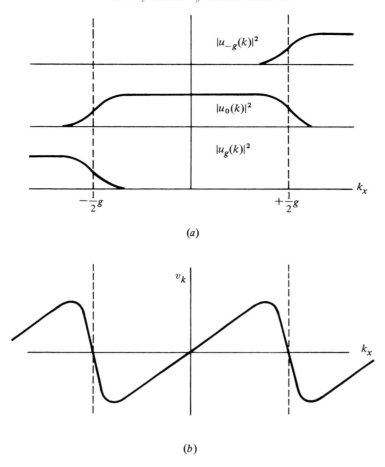

Figure 4.6.2. As the electron is accelerated by a weak electric field, its wave number k_x changes uniformly with time and the amplitudes (a) of the various plane wave components of $u_{\mathbf{k}}(\mathbf{r})$, the periodic part of the Bloch wave function, also change. Because the electron remains in the first Brillouin zone, the velocity (b) then changes periodically and not in the way shown in Fig. 4.6.1.

lattice potential. Thus the derivative $\partial (\ln u)/\partial \mathbf{k}$ is greatest when the lattice potential is weak, and it is then that the picture of the electron moving in a single band breaks down. This extra term that appears in (4.6.7) must be subtracted from the Hamiltonian if the electron is to remain in one band. This term is a function of $u_k(\mathbf{r})$, and is thus periodic with the period of the lattice. We may estimate its magnitude very simply by a glance at Fig. 4.6.3 which shows the band structure near the zone boundary. Since $u_1(\mathbf{k})$ is of order unity at the zone boundary, and has become very small by the time it

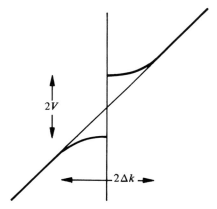

Figure 4.6.3. When the lattice potential is weak one may estimate with the aid of this diagram the range Δk of k_x-values over which the energy departs significantly from its free-electron value.

has reached a distance Δk away, where

$$V \simeq \frac{\partial \mathcal{E}}{\partial \mathbf{k}} \cdot \Delta \mathbf{k}$$

we may write

$$\frac{\partial u}{\partial \mathbf{k}} \sim \frac{1}{\Delta k} \sim \frac{\hbar^2 k}{mV},$$

and the extra term is of order $eE\hbar^2 k/mV$. When this term is of the order of the lattice potential it may cancel the lattice potential and allow the electron to make a transition to another band. The condition for this *not* to occur is then

$$eE \frac{\hbar^2 k}{mV} \ll V.$$

Since the Fermi energy $\mathcal{E}_F$ is roughly $\hbar^2 k^2/2m$ and k is of the order of $1/a$, where a is the lattice spacing, we may write

$$eEa \ll \frac{V^2}{\mathcal{E}_F}.$$

The condition for (4.6.5) to be valid is thus that the energy gained by the electron in being accelerated through one lattice spacing should be small

4.6 Dynamics of Bloch electrons

compared with $V^2/\mathcal{E}_F$. When this condition is not obeyed *Zener breakdown* is said to occur. While it is difficult to reach such high fields in homogeneous materials, the junction between n- and p-type semiconductors naturally contains a steep potential gradient which permits observation of these effects and as a result of which a variety of device applications are possible.

In the case of an applied magnetic field we might suppose that the Lorentz force would tend to change **k** in the same way as the force of the electric field, so that

$$\hbar \frac{d\mathbf{k}}{dt} = \frac{e}{c} \mathbf{v} \times \mathbf{H}. \tag{4.6.8}$$

This in fact turns out to be true within limitations similar to those imposed in the case of the electric field, although the demonstration of this result is a little more involved. Let us first choose the gauge so that the vector potential is

$$\mathbf{A} = \tfrac{1}{2}(\mathbf{H} \times \mathbf{r}), \tag{4.6.9}$$

and write the Hamiltonian (as in Eq. (3.10.8)) as

$$\mathcal{H} = \tfrac{1}{2} m\mathbf{v}^2 + V(\mathbf{r})$$

$$= \frac{1}{2m}\left(\mathbf{p} - \frac{e}{c}\mathbf{A}\right)^2 + V(\mathbf{r}),$$

where $V(\mathbf{r})$ is now the lattice potential. Then

$$i\hbar \frac{d\mathbf{v}}{dt} = [\mathbf{v}, \mathcal{H}]$$

$$= \tfrac{1}{2} m[\mathbf{v}, \mathbf{v}^2] + [\mathbf{v}, V(\mathbf{r})]$$

$$= i\hbar \frac{e}{mc} \mathbf{v} \times \mathbf{H} + [\mathbf{v}, V(\mathbf{r})] \tag{4.6.10}$$

as may be verified by substituting $(\mathbf{p} - e\mathbf{A}/c)$ for **v** and using the explicit form (4.6.9) for **A**.

If we had performed a similar manipulation in the case where an electric field was applied we should have found

$$i\hbar \frac{d\mathbf{v}}{dt} = i\hbar \frac{e\mathbf{E}}{m} + [\mathbf{v}, V(\mathbf{r})],$$

and since we were able to identify approximately $e\mathbf{E}$ with $\hbar\, d\mathbf{k}/dt$ we could then have written

$$[\mathbf{v}, V(\mathbf{r})] = i\hbar \left[\frac{d\mathbf{v}}{dt} - \frac{\hbar}{m} \frac{d\mathbf{k}}{dt} \right]. \qquad (4.6.11)$$

This relation describes the rate of change of velocity of a Bloch electron whose wavenumber is changing, but which is remaining in a single band. It does not discuss the agency that causes $\mathbf{k}$ to change, but merely states the consequent change in velocity. It is thus more general than the case of an applied electric field, and can be used in combination with (4.6.10) when a magnetic field is applied, the commutator $[\mathbf{A}(\mathbf{r}), V(\mathbf{r})]$ vanishing. Substitution of (4.6.11) in (4.6.10) then gives the expected result, (4.6.8).

While the electric field caused the wavenumber $\mathbf{k}$ to move in a straight line with uniform velocity in the repeated zone scheme, the effect of the magnetic field is more complicated. Equation (4.6.8) states that $d\mathbf{k}/dt$ is always perpendicular to both the electron velocity and the magnetic field. But since the velocity is proportional to $d\mathcal{E}/d\mathbf{k}$ the energy of the electron must remain constant, and $\mathbf{k}$ moves along an *orbit* in k-space which is defined by the two conditions that both the energy and the component of $\mathbf{k}$ in the direction of the magnetic field remain constant.

As an illustration we consider the possible orbits of an electron in sandwichium when the magnetic field is applied in the z-direction. For states of low energy the constant-energy surfaces are approximately spherical, and their intersections with the planes of constant k_z are nearly circular. The electrons thus follow *closed orbits* in k-space with an angular frequency, ω, close to the cyclotron frequency, ω_0, of a classical free electron, which is given by eH/mc. Such an orbit is labeled α in Fig. 4.6.4. In real space the path of such a classical electron would be a helix with its axis in the z-direction. For an electron of slightly higher energy the orbit passes closer to the zone boundary, and the electron velocity is reduced below the free-electron value. A circuit of the orbit labeled β in Fig. 4.6.4 thus takes a longer time than a circuit of α, and one says that the *cyclotron frequency* of the orbit β is less than ω_0, and would be the same as for a free particle of charge e and mass greater than m in the same magnetic field. One sometimes defines a *cyclotron mass* m^* in this way for a particular orbit by means of the relation

$$m^* = \frac{m\omega_0}{\omega}.$$

The cyclotron mass, which is a function of the electron velocity at all points on an orbit, must be distinguished from the inverse effective mass defined in

4.6 Dynamics of Bloch electrons

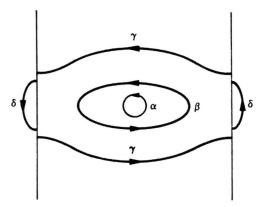

Figure 4.6.4. In a magnetic field in the z-direction an electron of low energy in sandwichium will travel the almost circular orbit α in k-space, while one of slightly larger energy will follow the distorted orbit β. At still higher energies the electron may either follow the second-zone orbit δ or the periodic open orbit γ that lies in the first Brillouin zone.

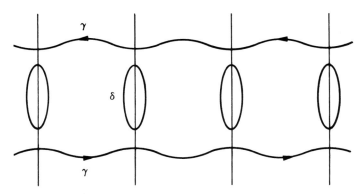

Figure 4.6.5. This diagram shows Fig. 4.6.4 replotted in the repeated zone scheme. The periodic open orbits γ carry a current that does not average to zero over a period of the motion.

Section 4.3 which characterized the band structure in the neighborhood of a single point in k-space.

If the electron energy is greater than $\mathcal{E}_g - V$ and k_z is sufficiently small there will be some orbits, such as γ in Fig. 4.6.4, that meet the zone boundary. The path of the electron in k-space is then a *periodic open orbit* in the repeated zone scheme, as shown in Fig. 4.6.5. Such orbits are particularly important in determining the conductivities of metals in magnetic fields in that the electron velocity does not average to zero over a period of the orbit. For energies greater than $\mathcal{E}_g + V$ there will also be orbits in the second band, such as those labeled δ in Fig. 4.6.4. In the repeated zone scheme these appear as the small closed orbits in Fig. 4.6.5. Because the velocity may be close to its free-electron value (in the extended zone scheme) over much of these orbits while their perimeter is much smaller, the time taken to complete an orbit may be very small. The cyclotron mass is then stated to be correspondingly small.

The range of validity of Eq. (4.6.8) may be deduced in a similar way to our estimate in the case of an electric field, and we find that

$$\frac{e}{c} \mathbf{v} \times \mathbf{H} \cdot \frac{\partial}{\partial \mathbf{k}} \ln u_\mathbf{k}(\mathbf{r})$$

must be small compared with the lattice potential. When we write

$$\left|\frac{\partial u}{\partial \mathbf{k}}\right| \sim \frac{\hbar^2 k}{mV}; \quad v \sim \frac{\hbar k}{m}; \quad \frac{eH}{mc} = \omega_0$$

we find the condition to be

$$\hbar \omega_0 \ll \frac{V^2}{\mathcal{E}_F}.$$

When this is violated *magnetic breakdown* is said to occur. The electron then has a finite probability of making a transition from a γ-orbit to a δ-orbit in Fig. 4.6.4, and the conductivity may be qualitatively affected.

4.7 Scattering by impurities

We have now seen how the application of an electric field causes the wavenumber of a Bloch electron to change, and hence how the electric current grows with time in a perfect periodic lattice. We know, however, that for moderate electric fields the current rapidly becomes constant and obeys Ohm's law in all normal metals. The current does not grow and then oscillate in the way that our simple dynamics predict, because the electron is scattered by some departure of the lattice from perfect periodicity. The two most important mechanisms that limit the magnitude the current attains in a particular field are scattering by lattice vibrations and scattering by impurities. The topic of the interaction of Bloch electrons with phonons is a major part of the theory of solids, and Chapter 6 is devoted to a discussion of such processes. The theory of alloys, in which the problem is to calculate the properties of *partially disordered* systems, is also a topic of some importance.

4.7 Scattering by impurities

For the present, however, we shall just consider the problem of a *single* impurity center in an otherwise periodic lattice. This avoids the statistical problems of the theory of alloys, but still allows us to formulate an expression for the probability per unit time that an electron is scattered from one Bloch state to another. We shall then have all the ingredients we need for the formulation of a simple theory of the conductivity of metals.

The customary approach to the scattering theory of a free particle involves the expansion of the wavefunction in spherical harmonics and the discussion of such quantities as phase shifts and cross sections. This approach is not so useful for Bloch electrons because of the reduced symmetry of the problem when the lattice potential is present. Instead we consider an electron initially in some Bloch state, $\phi_\mathbf{k}$, and then apply the perturbing potential, U. The wavefunction will then be transformed into some new function, $\psi_\mathbf{k}$. We interpret the scattering probability between the two Bloch states, $\phi_\mathbf{k}$ and $\phi_{\mathbf{k}'}$, as being proportional to the amount of $\phi_{\mathbf{k}'}$ contained in $\psi_\mathbf{k}$. That is, we form the integral $\langle\phi_{\mathbf{k}'}|\psi_\mathbf{k}\rangle$ to measure the amplitude that tells us how much of the state that was originally $\phi_\mathbf{k}$ has been transformed to $\phi_{\mathbf{k}'}$. The square of the modulus of this quantity will then be proportional to the probability $Q(\mathbf{k}, \mathbf{k}')$ that in unit time an electron is scattered between these states, i.e.,

$$Q(\mathbf{k}, \mathbf{k}') \propto |\langle\phi_{\mathbf{k}'}|\psi_\mathbf{k}\rangle|^2.$$

We may use the starting point of perturbation theory to rewrite this expression in a more useful form. We first write

$$\mathcal{H}_0\phi_\mathbf{k} = \mathcal{E}_\mathbf{k}\phi_\mathbf{k}$$

and

$$(\mathcal{H}_0 + U)\psi_\mathbf{k} = \mathcal{E}_\mathbf{k}\psi_\mathbf{k},$$

and note that the perturbed and unperturbed energies will be very close to each other provided no bound states are formed, since the impurity causing U only perturbs a negligible portion of our large volume Ω. We next note that these Schrödinger equations are satisfied by

$$|\psi_\mathbf{k}\rangle = |\phi_\mathbf{k}\rangle + (\mathcal{E}_\mathbf{k} - \mathcal{H}_0 + i\eta)^{-1} U|\psi_\mathbf{k}\rangle \qquad (4.7.1)$$

when $\eta \to 0$. Then because $\langle\phi_{\mathbf{k}'}|\phi_\mathbf{k}\rangle$ vanishes we find

$$Q(\mathbf{k}, \mathbf{k}') \propto |\langle\phi_{\mathbf{k}'}|(\mathcal{E}_\mathbf{k} - \mathcal{H}_0 + i\eta)^{-1} U|\psi_\mathbf{k}\rangle|^2$$
$$= [(\mathcal{E}_\mathbf{k} - \mathcal{E}_{\mathbf{k}'})^2 + \eta^2]^{-1}|\langle\phi_{\mathbf{k}'}|U|\psi_\mathbf{k}\rangle|^2.$$

Because the scatterer has no internal degrees of freedom in this model the energy of the electron must be conserved, and only *elastic scattering* can take place. This is expressed by the term in brackets. Since

$$[(\mathcal{E}_\mathbf{k} - \mathcal{E}_{\mathbf{k}'})^2 + \eta^2]^{-1} = \frac{1}{\eta} \frac{d}{d\mathcal{E}_\mathbf{k}} \arctan\left(\frac{\mathcal{E}_\mathbf{k} - \mathcal{E}_{\mathbf{k}'}}{\eta}\right)$$

and $\arctan[(\mathcal{E}_\mathbf{k} - \mathcal{E}_{\mathbf{k}'})/\eta]$ becomes a step function as $\eta \to 0$, we can interpret the derivative of the step function as a δ-function. The constant of proportionality can be found from time-dependent perturbation theory, which in lowest order gives the result

$$Q(\mathbf{k}, \mathbf{k}') \simeq \frac{2\pi}{\hbar} |\langle \phi_{\mathbf{k}'} | U | \phi_\mathbf{k} \rangle|^2 \delta(\mathcal{E}_\mathbf{k} - \mathcal{E}_{\mathbf{k}'}). \quad (4.7.2)$$

In order for our result to reduce to this when the potential is weak so that $\psi_\mathbf{k}$ may be replaced by $\phi_\mathbf{k}$ we must choose the same constant of proportionality, and write

$$Q(\mathbf{k}, \mathbf{k}') = \frac{2\pi}{\hbar} |\langle \phi_{\mathbf{k}'} | U | \psi_\mathbf{k} \rangle|^2 \delta(\mathcal{E}_\mathbf{k} - \mathcal{E}_{\mathbf{k}'}). \quad (4.7.3)$$

The approximation (4.7.2) is known as the *Born approximatiom*, and may be thought of as neglecting multiple scattering by the impurity. The exact formula (4.7.3) might be rewritten by repeatedly substituting for ψ from (4.7.1). We should then have a series of terms in which U appeared once, twice, three times, and so on. These could be interpreted as single, double, triple, and higher-order scattering by the impurity (Fig. 4.7.1).

It is sometimes useful to ask what the potential T would be that, if the Born approximation were exact, would give the scattering predicted by (4.7.3) for the potential U. That is, we ask for the operator T such that

$$\langle \phi_{\mathbf{k}'} | U | \psi_\mathbf{k} \rangle = \langle \phi_{\mathbf{k}'} | T | \phi_\mathbf{k} \rangle.$$

This operator is known as the *transition matrix* (or sometimes just as the *T*-matrix), and does not in general have the form of a simple potential. It can be seen from (4.7.1) that

$$T = U + U[\mathcal{E} - \mathcal{H}_0 + i\eta]^{-1} T.$$

Also

$$[\mathcal{E} - \mathcal{H}_0 + i\eta]^{-1} T = [\mathcal{E} - \mathcal{H} + i\eta]^{-1} U,$$

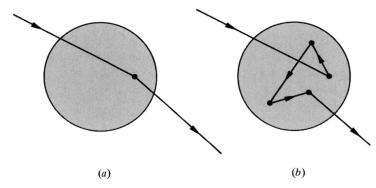

Figure 4.7.1. The Born approximation is a result of first-order perturbation theory, and can be diagrammaticaliy represented as a single scattering event (a). The T-matrix includes multiple scattering (b).

as may be seen by operating on both sides with $[\mathcal{E} - \mathcal{H} + i\eta]$. Thus

$$T = U + U[\mathcal{E} - \mathcal{H} + i\eta]^{-1}U.$$

Since the potential U is real, the only difference between T and its Hermitian conjugate $T\dagger$ will be that the term $i\eta$ will be replaced by $-i\eta$. We could thus have equally well used $T\dagger$ in calculating $Q(\mathbf{k}, \mathbf{k}')$. But since by the definition of the Hermitian conjugate

$$\langle \phi_{\mathbf{k}'} | T | \phi_{\mathbf{k}} \rangle^* = \langle \phi_{\mathbf{k}} | T\dagger | \phi_{\mathbf{k}'} \rangle$$

we see that the scattering probability must be the same in either direction, and

$$Q(\mathbf{k}, \mathbf{k}') = Q(\mathbf{k}', \mathbf{k}). \tag{4.7.4}$$

We could also argue this from the starting point of the principle of *microreversibility*, which states that the transition probability will be unaffected by time reversal. The time reversal of the state $\phi_{\mathbf{k}}$ will be $\phi_{-\mathbf{k}}$, and so

$$Q(\mathbf{k}, \mathbf{k}') = Q(-\mathbf{k}', -\mathbf{k}). \tag{4.7.5}$$

However, $\phi_{-\mathbf{k}} = \phi_{\mathbf{k}}^*$, as can be seen from the Schrödinger equation in the form (4.1.4), and we do not expect a real transition probability to depend on our convention as to complex numbers. We thus deduce (4.7.4) to be a consequence of (4.7.5).

We also note that the perturbation of the electron wavefunctions changes the density of electrons, and hence of electric charge, in the vicinity of an impurity in a metal. If the impurity represents an added electric charge the change in electron density will screen the field of the impurity. One can thus equate the excess charge of the impurity with the excess charge of the electrons that are in the process of being scattered. The formulation of this concept is rather complicated for Bloch electrons, but reduces to a simple form for free electrons, where it is known as the *Friedel sum rule*. It is a useful condition that all models of impurity potentials must approximately satisfy.

4.8 Quasicrystals and glasses

Our study of band structure so far has been built on the concept of the Bloch waves that we have proved to exist in perfectly periodic structures. In the real world, however, nothing is perfectly periodic, and so we should ask ourselves what the consequences are of deviations from perfect periodicity. In Chapter 6 we shall look at the effect of the weak deviations from perfect order that are introduced by phonons. There we shall see that this type of motion in a three-dimensional crystal does not destroy the long-range order. That is to say, when X-rays or neutrons are scattered by a thermally vibrating three-dimensional lattice there will still be sharp Bragg peaks, although in one or two dimensions this would not be the case. We now look at some other systems that lack perfect order, and examine whether the concept of band gaps will survive. The first of these is a remarkable family of structures known as *quasicrystals*. These are a form of not-quite-crystalline solid that was discovered experimentally as recently as 1984, although similar structures had been studied as mathematical constructs much earlier.

An example of a quasicrystal in two dimensions is given in Fig. 4.8.1. It clearly depicts an ordered array, but closer inspection shows it not to be a Bravais lattice. The telltale sign is the fact that it has a five-fold rotational symmetry. This is forbidden for Bravais lattices in two dimensions, as one cannot completely cover a plane using pentagonal tiles. One can, however, tile a plane using two types of diamond-shaped *Penrose tile*, one of which has an acute angle of $\pi/5$, the other tile having an angle of $2\pi/5$ (Fig. 4.8.2). In three dimensions the task becomes much harder to accomplish, and nearly impossible to illustrate. Nevertheless, experiment shows that if a molten mixture of aluminum and manganese in an atomic ratio of 4:1 is cooled ultrarapidly (~ 1 megakelvin/second!) then small pieces of solid are produced that give diffraction patterns having the five-fold symmetry characteristic of an icosahedron. These materials are thus clearly not crystalline (this is deduced from the five-fold symmetry) but do have long-range order (deduced from the existence of sharp Bragg peaks).

4.8 Quasicrystals and glasses

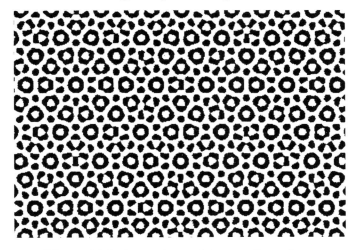

Figure 4.8.1. A quasicrystal in two dimensions.

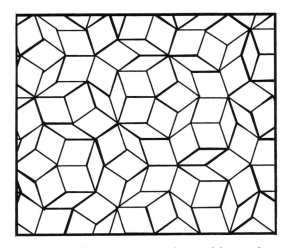

Figure 4.8.2. Two types of tile can cover a plane with quasicrystalline symmetry.

Figure 4.8.3. A Fibonacci chain is built from atoms separated by either long or short spacers placed in a special order.

We can gain some insight into the nature of quasicrystals by looking at the one-dimensional chain of atoms shown in Fig. 4.8.3. The spacing between atoms is either long (L) or short (S), with L/S an irrational number. If the arrangement of L and S spacings were random, then the chain would have no long-range order, and would give rise to no sharp Bragg diffraction peaks.

But it is not random. It is a *Fibonacci chain*, built according to the following prescription. We start with a single spacing S, and then repeatedly apply the operation that each S is turned into L and each L is turned into the pair LS. In this way $S \to L \to LS \to LSL \to LSLLS \to LSLLSLSL$ and so on. (An important special case occurs when $L/S = 2\cos(\pi/5) = \frac{1}{2}(1 + \sqrt{5})$, a number known as the golden mean.) Although this sequence does not at first sight appear to have any long-range order, one can, with the aid of some ingenious arguments, calculate the Fourier transform of the atomic density exactly. One finds that there are large, sharp, Bragg peaks at various wavenumbers. The chain is clearly not periodic in the sense of a Bravais lattice, but it does have some sort of long-range order. Evidently there are some hidden repeat lengths that are disguised by local deviations from periodicity. An invisible hand is placing the L and S segments in just such a way as to retain the Bragg peaks.

We find a clue to what is happening by looking at a strip cut from a true Bravais lattice in a higher dimension. In Fig. 4.8.4 we see a square lattice across which two parallel lines have been drawn with a slope equal to the reciprocal of the golden mean and passing through the opposite corners of one unit cell. We then project all the lattice points included in this strip onto the lower line to form a one-dimensional array. This array turns out to be precisely the special-case Fibonacci chain. We have thus made a connection

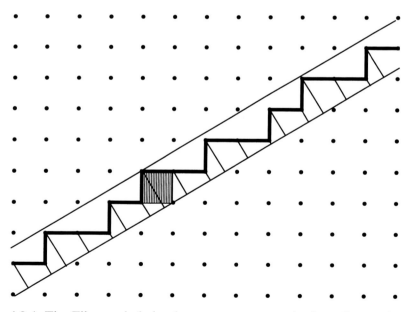

Figure 4.8.4. The Fibonacci chain also appears as a projection of a regular square lattice.

4.8 Quasicrystals and glasses

between a quasiperiodic array in one dimension and a Bravais lattice in a higher dimension. This idea may be extended to show that five-fold rotational symmetry may be found in spaces of six or more dimensions. In particular, icosahedral symmetry may be found in a cubic lattice in six dimensions. An icosahedron has 20 identical faces, each of which is an equilateral triangle. Five of these faces meet at each of the 12 vertices, and so there are six five-fold symmetry axes. This symmetry is clearly seen experimentally in single grains of some quasicrystals which form beautiful structures resembling five-petaled flowers. It is truly remarkable that this obscure crystallographic niche is actually occupied by real materials.

In one dimension, the existence of sharp Bragg peaks will always lead to gaps in the electronic density of states, but in three dimensions this is not assured. Thus the density of states for electrons in the potential due to a Fibonacci chain of atoms will always have band gaps. These chains would then be good insulators if there were two electrons per atom. In three dimensions the long-range order characteristic of quasicrystals will not necessarily cause gaps in the density of states, so that even if the number of electrons were two per atom, the material might still be a metallic conductor.

As we moved from considering crystalline lattices to the less-ordered quasicrystals, we have found that the continued existence of long-range order was the factor that made plausible the sustained presence of band gaps. If we move further in this direction we find amorphous or glassy solids, in which no long-range order remains. The structure factor revealed by X-ray scattering shows no sharp peaks, but only broad maxima. Surely these materials should not have band gaps in their electronic density of states? Surprisingly, band gaps persist in amorphous materials. In silicon, the effective band gap is even greater in amorphous material than it is in a crystal.

It was only in 1966 that a demonstration was given of how band gaps could be proved to persist in one simple model of an amorphous solid. In this model the potential has the muffin-tin form, in which identical spherically symmetric attractive potential wells are separated by regions of constant potential V_0. No two wells overlap or have their centers closer together than a distance we define as 2λ. We consider the real wavefunction ψ describing an eigenstate of energy $\mathcal{E} < V_0$. In units in which $\hbar = 2m = 1$, the Schrödinger equation is

$$\nabla^2 \psi = (V - \mathcal{E})\psi. \quad (4.8.1)$$

If we multiply by ψ and integrate over the volume of the container we find

$$\int \{(V - \mathcal{E})\psi^2 + (\nabla \psi)^2\} \, d\tau = 0, \quad (4.8.2)$$

provided ψ is equal to zero over the surface of the box. Let us define a cell as the region closer to one particular well than to any other (this is sometimes known as a *Voronoy polyhedron*). Then we can certainly find a cell such that

$$\int_{\text{cell}} \{(V - \mathcal{E})\psi^2 + (\nabla\psi)^2\} \, d\tau \leq 0 \tag{4.8.3}$$

when the integrations are confined to the volume of the cell. Because both parts of the integrand are positive at distances greater than λ from the center of the well the integral will furthermore be negative when the integration is restricted to a sphere of radius λ. If S is the surface of this sphere it then follows that

$$\int_S \psi \nabla \psi \cdot \mathbf{dS} < 0. \tag{4.8.4}$$

Taking spherical polar coordinates with the center of this well as the origin, we expand ψ in spherical harmonics, writing

$$\psi = \sum_{l,m} c_{l,m} Y_{l,m}(\theta, \phi) R_l(r) \tag{4.8.5}$$

and substitute in the inequality to obtain

$$\sum_{l,m} c_{l,m}^2 \frac{d}{dr} \{R_l(r)\}^2 |_{r=\lambda} < 0. \tag{4.8.6}$$

If this inequality holds for the sum of terms, it must also be true for at least one term of the sum, and so an l must exist for which

$$\frac{d}{dr} \{R_l(r)\}^2 |_{r=\lambda} < 0. \tag{4.8.7}$$

If there are bands of energy for which no l can be found such that this inequality is satisfied, then the existence of gaps in the density of states is proved.

The presence of band gaps in the electronic structure is central to many of the most important properties of solids. It is thus satisfying that we can calculate band structures and band gaps in a variety of structures provided that the one-electron model is a satisfactory approximation. Our next step must be a more careful look at this assumption, and an exploration of the elegant analysis with which it can be justified.

Problems

4.1 In sandwichium metal the lattice potential is $2V \cos gx$. Investigate, in the nearly-free-electron model, the electron velocity in the neighborhood of the point $(g/2, 0, 0)$ in reciprocal space.

4.2 Investigate qualitatively the density of states of the sandwichium defined in Problem 4.1 in the regions near $\mathcal{E} = \hbar^2 g^2/8m \pm V$, and sketch the overall density of states.

4.3 Another type of sandwichium has a lattice potential

$$V(\mathbf{r}) = \sum_{n=-\infty}^{\infty} Va\,\delta(x - na).$$

Investigate its band structure in the nearly-free-electron model, using two plane waves.

4.4 Apply the nearly-free-electron approach using four plane waves to the band structure of a two-dimensional crystal whose lattice potential is

$$V(\mathbf{r}) = 2V[\cos gx + \cos gy].$$

Under what conditions will this crystal be an insulator if there are two electrons per "atom"? (An "atom" is assumed to occupy one unit cell of dimensions $2\pi/g \times 2\pi/g$.)

4.5 What are the possible forms of the inverse-effective-mass tensor in the model of Problem 4.4 at the point $(\tfrac{1}{2}g, \tfrac{1}{2}g)$ in k-space?

4.6 In the Kronig–Penney model a one-dimensional electron moves in a potential

$$V(x) = -\sum_{n=-\infty}^{\infty} Va\,\delta(x - na).$$

Contrast the exact solution for the width of the lowest band with that given by the method of tight binding when V is very large. Assume overlap only of nearest neighbors in the tight-binding approach.

4.7 Examine the inverse effective mass of the states at the bottom of the third band of the model in Problem 4.6, again assuming V to be large. Solve this problem in the following ways.

(1) Exactly.
(2) In the two-plane-wave NFE approximation.
(3) In the OPW method, treating the first band as core states in the tight-binding approximation. [Use two OPW's, and neglect the k-dependence of $\mathcal{E}_c^{\mathbf{k}}$ – i.e., take $\mathcal{E}_c^{\mathbf{k}}$ as the energy of the "atomic" bound state.]

4.8 Evaluate the Korringa–Kohn–Rostoker $G_{\mathbf{k}}(\mathbf{r} - \mathbf{r}')$ for the sandwichium of Problem 4.1. [Hint: $\int_C \operatorname{cosec} \pi z f(z)\, dz$ may be a helpful integral to consider.]

4.9 Calculate an approximate value for the energy discontinuity and effective inverse masses in the neighborhood of $\mathbf{k} = (g, 0, 0)$ in sandwichium by using the nearly-free-electron approximation with three plane waves.

4.10 Draw the Jones zone for a square lattice of side a with four identical atoms in each cell at the points $\pm(a/8, -a/8)$ and $\pm(3a/8, 3a/8)$.

4.11 In the limit of vanishingly small size of an orbit the cylotron mass m^* and the inverse-effective-mass tensor $(M^{-1})_{ij}$ are related. What is this relationship between m^*, (M^{-1}), and the direction $\hat{\omega}$ of the applied magnetic field? [It is helpful to consider the area $\mathcal{A}$ of an orbit, and its variation with energy, $d\mathcal{A}/d\mathcal{E}$.]

4.12 A magnetic field is applied in the z-direction to sandwichium. How, qualitatively, does m^* vary for orbits with $k_z = 0$ as $\mathcal{E} \to \mathcal{E}_g - V$?

4.13 A Bloch electron in sandwichium is scattered from (k_x, k_y, k_z) to $(-k_x, k_y, k_z)$ by the potential $U \exp[-(gr/4)^2]$. Investigate qualitatively how the transition probability for this process varies with k_x. [Use the Born approximation for $Q(\mathbf{k}, \mathbf{k}')$ and the two-plane-wave approximation for $\phi_{\mathbf{k}}$ and $\phi_{\mathbf{k}'}$. Sketch the variation of Q as k_x varies from 0 to $\frac{1}{2}g$.]

4.14 When the Coulomb interaction is included in the Hamiltonian of an insulator it becomes possible for an electron in the conduction band and a hole in the valence band to form a bound state together; this elementary excitation of the crystal is known as an *exciton*. In the simple model of an insulator in which the lattice potential is $2V(\cos gx + \cos gy + \cos gz)$ such a state can be formed if we allow an interaction $e^2/\epsilon |\mathbf{r}_e - \mathbf{r}_h|$ to exist between the electron and hole states at the corner of the first Brillouin zone. Investigate the possible energies of such an excitation by solving a Schrödinger equation analogous to that describing a hydrogen atom, but in which the proton and electron are replaced by an electron and a hole having the appropriate effective masses.

4.15 Verify that the constant a of Section 4.5 vanishes, as claimed in the sentence preceding Eq. (4.5.1).

4.16 In the model illustrated in Fig. 4.5.3 it was shown that spin–orbit coupling introduces energy discontinuities at the zone boundaries shown as dashed lines in Fig. 4.5.4. Does (a) the mass–velocity term or (b) the Darwin term cause a similar effect?

Chapter 5
Density functional theory

5.1 The Hohenberg–Kohn theorem

In Chapter 2 we explored some of the consequences of electron–electron interactions, albeit in some simple perturbative approaches and within the random phase approximation. There we found that the problem of treating these interactions is exceedingly difficult, even in the case where there is no external one-particle potential applied to the system. We have also explored some of the properties of *noninteracting* electrons in an external potential, in this case the periodic lattice potential. This led to the concepts of electron bands and band structure, subjects of fundamental importance in understanding the physics of metals, insulators, and semiconductors. Of course, in the real world, electrons in matter are subjected both to electron–electron interactions and to external potentials. How to include systematically and correctly the electron–electron interactions in calculations of real systems is truly a formidable problem.

Why that is so is easily demonstrated. Suppose that we want to solve the problem of N electrons interacting in some external potential. The N-electron wavefunction can be expanded in Slater determinants of some suitable single-particle basis such as plane waves. We can describe the Slater determinants by occupation numbers in our second-quantized notation. Suppose furthermore that we have a basis of a total of N_k plane wave states at our disposal. Here N_k must be large enough that all reasonable "wiggles" of the many-body wavefunction can be included. The size of our Hilbert space and hence the size of the Hamiltonian matrix to be diagonalized can then be found by using combinatorics: the size of the Hilbert space is given by the number of ways that we can put N "balls" in N_k "boxes," with only one ball per box. This number is a binomial factor, $N_k!/N!(N_k - N)!$, which has the unfortunate property that it grows factorially. Careful use of symmetry may help us reduce the size of the Hamiltonian by a factor of ten or so, and the increasing power of computers allows us to consider ever-larger systems, but it remains stubbornly the case that current state-of-the-art exact numerical diagonalizations have difficulty handling more than a few tens of electrons. Also, even though the computer power at our disposal grows exponentially with time, the size of the Hilbert space of our N-electron problem grows much faster than exponentially with N. We may therefore, somewhat pessimistically, conclude that we may never have enough computer resources available to solve a problem with a macroscopic number of electrons.

This draws attention to the urgent need for some alternative way to include electron–electron interactions in our calculations. Virtually the only way to do so in realistic calculations is provided by density functional theory (DFT). Since its formulation in the mid 1960s and early 1970s, DFT has been used extensively in condensed matter physics in almost all band-structure and electronic structure calculations. It has also been widely adopted in the quantum chemistry community, and has led to a computational revolution in that area. Density functional theory was conceived by Walter Kohn, who also led many of the successive developments in this field.

What makes density functional theory so powerful to use is a deceptively simple-looking theorem, the Hohenberg–Kohn theorem, which has profound implications. This theorem allows for the systematic formulation of a many-body problem – interacting electrons in an external potential – in terms of the electron density as the basic variable. It is worth spending a moment to reflect on this. Consider the Schrödinger equation for N interacting electrons. This is a differential equation for a complex quantity, the Schrödinger wavefunction, which in three dimensions is a function of $3N$ variables. This large number makes it impractical to solve even for just the ground-state wavefunction, which will generally be insufficient, as we also need information about the excited states. Finally, the physical quantities in which we are interested have to be extracted from the wavefunctions that we have laboriously obtained. This in itself may be technically very difficult. It is clear that if we can instead work with just the electron density as the basic variable, this will lead to an enormous simplification, since the density of a three-dimensional system is a scalar field of only three variables. What is truly remarkable is, as we shall see, that all physical properties of the system can in principle be determined with knowledge only of the ground-state density! That is precisely the statement of the Hohenberg–Kohn theorem, as we now prove for systems with nondegenerate ground states.

Let

$$\mathcal{H} = T + V_{\text{ext}} + V$$

be the nonrelativistic, time-independent Hamiltonian of a system of N electrons. Here, T is the kinetic energy, V_{ext} is an external potential which couples to the density (an example being that from the nuclei in a solid), and V is the two-body electron–electron interaction (usually the Coulomb interaction). In second-quantized notation we write

$$\mathcal{H} = \sum_{\mathbf{k},s} \frac{\hbar^2 k^2}{2m} c^\dagger_{\mathbf{k},s} c_{\mathbf{k},s} + \sum_{\mathbf{k},\mathbf{q},s} V_{\text{ext}}(q) c^\dagger_{\mathbf{k},s} c_{\mathbf{k}+\mathbf{q},s} + \frac{1}{2} \sum_{\mathbf{k},\mathbf{k}',\mathbf{q},s,s'} V_\mathbf{q} c^\dagger_{\mathbf{k}-\mathbf{q},s} c^\dagger_{\mathbf{k}'+\mathbf{q},s'} c_{\mathbf{k}',s'} c_{\mathbf{k},s}.$$

The Hohenberg–Kohn theorem then states that the expectation value O of any operator $\mathcal{O}$ is a unique *functional* $O[n_0(\mathbf{r})]$ of the ground-state density $n_0(\mathbf{r})$, by which we mean that the value of O depends on the value of $n_0(\mathbf{r})$ at all points $\mathbf{r}$.

What does this imply? Well, we already know that if we could solve the Schrödinger equation for the Hamiltonian $\mathcal{H}$ and find all the many-body eigenstates Ψ_α, we could then calculate the expectation value of any operator. The Hamiltonian therefore determines the expectation value of any operator, and, in particular, the Hamiltonian determines the ground-state density, since this is just the ground-state expectation value of the density operator. We can be even more specific: since the kinetic energy operator T and the interaction V are universal, meaning that they are the same for all nonrelativistic interacting N-electron systems, it is really only the external potential V_{ext} that characterizes the Hamiltonian, and thus the eigenstates and the ground-state density. This is straightforward. What the Hohenberg–Kohn theorem states is that this mapping from external potential to ground-state density is invertible. Given any density $n(\mathbf{r})$, which is specified to be the ground-state density for *some* N-electron system, the Hamiltonian of that system is then uniquely determined, and so then are all the eigenstates and the expectation value of any operator. So with knowledge of only the ground-state density of an N-electron system, we can (in principle, at least) determine *everything* about that system, including excited states, excitation energies, transport properties, etc.

The proof of this theorem is simple. We first show that two potentials, V_{ext} and V'_{ext}, that differ by more than a trivial constant (a constant is unimportant since we can always shift the reference point of the potential energy),

5.1 The Hohenberg–Kohn theorem

necessarily lead to different ground states Ψ_0 and Ψ_0'. The Schrödinger equations for Ψ_0 and for Ψ_0' are

$$(T + V + V_{\text{ext}})\Psi_0 = \mathcal{E}_0 \Psi_0 \tag{5.1.1}$$

$$(T + V + V_{\text{ext}}')\Psi_0' = \mathcal{E}_0' \Psi_0', \tag{5.1.2}$$

where $\mathcal{E}_0$ and $\mathcal{E}_0'$ are the respective ground-state energies. We prove the first part of the theorem by contradiction. Suppose now that Ψ_0 and Ψ_0' are the same. We then subtract Eq. (5.1.1) from Eq. (5.1.2) to obtain

$$(V_{\text{ext}} - V_{\text{ext}}')\Psi_0 = (\mathcal{E}_0 - \mathcal{E}_0')\Psi_0.$$

But $\mathcal{E}_0$ and $\mathcal{E}_0'$ are just real numbers, so this means that the two potentials V_{ext} and V_{ext}' can differ at most by a constant, in contradiction to our hypothesis. We have thus shown that if $V_{\text{ext}} \neq V_{\text{ext}}'$ then $\Psi_0 \neq \Psi_0'$.

At this point we pause to note the relation between $n_0(\mathbf{r})$, $V_{\text{ext}}(\mathbf{r})$, and $\langle \Psi_0 | V_{\text{ext}} | \Psi_0 \rangle$. We recall that

$$n_0(\mathbf{r}) = \int \Psi_0^*(\mathbf{r}_1, \mathbf{r}_2, \ldots) \sum_i^N \delta(\mathbf{r} - \mathbf{r}_i) \Psi_0(\mathbf{r}_1, \mathbf{r}_2, \ldots) d\mathbf{r}_1, d\mathbf{r}_2, \ldots,$$

which allows us to write

$$\langle \Psi_0 | V_{\text{ext}} | \Psi_0 \rangle$$

$$= \int \Psi_0^*(\mathbf{r}_1, \mathbf{r}_2, \ldots) \sum_i^N V_{\text{ext}}(\mathbf{r}_i) \Psi_0(\mathbf{r}_1, \mathbf{r}_2, \ldots) d\mathbf{r}_1, d\mathbf{r}_2, \ldots, d\mathbf{r}_N$$

$$= \int \Psi_0^*(\mathbf{r}_1, \mathbf{r}_2, \ldots) \sum_i^N \delta(\mathbf{r}_p - \mathbf{r}_i) V_{\text{ext}}(\mathbf{r}_p) \Psi_0(\mathbf{r}_1, \mathbf{r}_2, \ldots) d\mathbf{r}_1, d\mathbf{r}_2, \ldots, d\mathbf{r}_N, d\mathbf{r}_p$$

$$= \int n_0(\mathbf{r}) V_{\text{ext}}(\mathbf{r}) d\mathbf{r}.$$

Now we can prove that if $V_{\text{ext}} \neq V_{\text{ext}}'$ (so that consequently $\Psi_0 \neq \Psi_0'$), then we must also have $n_0(\mathbf{r}) \neq n_0'(\mathbf{r})$. Again, we prove this assertion by contradiction. Assume that $n_0(\mathbf{r}) = n_0'(\mathbf{r})$, and that $\mathcal{H}$ and $\mathcal{H}'$ are the two Hamiltonians corresponding to V_{ext} and V_{ext}', respectively. According to the Rayleigh–Ritz variational principle, we have

$$\mathcal{E}_0 = \langle \Psi_0 | \mathcal{H} | \Psi_0 \rangle < \langle \Psi_0' | \mathcal{H} | \Psi_0' \rangle,$$

and

$$\langle\Psi_0'|\mathcal{H}|\Psi_0'\rangle = \langle\Psi_0'|\mathcal{H}' + V_{\text{ext}} - V_{\text{ext}}'|\Psi_0'\rangle = \mathcal{E}_0' + \int n_0'(\mathbf{r})[V_{\text{ext}}(\mathbf{r}) - V_{\text{ext}}'(\mathbf{r})]\,\mathbf{dr},$$

so that

$$\mathcal{E}_0 < \mathcal{E}_0' + \int n_0'(\mathbf{r})[V_{\text{ext}}(\mathbf{r}) - V_{\text{ext}}'(\mathbf{r})]\,\mathbf{dr}. \tag{5.1.3}$$

An analogous argument, obtained by interchanging primed and unprimed quantities, yields

$$\mathcal{E}_0' < \mathcal{E}_0 + \int n_0(\mathbf{r})[V_{\text{ext}}'(\mathbf{r}) - V_{\text{ext}}(\mathbf{r})]\,\mathbf{dr}. \tag{5.1.4}$$

Adding Eqs. (5.1.3) and (5.1.4), and using our assumption that $n_0(\mathbf{r}) = n_0'(\mathbf{r})$ then leads to the expression

$$\mathcal{E}_0 + \mathcal{E}_0' < \mathcal{E}_0 + \mathcal{E}_0',$$

which appears unlikely. We have thus established that two different, nondegenerate ground states necessarily lead to different ground-state densities. It follows that two identical ground-state densities must stem from identical external potentials, and with that our proof of the Hohenberg–Kohn theorem is complete.

There is also an important variational principle associated with the Hohenberg–Kohn theorem. Since the expectation value of any operator $\mathcal{O}$ of a system is a unique functional of the ground-state density $n_0(\mathbf{r})$, this certainly applies to the ground-state energy. We write this functional as

$$\mathcal{E}[n] \equiv \langle\Psi_0[n]|T + V_{\text{ext}} + V|\Psi_0[n]\rangle, \tag{5.1.5}$$

where V_{ext} is the specific external potential of a system with ground-state density $n_0(\mathbf{r})$ and ground-state energy $\mathcal{E}_0$. For the case where the density $n(\mathbf{r})$ equals the ground-state density $n_0(\mathbf{r})$ corresponding to the external potential V_{ext}, the functional $\mathcal{E}[n]$ then takes on the value $\mathcal{E}_0$. Since the ground-state energy is uniquely determined by $n_0(\mathbf{r})$, the Rayleigh–Ritz principle establishes that

$$\mathcal{E}_0 < \mathcal{E}[n] \quad \text{for } n \neq n_0.$$

We shall find that this is a very useful property. The ground-state energy can be found by varying the density to minimize the energy, provided we know the form of the functional $\mathcal{E}[n]$, or at least have a good approximation for it. In fact, we can write the ground-state energy functional as

$$\mathcal{E}[n] = F_{\text{HK}}[n] + \int V_{\text{ext}}(\mathbf{r})n(\mathbf{r})\,\mathbf{dr}, \tag{5.1.6}$$

where $F_{\text{HK}}[n] = \langle \Psi[n] | T + V | \Psi[n] \rangle$ is a unique functional. By that we mean that $F_{\text{HK}}[n]$ is the same functional of the density $n(\mathbf{r})$ for *all* interacting N-electron systems. We thus need to determine it only once, and can then apply it to all systems.

We have here discussed the Hohenberg–Kohn theorem only for nondegenerate ground states. The theorem can also be extended to include the case of degenerate ground states, which is formally very important. There are also many other extensions that are important for practical calculations, such as extensions to polarized systems, and to systems at finite temperatures. For example, we might consider a spin-polarized system with a fixed quantization axis, which we take to be the z-axis. The system may then have a net magnetization along this axis. In this case, we can define up- and down-spin densities $n_\uparrow$ and $n_\downarrow$, or, equivalently, total density n and polarization ξ, with

$$n = n_\uparrow + n_\downarrow$$
$$\xi = \frac{n_\uparrow - n_\downarrow}{n_\uparrow + n_\downarrow}.$$

A Hohenberg–Kohn theorem can then be formulated in terms of $n_\uparrow$ and $n_\downarrow$ (or in terms of n and ξ). It turns out that calculations formulated in this way are usually much more accurate than calculations cast in terms of density alone, even if the system itself has no net polarization.

5.2 The Kohn–Sham formulation

While the Hohenberg–Kohn theorem rigorously establishes that we may use the density, and the density alone, as a variable to find the ground-state energy of an N-electron problem, it does not provide us with any useful computational scheme. This is provided by the Kohn–Sham formalism. The idea here is to use a noninteracting "reference," or auxiliary, system, and to look for an external potential V_s such that the noninteracting system has the same ground-state density as the real, interacting system. Once we have obtained this density, we can use it in the energy functional Eq. (5.1.5), or in some approximation of it. The ground-state of a noninteracting system of N electrons is given by a Slater determinant of the N lowest-lying single-particle states. Since we can much more readily solve for these, the Kohn–Sham scheme provides us with a route to practical calculations. But there is no free lunch – the price we pay in the Kohn–Sham scheme is that the equations have to be solved self-consistently. The effective potential V_s will turn out to depend on the electron density. In practical calculations, one then typically starts by assuming an initial density. This gives an input potential V_s, which can then be used to solve for the single-particle states. From these a new density is obtained, which gives a new V_s. The equations are then solved again, and this process is repeated until self-consistency is obtained, i.e., until the input and output density in one iteration are sufficiently close to one another. Much effort has been spent over the years to come up with efficient schemes for such self-consistent calculations.

Let us then start by considering a noninteracting N-electron system in an external potential V_s. The Hamiltonian $\mathcal{H}_s$ of this system is given by

$$\mathcal{H}_s = T + V_s.$$

We then apply the Hohenberg–Kohn theorem to this system. Accordingly, there exists a unique energy functional

$$\mathcal{E}_s[n] = T_s[n] + \int V_s(\mathbf{r}) n(\mathbf{r}) \, d\mathbf{r}. \tag{5.2.1}$$

We note here that $T_s[n]$ is the kinetic energy functional of a system of N *noninteracting* electrons, and is consequently a different functional from the $T[n]$ that forms part of $F_{HK}[n]$ in Eq. (5.1.6).

The ground-state density of this system is easily obtained. It is simply

$$n_s(\mathbf{r}) = \sum_{i=1}^{N} |\phi_i(\mathbf{r})|^2, \tag{5.2.2}$$

where we have occupied the N single-particle states, or orbitals, that satisfy the Schrödinger-like equation

$$\left[\frac{-\hbar^2}{2m} \nabla^2 + V_s(\mathbf{r}) \right] \phi_i(\mathbf{r}) = \mathcal{E}_i \phi_i(\mathbf{r}), \quad \mathcal{E}_1 \leq \mathcal{E}_2 \leq \cdots, \tag{5.2.3}$$

5.2 The Kohn–Sham formulation

and have the N lowest eigenvalues $\mathcal{E}_i$. But we are really interested in a system of N *interacting* electrons in an external potential V_{ext}, so the question we would like to answer is the following: can we determine the form that V_s (the external potential of the noninteracting system) must take in order for the noninteracting system to have the same ground-state density as the interacting system in the external potential V_{ext}? The strategy we use is to solve for the density using the auxiliary noninteracting system, and then insert this density (which by construction is the same as that for the interacting system) into an approximate expression for the total energy of the interacting system.

The first step in this process is to rewrite the energy functional $\mathcal{E}[n]$ of the interacting system, which was given in Eq. (5.1.5), as

$$\mathcal{E}[n] = T_s[n] + \left\{ T[n] - T_s[n] + V[n] - \frac{e^2}{2} \iint \frac{n(\mathbf{r})n(\mathbf{r}')}{|\mathbf{r}-\mathbf{r}'|} \, d\mathbf{r} \, d\mathbf{r}' \right\}$$
$$+ \frac{e^2}{2} \iint \frac{n(\mathbf{r})n(\mathbf{r}')}{|\mathbf{r}-\mathbf{r}'|} \, d\mathbf{r} \, d\mathbf{r}' + \int n(\mathbf{r}) V_{\text{ext}}(\mathbf{r}) \, d\mathbf{r}$$
$$\equiv T_s + \frac{e^2}{2} \iint \frac{n(\mathbf{r})n(\mathbf{r}')}{|\mathbf{r}-\mathbf{r}'|} \, d\mathbf{r} \, d\mathbf{r}' + \int n(\mathbf{r}) V_{\text{ext}}(\mathbf{r}) \, d\mathbf{r} + \mathcal{E}_{\text{xc}}[n]. \quad (5.2.4)$$

Here we have added and subtracted both the kinetic energy functional $T_s[n]$ of a noninteracting system and the direct, or Hartree, term in the electrostatic energy. We have then defined the sum of the terms in braces to be the exchange-correlation energy functional $\mathcal{E}_{\text{xc}}[n]$. From Eq. (5.1.6), this functional is

$$\mathcal{E}_{\text{xc}}[n] \equiv F_{\text{HK}}[n] - \frac{e^2}{2} \iint \frac{n(\mathbf{r})n(\mathbf{r}')}{|\mathbf{r}-\mathbf{r}'|} \, d\mathbf{r} \, d\mathbf{r}' - T_s[n]. \quad (5.2.5)$$

We have thus swept all our ignorance about electron interactions beyond the Hartree term under the rug that we call $\mathcal{E}_{\text{xc}}[n]$. What we gain in writing $\mathcal{E}[n]$ in this way is that we can eventually focus on developing reasonable approximations for $\mathcal{E}_{\text{xc}}[n]$.

According to the Hohenberg–Kohn theorem, the density n that minimizes the functional $\mathcal{E}[n]$ is the ground-state density. Thus by taking the variation of Eq. (5.2.4) with respect to the particle density we obtain

$$\frac{\delta \mathcal{E}[n]}{\delta n(\mathbf{r})} = \frac{\delta T_s[n]}{\delta n(\mathbf{r})} + e^2 \int \frac{n(\mathbf{r}')}{|\mathbf{r}-\mathbf{r}'|} \, d\mathbf{r}' + V_{\text{ext}}(\mathbf{r}) + v_{\text{xc}}[n(\mathbf{r})] = 0, \quad (5.2.6)$$

where we have formally defined the *exchange-correlation potential* as

$$v_{xc}[n(\mathbf{r})] \equiv \frac{\delta \mathcal{E}_{xc}[n]}{\delta n(\mathbf{r})}.$$

We now use the auxiliary noninteracting system and its Schrödinger equation, from which we can similarly show that

$$\frac{\delta T_s[n]}{\delta n(\mathbf{r})} + V_s(\mathbf{r}) = 0.$$

By comparing this result with Eq. (5.2.6) we see that this effective potential $V_s(\mathbf{r})$ must satisfy

$$V_s(\mathbf{r}) = V_{ext}(\mathbf{r}) + e^2 \int \frac{n(\mathbf{r}')}{|\mathbf{r}-\mathbf{r}'|}\, d\mathbf{r}' + v_{xc}(\mathbf{r}). \tag{5.2.7}$$

We are now in a position to implement the self-consistent Kohn–Sham scheme. We first choose an initial trial form of the function $n(\mathbf{r})$ and substitute into Eq. (5.2.7) to find a trial form of V_s. We then solve Eq. (5.2.3) for the single-particle wavefunctions $\phi_i(\mathbf{r})$, and use Eq. (5.2.2) to find the next iteration for $n(\mathbf{r})$. When this procedure has been repeated a sufficient number of times that no further changes occur, then a solution for $n(\mathbf{r})$ has been found that not only satisfies the Schrödinger equation for the reference noninteracting electrons, but also is the correct density for the interacting system.

We close this section by highlighting a few points about the Kohn–Sham formalism. First of all, it is formally *exact*, supposing that we can find the exact exchange-correlation potential $v_{xc}(\mathbf{r})$. Second, we have cast the solution of the interacting N-electron problem in terms of noninteracting electrons in an external potential $V_s(\mathbf{r})$. This is of great practical importance. The ground-state wavefunction of the noninteracting system is just a Slater determinant of the N orbitals, the so-called Kohn–Sham orbitals, with the lowest eigenvalues $\mathcal{E}_\alpha$. It is relatively easy to solve for these single-particle orbitals even for as many as a few hundred electrons. The Kohn–Sham equations formally look very much like self-consistent Hartree equations, the only difference being the presence of the exchange-correlation potential. This makes them much simpler to solve than the Hartree–Fock equations, in which the potential is orbital-dependent. In the Kohn–Sham and Hartree equations, the effective potential is the same for every orbital.

5.3 The local density approximation

Before we can actually implement the Kohn–Sham formalism, we have to devise some workable approximation for the exchange-correlation potential $v_{xc}(\mathbf{r})$. The first such approximation to be suggested was the Local Density Approximation, or LDA. The idea behind the LDA is very simple; it just ignores the nonlocal aspects of the functional dependence of v_{xc}. The true form of $v_{xc}(\mathbf{r})$ will depend not only on the local density $n(\mathbf{r})$ but also on n at all other points $\mathbf{r}'$, and this functional dependence is in general not known. This difficulty is avoided with the assumption that v_{xc} depends only on the local density $n(\mathbf{r})$, and that $\mathcal{E}_{xc}[n]$ can thus be written as

$$\mathcal{E}_{xc}[n] = \int \epsilon_{xc}[n] \, n(\mathbf{r}) \, d\mathbf{r},$$

where $\epsilon_{xc}[n]$ is the exchange-correlation energy per particle of a homogeneous system of density n. In the LDA, we assume that the density of our (*inhomogeneous*) system varies very slowly, so that the exchange-correlation energy is *locally* that of a homogeneous system at the local density.

For practical calculations, we must then determine what $\epsilon_{xc}[n]$ is. Although no general form is known exactly, the low-density and high-density limits can be calculated analytically. Usually, the density is expressed in terms of the dimensionless parameter r_s, which is the radius of the sphere that can be assigned to each electron (so that the volumes of all these spheres add up to the total volume of the system), measured in units of the Bohr radius a_0. That is,

$$r_s = (4\pi n a_0^3 / 3)^{-1/3}.$$

In the low-density limit ($r_s \gg 1$), the electrostatic potential energy dominates, and the electrons condense into what is known as a Wigner crystal, the energy of which can be calculated. While the density of the Wigner crystal is not strictly uniform, we can still use the energy per electron of this system to develop an estimate for $\epsilon_{xc}[n]$ at low densities. In the high-density limit ($r_s \ll 1$), the kinetic energy dominates, and the random-phase approximation becomes exact. Unfortunately, real metals have r_s of the order of unity. Usually, one then uses one of several interpolation schemes that join the low- and high-density limits of ϵ_{xc}. The most commonly used approximations combine numerical calculations with Padé approximants, which are ratios of polynomials, for the interpolations. In local spin density functional calculations, the exchange-correlation energy is frequently separated into exchange and correlation parts. The exchange part is in this context just the exchange energy that we discussed in Chapter 2, and the correlation energy is everything else. These two quantities are then calculated at zero and unit polarizations ($\xi = 0$ and $\xi = 1$). The exchange energy is straightforward to calculate, and the result is

$$\epsilon_x(n, \xi = 0) = -\frac{3}{4\pi}\left(\frac{9\pi}{4}\right)^{1/3} \frac{e^2}{r_s a_0},$$

and

$$\epsilon_x(n, \xi = 1) = 2^{1/3} \epsilon_x(n, \xi = 0).$$

A popular approximation for the correlation energy ϵ_c is based on Monte Carlo calculations by Ceperley and Alder for certain values of r_s for $\xi = 0$ and $\xi = 1$. These are then parametrized. In one example, Perdew and Zunger use a Padé approximant in $r_s^{1/2}$ for $r_s \geq 1$:

$$\epsilon_c(n, \xi = 0, 1) = \frac{\gamma(\xi)}{1 + \beta_1(\xi)\sqrt{r_s} + \beta_2(\xi)r_s},$$

with

$$\begin{aligned}\gamma(0) &= -0.1423 e^2/a_0, & \beta_1(0) &= 1.0529, & \beta_2(0) &= 0.3334 \\ \gamma(1) &= -0.0843 e^2/a_0, & \beta_1(1) &= 1.3981, & \beta_2(1) &= 0.2611.\end{aligned} \quad (5.3.1)$$

This form is then joined smoothly to the high-density form of ϵ_c (for $r_s \leq 1$), which is

$$\epsilon_c(n, \xi = 0, 1) = A(\xi) \ln r_s + B(\xi) + C(\xi) r_s \ln r_s + D(\xi) r_s. \quad (5.3.2)$$

For $\xi = 0$, the result is matched to the classical random-phase approximation result by Gell-Mann and Brueckner, and for $\xi = 1$ there exists a scaling relation (also obtained from the random-phase approximation), which states that

$$\epsilon_c(r_s, \xi = 1) = \tfrac{1}{2}\epsilon_c(r_s/2^{4/3}, \xi = 0). \quad (5.3.3)$$

Finally, by requiring that the correlation energy and the resulting correlation potential, $v_c(n, \xi = 0, 1) = (d/dn)(n\epsilon_c(n, \xi = 0, 1))$ be continuous at $r_s = 1$, the parameters $A(\xi)$, $B(\xi)$, $C(\xi)$, and $D(\xi)$ can be fixed. The result (with all parameters given in units of e^2/a_0) is

$$A(0) = 0.0311, \quad B(0) = -0.048, \quad C(0) = 0.0020, \quad D(0) = -0.0166$$
$$A(1) = 0.01555, \quad B(1) = -0.0269, \quad C(1) = 0.0007, \quad D(1) = -0.0048. \tag{5.3.4}$$

It then remains to interpolate exchange and correlation energies to arbitrary polarizations. For the exchange energy alone, one can show that there is an exact expression for the exchange energy of a homogeneous system at arbitrary polarizations. This expression is usually written

$$\epsilon_x(n, \xi) = \epsilon_x(n, \xi = 0) + [\epsilon_x(n, \xi = 1) - \epsilon_x(n, \xi = 0)] f(\xi), \tag{5.3.5}$$

where the dimensionless function $f(\xi)$ is

$$f(\xi) = \frac{(1+\xi)^{4/3} + (1-\xi)^{4/3} - 2}{2(2^{1/3} - 1)}. \tag{5.3.6}$$

This function has the value zero at $\xi = 0$ and unity at $\xi = 1$. In the LDA, one then just replaces n and ξ in Eq. (5.3.5) by their local values $n(\mathbf{r})$ and $\xi(\mathbf{r})$. The correlation energy is again a little trickier. There exists no exact interpolation to arbitrary polarization, even for the homogeneous system, as there does for the exchange energy. In lieu of better approximations, the first thing that comes to mind is to use the *same* scaling relation for the correlation energy as for the exchange energy, and write

$$\epsilon_c(n(\mathbf{r}), \xi(\mathbf{r})) = \epsilon_c(n(\mathbf{r}), \xi(\mathbf{r}) = 0) + [\epsilon_c(n(\mathbf{r}), \xi(\mathbf{r}) = 1) - \epsilon_c(n(\mathbf{r}), \xi(\mathbf{r}) = 0)] f(\xi(\mathbf{r})),$$

with the same scaling function $f(\xi)$ as for the exchange energy, Eq. (5.3.6). This turns out to be surprisingly accurate: the error given by this for the Perdew–Zunger parametrization of the exchange and correlation energies is at most 3.5% when compared with values calculated by more laborious numerical schemes.

The first real calculation using DFT was by Lang and Kohn, who found the work functions of simple metals in the LDA. Before their calculation, this had been a difficult problem. The Hartree approximation typically gave at least the right sign, but wrong values, while "improvements" using Hartree–Fock theory often would yield the wrong sign, meaning that according to Hartree–Fock theory, metals were not stable! The calculations by Lang and Kohn not only showed metals to be stable, but gave very good quantitative agreement with experiments. These calculations thus showed the importance of the correlation energy. The exchange energy is typically attractive, but too large to give good results for metals without further corrections. The inclusion of correlation energy compensates for the large exchange energy.

It may seem surprising that the LDA gives such good results for work functions. After all, the calculation of work functions involves metal *surfaces*. While the LDA is based on the assumption of a slowly varying electron density, the density near metal surfaces varies very rapidly, and so metal surfaces should perhaps not be within the region of applicability of the LDA. Other cases in point are also given by DFT–LDA calculations of atomic systems, in which the density again varies rapidly on atomic length scales. Yet, the LDA yields very good results (of the order of one to ten percent error depending on the quantity). Why is that?

First of all, the LDA is by construction the exchange-correlation energy of a physical system (an infinite homogeneous system). As such, it then satisfies many of the relations, the so-called sum rules, that are required of the exchange-correlation energy and potential. There is also another reason that has a deep and physical origin, and for which we need to know a little about something called the *exchange hole*. Consider an electron system interacting through Hartree and exchange terms alone. While the Hartree term is blind to the spin of electrons, the exchange term is not. This term, which is really a manifestation of the Pauli Exclusion Principle, acts to keep electrons of like spins apart. As a consequence, if we put an up-spin electron at the origin, there will be a deficit of other up-spin electrons in a neighborhood around the origin. One can actually write down a precise expression for this deficit, which, if integrated out, adds up to precisely one electron. We can therefore think of each up-spin (or down-spin) electron as it is moving through the system as being surrounded by a little bubble of deficit of up-spin (or down-spin) electrons. This "bubble," which moves with the electron, is the exchange hole. Now, if we go beyond Hartree and Fock terms, and add electron correlations, there is still a deficit of up-spin (or down-spin) charge around the electron. The correlation effects counteract the exchange term locally to some extent, but the net deficit is still precisely one electron. This net deficit is, quite naturally, called the exchange-correlation hole. If we consider an electron at a point $\mathbf{r}$, we can write the exchange-correlation hole density at $\mathbf{r}'$ as $\rho_{xc}(\mathbf{r}, \mathbf{r}')$, and one can rigorously show that

$$\mathcal{E}_{xc} = \frac{1}{2} \iint n(\mathbf{r}) V(\mathbf{r} - \mathbf{r}') \rho_{xc}(\mathbf{r}, \mathbf{r}') \, d\mathbf{r} \, d\mathbf{r}'.$$

For translationally invariant interactions, $\rho_{xc}(\mathbf{r}, \mathbf{r}') = \rho_{xc}(|\mathbf{r} - \mathbf{r}'|)$. The exchange-correlation energy is consequently determined by the form of this spherically symmetric exchange-correlation hole. The LDA amounts to using a particular approximation for the exchange-correlation hole, and this approximation is successful in representing the form of the true exchange-correlation hole. Part of this is due to the fortunate fact that errors from exchange and correlation terms in the LDA cancel each other to some degree.

5.4 Electronic structure calculations

Density functional theory is the approach that is now almost universally used in performing electronic structure calculations in condensed matter physics, and the same is fast becoming the case in quantum chemistry. While many of the earlier calculations used the Kohn–Sham scheme, there is now also a large body of work, in particular for large systems, in which the approach is to focus on minimizing the DFT–LDA expression for the ground-state energy *directly*. There are in common use two different kinds of DFT ground-state calculation. One is the so-called all-electron calculation, in which the Coulomb potentials of the fully ionized atoms, i.e., of the bare nuclei, are used for the external potential. All the electrons are then added to the system and are considered in the calculations. This approach, however, is frequently not practical for large systems consisting of hundreds of atoms, as there are then many thousands of electrons.

This difficulty was resolved in the previous chapter for the case of non-interacting electrons when we introduced the concept of the pseudopotential in Section 4.4. There we argued that the bare Coulomb potential is very strong, and thus tightly binds the core electrons, which consequently have little effect on the properties of the system. We developed an analysis in which the comparatively weak pseudopotential $U_\mathbf{k}$ given in Eq. (4.4.4) acted on a comparatively smooth pseudo-wavefunction $\chi_\mathbf{k}$. We can proceed to apply a similar technique within DFT, but at a cost of further loss of rigor, since there is in this case no Hohenberg–Kohn theorem that formally allows us to cast the problem in terms of only the ground-state density. This difficulty stems from the fact that pseudopotentials are operators, and hence are non-local. That is, a pseudopotential from an ion at position $\mathbf{r}$ acting on a valence electron at $\mathbf{r}'$ depends on $\mathbf{r}$ and $\mathbf{r}'$ *separately*, and not just on the difference $\mathbf{r} - \mathbf{r}'$. A DFT can be constructed for pseudopotentials, but the basic variable here is the so-called single-particle density–density matrix $\gamma(\mathbf{r}, \mathbf{r}')$, which introduces another layer of complexity. For reasons of computational convenience, many users of pseudopotential methods within DFT typically ignore the inconsistency posed by the nonlocality of U_k. A powerful justification is that pseudopotential calculations frequently give results that are as accurate as any all-electron calculations. This may well be due to the fact that it is in the end necessary to resort to approximations for the exchange-correlation energy and potential, and these approximations are probably responsible for most of the errors in the calculations.

Density functional theory, as we have described it here, is a theory for the ground state of an interacting electron system, although there do exist extensions to thermodynamic equilibria and to excited states. While it is formally exact and can in principle be used to determine the expectation value of any observable of the system, practical calculations focus on the exact formulation for the ground-state energy and electron density, and are constructed to give good approximations for these quantities. It is therefore not surprising that quantities extracted directly from the ground-state energy and electron density tend to be more accurate than quantities found by methods for which there exists no formal justification. But even if we restrict ourselves to only the former class, we can, with a little ingenuity, calculate a host of quantities.

First of all, we have the total energy. By calculating this for a variety of different possible unit cells for a solid, one can predict the crystalline structure at various macroscopic densities. Calculated values for the total energy at different lattice parameters for each lattice structure for a given material can then be fitted to an approximate equation of state, from which the bulk modulus can be calculated. By analyzing these equations of state, predictions can be made about phase transitions as a function of pressure. In fact, a high-pressure phase (a so-called β-tin phase) of silicon was first predicted from total-energy DFT calculations, and later found experimentally. Most common semiconductors, such as Si, Ge, GaAs, GaP, InP, InAs, and InSb, have been studied extensively in total-energy calculations of this sort. Unit cell volumes and bond distances can be determined directly from the structure having the minimum energy, with an accuracy of a few percent for the unit cell volumes, and up to about one third of one percent (less than 0.01 Å) for bond distances. Total energies (for nonmagnetic crystals) are also well predicted, with energies accurate to within ten percent of experimental values. The bulk modulus is found as a second derivative of the energy with respect to volume, and is correspondingly less accurately predicted, with errors typically being some tens of percent.

Surfaces have also been the subject of many calculations. For example, the Lang–Kohn calculations mentioned earlier were the first DFT calculations of simple surface properties. Compared with exact calculations of surface exchange-correlation energies for jellium systems, the LDA performs acceptably well, with errors of the order of two percent. Surface reconstructions, in particular of semiconductors, have been particularly extensively studied. In these calculations, a perfect bulk crystal is terminated at a particular plane surface, and then the atoms are allowed to move to establish new positions of lower energy. Since the atoms at or near the surface have fewer near neighbors than those in the bulk, they experience a net force. As these atoms move to new equilibrium positions, the symmetry of their arrangement at the surface is typically altered from the bulk symmetry. Perhaps the most celebrated example is the reconstruction of Si atoms on a (111) surface, the so-called 7×7 reconstruction.

Phonon energies can be determined from "frozen phonon" calculations, in which the lattice is given a static deformation corresponding to a phonon of a particular branch of a chosen wavevector **q** and polarization **s**, and the total energy is calculated and compared with the ground-state energy. These "frozen phonon" calculations use the Born–Oppenheimer approximation, which is based on the fact that electrons are very much lighter than the ions, and so move much faster. The electrons can thus adjust very quickly to a change in the ionic positions and at every instant form the ground state of the system defined by whatever positions the ions are occupying at that time. These calculations are easiest for special phonon wavevectors **q** at high symmetry points in the Brillouin zone, and give results that are typically accurate to within a few percent.

For atoms and simple molecules, the LDA (or alternatively the Local Spin Density Approximation, LSDA) gives good results for geometrical quantities, such as bond lengths, and for electron densities, vibrational frequencies, and energy differences such as ionization potentials. These results are often an improvement over results obtained using the Hartree–Fock approximation. We remember also that the Hartree–Fock approximation is difficult and time-consuming to use in practical calculations, since the effective potential is orbital-dependent. For open-shell atoms the LSDA tends to overestimate the ground-state energy in comparison with the best experimental values, and with the Hartree–Fock approximation. Another important quantity for molecules is the bond dissociation energy, or the atomization energy. This is the energy required to break bonds and dissociate a simple molecule into its atomic constituents. The LDA typically does rather poorly for this quantity, overestimating it with errors around 20%, and sometimes over 100%.

The band structures of solids are usually calculated by interpreting the Kohn–Sham eigenvalues $\mathcal{E}_{\mathbf{k},n}$ for Bloch states of wavevector **k** in band n as being the band energies. Although there is no formal justification for this interpretation, it usually works remarkably accurately. However, if one tries to calculate the band gap of insulators and semiconductors by taking the difference between the highest occupied LDA Kohn–Sham eigenvalue and the lowest unoccupied one, the result typically underestimates the real band gap by as much as 50%. The error has two distinct sources. One is the LDA approximation, which introduces errors due to the inexact nature of the LDA exchange-correlation potential. The other is of a more subtle kind. One can rigorously show that the exchange-correlation potential must have discontinuities in its functional derivative at integer particle numbers. In order to evaluate the band gap of an N-electron system, one must include this discontinuity at N. The necessity of doing this adds such complexity to the calculations that this correction is often neglected, sometimes with unfortunate consequences. It is known that for some systems, such as Si, omission of the discontinuity in the derivative of the exchange-correlation potential is responsible for over 80% of the error in the calculated band gap.

5.5 The Generalized Gradient Approximation

As we have mentioned, the local density approximation (or its extension to spin-polarized systems) has been the most commonly used approximation for the exchange-correlation energy. It is simple to implement in calculations, it gives very reasonable results, and it has the appeal of actually being the exchange-correlation energy of something real (an infinite homogeneous electron gas) and thus satisfies sum rules and other constraints. It is, however, tempting to regard the LDA as only the lowest-order term in an expansion of the exchange-correlation energy in powers of the first- and higher-order gradients of the density.

For a homogeneous electron gas, there is really only one intrinsic length scale, and that is k_F^{-1}, which is proportional to $n^{-1/3}$. For the expansion to be justified the length scale of variation of the density must thus be large compared with k_F^{-1}. We can then formally expand the exchange-correlation energy in density variations about the constant density of the homogeneous gas by writing

$$\mathcal{E}_{xc}[n] = \int [g_{00}(n) + g_{22}(n)(\nabla n)^2 + g_{42}(n)(\nabla^2 n)^2$$
$$+ g_{43}(n)(\nabla^2 n)(\nabla n)^2 + g_{44}(n)(\nabla n)^4 + \cdots] \, d\mathbf{r}. \quad (5.5.1)$$

This form satisfies some constraints on rotational invariance, and the first term in this expression, $g_{00}(n(\mathbf{r}))$, constitutes the LDA. As terms of higher

5.5 The Generalized Gradient Approximation

order bring in more powers of ∇n, the coefficients that accompany them must bring in an equal number of powers of $(k_F n)^{-1}$ if the dimensions of the expression are to be preserved. Because $k_F \propto n^{1/3}$ we conclude that Eq. (5.5.1) is an expansion in powers of $|\nabla n|/n^{4/3}$.

Early attempts to incorporate the corrections in Eq. (5.5.1) were focused almost exclusively on the term $g_{22}(n(\mathbf{r}))$, mainly because it has the attractive property of being quadratic in ∇n. As a result, any variation in the energy produces a linear response that can be calculated more or less straightforwardly. However, the results obtained were discouraging in that they did not improve on the LDA. In fact, the converse was often true, and inclusion of gradient corrections gave *worse* results than use of just the LDA. Blindly expanding the exchange-correlation energy in gradients of the density leads to functionals that violate some sum rules that the LDA has to satisfy. For example, such expansions often lead to an incorrect behavior for the long-wavelength contributions to the exchange-correlation energy, and diverge or increase sharply as $k \to 0$, whereas the long-wavelength contributions to the true exchange-correlation energy vanish as $k \to 0$. The incorrect behavior of this expansion is particularly easy to see for finite systems, such as atoms and molecules. For such systems, the density vanishes exponentially at distances far from the nuclei, and this makes $|\nabla n|/n^{4/3}$ diverge.

Further analysis of the gradient expansion shows that the resulting exchange-correlation hole is not negative definite. The question then arises of whether one can construct a gradient expansion that avoids these shortcomings, so that the resulting exchange-correlation hole satisfies the most important sum rules. Such analyses and "fixes" have been performed, most notably by Perdew and co-workers. The resulting gradient corrections with exchange-correlation hole constraints restored are referred to as *Generalized Gradient Approximations* (GGAs). Carefully constructed GGAs also satisfy other physical limits, such as giving the correct exchange-correlation energy in the homogeneous electron gas limit ($\nabla n(\mathbf{r}) \to 0$), coordinate scaling relations, global bounds on the exchange-correlation energy, and correctly giving the lowest-order term $g_{22}(n(\mathbf{r}))$ in the expansion Eq. (5.5.1). The exchange-correlation energy in the GGA is thus written as

$$\mathcal{E}_{xc}^{GGA} = \int f(n_\uparrow, n_\downarrow, \nabla n_\uparrow, \nabla n_\downarrow)\, d\mathbf{r}, \tag{5.5.2}$$

where the function f is a universal function of the spin-up and spin-down densities and their gradients. There exist several versions of the GGA, all of which are parametrized somewhat differently.

The great strength of the GGA lies in the dramatic improvement it gives over the LDA in calculating such properties as bond dissociation energies, which the LDA may overestimate by as much as 100%, while the GGA gives errors typically of the order of ten percent or less. The GGA also gives a great improvement over the LDA for bulk moduli of solids, with an error of around ten percent, compared with around 20% for the LDA (or LSDA). With modern GGA exchange-correlation potentials, atomic and molecular quantities can now be calculated with chemical accuracy. Before the GGA, this was only possible using so-called configuration-interaction schemes, in which the true many-body wavefunction is expanded in some small set of Slater determinants composed of suitable atomic basis functions. This is an extremely arduous and computationally intensive approach. Density functional theory with the GGA, on the other hand, is essentially no more complicated than Hartree or LDA calculations. The exchange-correlation potential is only slightly more complicated than in the LDA, and (more importantly) it is still a *multiplicative* potential, leading to simple, effective single-particle equations. The combined advantages of simple equations and high accuracy have resulted in a revolution in quantum chemistry.

We close this section with a cautionary note. While mean errors, averaged over many compounds, atoms, or molecules, tend to be improved in the GGA compared with the LDA, this is no guarantee that the GGA will be better for some specific calculation. For example, while the LDA on average gives poorer results for the bulk modulus, it has been found better for GaAs than some GGA calculations. New forms for the functions used in Eq. (5.5.2) are continually being suggested, each of which has its own advantages. The so-called PBE (Perdew–Burke–Ernzerhof) GGA is one of those constructed to satisfy the constraints and limits mentioned earlier. There are other GGAs not constructed this way, but obtained by brute-force fitting to a large data set. While such GGAs may give excellent results for quantities included in the fitting procedure, they can yield large errors for other quantities. Because these kinds of fitting procedure tend not to satisfy physical constraints and limits and often violate, for example, the uniform electron gas limit, their use is limited to specialized applications.

5.6 More acronyms: TDDFT, CDFT, and EDFT

We finish our discussion of density functional theory with a brief tour of some later developments. We start with time-dependent density functional theory (TDDFT), which extends the reach of stationary density functional theory in a very powerful way. Not only are strongly time-dependent phenomena accessible to computations, but TDDFT also provides a natural way to calculate the excitation energies of a system.

We have previously seen how the Hohenberg–Kohn theorem establishes a one-to-one correspondence between external potentials and electron densities for time-independent systems. It is natural to ask under what circumstances something similar holds true for systems in which the external potential depends on time, so that $V_{ext} = V_{ext}(\mathbf{r}, t)$. The answer is given by the Runge–Gross theorem, which we will here only state and not prove. This theorem establishes a one-to-one correspondence between density $n(\mathbf{r}, t)$ and its governing time-dependent external potential $v(\mathbf{r}, t)$, but there is a catch: the correspondence can be established only for a specified initial many-body state Ψ_0, and consequently the functional relationships between density and potential implicitly depend on Ψ_0. However, this will not cause any problem if Ψ_0 is a nondegenerate ground state. To state the theorem, we begin with a system initially in an eigenstate Ψ_0 of the usual homogeneous N-electron Hamiltonian,

$$\mathcal{H}_0 = T + V_{int}.$$

At some time t_0 we turn on a time-dependent, scalar potential

$$V_{ext}(t) = \sum_i^N v_{ext}(\mathbf{r}_i, t).$$

Here the only requirements we put on $v_{ext}(\mathbf{r}, t)$ are that it be finite and Taylor-expandable about t_0. We do not require Ψ_0 to be an eigenstate of $V_{ext}(t_0)$. The theorem then states that the densities $n(\mathbf{r}, t)$ and $n'(\mathbf{r}, t)$ that evolve from a common initial state Ψ_0 under the influence of two potentials $v_{ext}(\mathbf{r}, t)$ and $v'_{ext}(\mathbf{r}, t)$ are different provided that the potentials differ by more than a purely time-dependent but spatially uniform function, so that

$$v_{ext}(\mathbf{r}, t) \neq v'_{ext}(\mathbf{r}, t) + c(t).$$

The proof of the theorem proceeds by first demonstrating that the current densities resulting from the application of $v_{ext}(\mathbf{r}, t)$ and $v'_{ext}(\mathbf{r}, t)$ necessarily must differ, given the property of Taylor-expandability of the potentials. Once this is established, one can then use the continuity equation to show that the densities $n(\mathbf{r}, t)$ and $n'(\mathbf{r}, t)$ must also differ. Therefore, the density determines the potential, and so the Hamiltonian, and thus the expectation value of any operator, is uniquely determined by the time-dependent density.

However, we must bear in mind that all these relationships implicitly depend on the choice of initial state Ψ_0.

Once the Runge–Gross theorem is established, one can then continue and derive Kohn–Sham equations, just as for the time-independent case, by considering a noninteracting system with the same density as the interacting one. Formally, the equations look just as one would expect for the time-dependent Kohn–Sham orbitals $\varphi(\mathbf{r}, t)$:

$$i\hbar \frac{\partial \varphi_j(\mathbf{r}, t)}{\partial t} = \left[-\frac{\hbar^2}{2m} \nabla^2 + v_s[n](\mathbf{r}, t) \right] \varphi_j(\mathbf{r}, t).$$

Here, the effective potential $v_s(\mathbf{r}, t)$ is given by

$$v_s(\mathbf{r}, t) = v_{xc}(\mathbf{r}, t) + v_{ext}(\mathbf{r}, t) + v_H(\mathbf{r}, t),$$

with the exchange-correlation and Hartree potentials now being time-dependent.

Once the time-dependent formalism is established, one can derive formally *exact* expressions for the density–density response functions, which describe how the density of the system changes in response to an external potential that couples to the density. As an example we consider the linear response function obtained by expanding the density response to first order in the applied potential. We suppose an external potential of the form

$$v_{ext}(\mathbf{r}, t) = v_0(\mathbf{r}) + v_1(\mathbf{r}, t)\theta(t - t_0),$$

where $\theta(x)$ is the Heaviside step function, which vanishes for $x < 0$ and is unity for $x \geq 0$. This form of external potential is allowed by the Runge–Gross theorem. For $t < t_0$, the external potential is $v_0(\mathbf{r})$, which we take to be the potential of the ion cores of the system, and the system is in its ground state corresponding to $v_0(\mathbf{r})$ with density $n_0(\mathbf{r})$. At a time t_0 an additional, time-dependent perturbation $v_1(\mathbf{r}, t)$ is applied. According to the Runge–Gross theorem, the time-dependent density $n(\mathbf{r}, t)$ is then a functional of the external potential (since we had fixed the state of the system prior to turning on the time-dependent perturbation). We can then expand the density in a Taylor series about $n_0(\mathbf{r})$:

$$n(\mathbf{r}, t) = n_0(\mathbf{r}) + n_1(\mathbf{r}, t) + \cdots$$

5.6 More acronyms: TDDFT, CDFT, and EDFT

The first-order response of the density is by definition linear in $v_1(\mathbf{r}, t)$, and thus can be written

$$n_1(\mathbf{r}, t) = \int \chi(\mathbf{r}, t; \mathbf{r}', t') v_1(\mathbf{r}', t') \, d\mathbf{r}' \, dt.$$

The response function $\chi(\mathbf{r}, t; \mathbf{r}', t')$ is the amount by which the density $n(\mathbf{r}, t)$ varies when a change in external potential is applied that is localized in both time and space, and of the form $\delta(\mathbf{r} - \mathbf{r}')\delta(t - t')$. It can be expressed as a functional derivative, but of a type more complicated than those encountered in Section 5.2, and written as

$$\chi(\mathbf{r}, t; \mathbf{r}', t', n_0) \equiv \frac{\delta n[v_{\text{ext}}](\mathbf{r}, t)}{\delta v_{\text{ext}}(\mathbf{r}', t')},$$

where we have made the dependence of χ on the initial density n_0 explicit. The formal definition of the functional derivative $\delta F[\eta(\mathbf{r})]/\delta \eta(\mathbf{r}')$ of the scalar functional $F[\eta(\mathbf{r})]$ with respect to the scalar function $\eta(\mathbf{r}')$ is a function of $\mathbf{r}$ and $\mathbf{r}'$, and is defined as

$$\frac{\delta F[\eta(\mathbf{r})]}{\delta \eta(\mathbf{r}')} \equiv \lim_{\epsilon \to 0} \frac{F[\eta(\mathbf{r}) + \epsilon \delta(\mathbf{r} - \mathbf{r}')] - F[\eta(\mathbf{r})]}{\epsilon}.$$

We note that the dimensions of functional derivatives are not what they at first appear to be, since ϵ has the dimensions of η multiplied by a volume. The functional derivative $\delta n[v_{\text{ext}}](\mathbf{r}, t)/\delta v_{\text{ext}}(\mathbf{r}', t')$ thus has dimensions of (number density)/(energy × volume × time).

Similarly, we can consider the response of the noninteracting auxiliary system to the external potential $v_{\text{ext}}(\mathbf{r}, t)$, which yields its response function

$$\chi_s(\mathbf{r}, t; \mathbf{r}', t') \equiv \frac{\delta n[v_s](\mathbf{r}, t)}{\delta v_{\text{ext}}(\mathbf{r}', t')}. \tag{5.6.1}$$

It is possible to show that when this response function is Fourier transformed with respect to the time difference $(t - t')$ and is given in terms of the *static, unperturbed* Kohn–Sham orbitals $\varphi_i(\mathbf{r})$ one finds

$$\chi_s(\mathbf{r}, t; \mathbf{r}', \omega) = \sum_{i,j} (f_j - f_i) \frac{\varphi_i(\mathbf{r}) \varphi_j^*(\mathbf{r}) \varphi_i^*(\mathbf{r}') \varphi_j(\mathbf{r}')}{\hbar \omega - (\epsilon_i - \epsilon_j) + i\eta}. \tag{5.6.2}$$

Here the ϵ_i are the Kohn–Sham eigenvalues, η is a positive infinitesimal to fix the poles of χ_s, and f_i are the occupation numbers (typically 0 or 1) of the Kohn–Sham orbitals. The sum in Eq. (5.6.2) vanishes unless one of the orbitals is occupied and one is empty. This can be understood in terms of electrons making virtual transitions from occupied to unoccupied states. This means that any practical calculations of the response function must include enough of the unoccupied orbitals to capture the principal contributions to χ_s.

What is now left is to relate the response function χ of the interacting system to that of the noninteracting system. First, we note that, by assumption, the densities of the interacting and the noninteracting systems are identical. Therefore, we can think of $n(\mathbf{r}, t)$ in Eq. (5.6.1) as the density of the noninteracting system. This density is a functional of the effective potential $v_s(\mathbf{r}, t)$ of the noninteracting system, and v_s is in its turn a functional of the external potential $v_{\text{ext}}(\mathbf{r}, t)$. Putting that all together using the chain rule for functional differentiation, we obtain

$$\chi(\mathbf{r}, t; \mathbf{r}', t') = \int \frac{\delta n(\mathbf{r}, t)}{\delta v_s(\mathbf{r}'', t'')} \frac{\delta v_s(\mathbf{r}'', t'')}{\delta v_{\text{ext}}(\mathbf{r}', t')} \, d\mathbf{r}'' \, dt''. \tag{5.6.3}$$

If we then write the exchange-correlation potential as

$$v_{\text{xc}}[n](\mathbf{r}, t) = v_s[n](\mathbf{r}, t) - v_{\text{ext}}(\mathbf{r}, t) - v_H[n](\mathbf{r}, t)$$

we can use this relation between v_s, v_{ext}, v_H, and v_{xc} to rewrite the second functional derivative in Eq. (5.6.3) as

$$\frac{\delta v_s(\mathbf{r}, t)}{\delta v_{\text{ext}}(\mathbf{r}', t')} = \delta(\mathbf{r} - \mathbf{r}')\delta(t - t') + \int \left[\frac{\delta(t - t'')}{|\mathbf{r} - \mathbf{r}''|} + \frac{\delta v_{\text{xc}}(\mathbf{r}, t)}{\delta n(\mathbf{r}'', t'')} \right] \frac{\delta n(\mathbf{r}'' t'')}{\delta v_{\text{ext}}(\mathbf{r}' t')} \, d\mathbf{r}'' \, dt''.$$

We then insert this into Eq. (5.6.1), and use Eq. (5.6.2) to arrive finally at

$$\chi(\mathbf{r}, t; \mathbf{r}', t') = \chi_s(\mathbf{r}, t; \mathbf{r}', t') + \iint \chi_s(\mathbf{r}, t; \mathbf{r}'', t'') \left[\frac{\delta(t'' - t''')}{|\mathbf{r}'' - \mathbf{r}'''|} + \frac{\delta v_{\text{xc}}(\mathbf{r}'', t'')}{\delta n(\mathbf{r}''', t''')} \right]$$
$$\times \chi(\mathbf{r}''', t'''; \mathbf{r}', t') \, d\mathbf{r}'' \, dt'' \, d\mathbf{r}''' \, dt'''. \tag{5.6.4}$$

Equation (5.6.4) is a formally *exact* equation for the density–density response function of the interacting system. Note that this expression (being a linear response quantity) depends only on the ground-state properties of the system, and can be calculated if the *ground-state* Kohn–Sham orbitals and their eigenvalues have been calculated. By multiplying Eq. (5.6.4) by the

5.6 More acronyms: TDDFT, CDFT, and EDFT

perturbing potential under consideration, and integrating over $\mathbf{r}'$ and t', we can also derive a formally exact expression for the linear density response to the time-dependent potential of our choosing. Moreover, after a little thought we realize that the exact density–density response function $\chi(\mathbf{r}, \mathbf{r}'; \omega)$ has poles at the exact excitation energies of the interacting system. Thus, the time-dependent linear response theory gives us a practical way to calculate the excitation energies of an interacting system. This calculation would be exact if we were fortunate enough to have an exact expression for the quantity

$$f_{xc}(\mathbf{r}, t; \mathbf{r}', t') \equiv \frac{\delta v_{xc}(\mathbf{r}, t)}{\delta n(\mathbf{r}', t')}.$$

This quantity is called the exchange-correlation kernel. Perhaps the most natural, and also the most frequently used, approximation for it is one made in the spirit of the LDA, and is called the adiabatic local density approximation (ALDA). Here, we consider a homogeneous electron gas with a time-dependent perturbation that varies sufficiently slowly that the system is "adiabatic," meaning that the exchange-correlation energy per particle at $\mathbf{r}$ at any instant t is that of a homogeneous electron gas in an external potential $v_{ext}(\mathbf{r}, t')|_{(t'=t)}$. This implies that there is no frequency dependence of f_{xc} in the ALDA. Just as for the LDA, it is found that the ALDA works reasonably well, especially for low-frequency perturbations, and, just as for the LDA, attempts to improve on it turn out to be difficult. The reason for this is very deep: any simple attempt to include a frequency dependence in a local approximation will violate something called the Harmonic Potential Theorem, which is related to Galilean invariance. The conclusion is, somewhat discouragingly, that there is no local, frequency-dependent exchange-correlation kernel. Instead, to build a theory that does not run into trouble with Galilean invariance, one has to consider the current density as the fundamental variable, not the particle density. Such a theory is called current density functional theory (CDFT).

Current density functional theory was developed to deal with magnetic systems for which the external magnetic field is strong enough that the full canonical kinetic energy

$$T = \frac{1}{2m}\left[\mathbf{p} - \frac{e}{c}\mathbf{A}\right]^2$$

has to be considered. As a consequence, there will in general be diamagnetic currents in response to the external magnetic field. The Hohenberg–Kohn theorem remains perfectly valid, even in the presence of a magnetic field, so we can *formally* still extract everything we want from "standard" density functional theory. However, experience has shown that practical calculations are usually more accurate if the broken symmetry can be explicitly included in the formalism. This is, for example, the case with spin density functional theory. While we could *in principle* use regular density-only functional theory, the approximations we make when we consider the spin densities separately lead to more accurate calculations. It was thought, therefore, that building the current density into the formulation would likewise lead to more accurate computational approaches. While this essentially turned out *not* to be the case, current density functional theory has given us a way to formulate more refined approximations for the exchange-correlation kernel than can be done using time-dependent density functional theory, and which are not encumbered by flagrant violations of Galilean invariance.

Finally, in this discourse on more acronyms, we briefly discuss ensemble density functional theory (EDFT), which helps us deal with degeneracies. The origin of EDFT is simple enough to relate. Consider a time-independent system, for which we are solving the Kohn–Sham equations. It does not matter whether we are using the LDA or GGA, or even some exact exchange-correlation energy. According to the Kohn–Sham scheme, we fill all the orbitals having the lowest energy eigenvalues. The occupation numbers f_i for the N lowest-lying orbitals are then unity while $f_i = 0$ for all the others. Suppose now that, while iterating our equations to reach self-consistency, we find that as we fill the orbitals in order of ascending energy, p eigenvalues at the highest occupied level ϵ_F are degenerate. Then what are we supposed to do? The answer is formally given by ensemble density functional theory. Here, the density is not constructed from a single Slater determinant of Kohn–Sham orbitals, but by an ensemble of Slater determinants Φ_k. We do this by writing the density matrix operator as

$$D_s = \sum_{k=1}^{p} d_k |\Phi_k\rangle\langle\Phi_k|,$$

where the weights d_k of the Slater determinants Φ_k are positive numbers that sum to unity, so that $\sum_k d_k = 1$. This leads to a ground-state ensemble density

$$n_s(\mathbf{r}) = \mathrm{Tr}\{D_s n(\mathbf{r})\} = \sum_{k=1}^{p} d_k \langle\Phi_k|n(\mathbf{r})|\Phi_k\rangle = \sum_{i:\epsilon_i<\epsilon_F} |\varphi_i(\mathbf{r})|^2 + \sum_{i:\epsilon_i=\epsilon_F} f_i|\varphi_i(\mathbf{r})|^2,$$

where now the occupation numbers f_i of the orbitals on the "Fermi surface" are fractional, $0 \leq f_i < 1$. Formally, these fractional occupation numbers are given as follows. Consider a particular orbital φ_j. This orbital does or does not appear in a particular Slater determinant Φ_k, and we can assign an occupation number $\theta_{kj} = 0, 1$ to this orbital in that Slater determinant. The occupation number f_j for this orbital is then given by a weighted average of the θ_{kj}, with the weights given by the ensemble weights d_k:

$$f_j = \sum_{k=1}^{q} d_k \theta_{kj}.$$

One can show that EDFT has very general validity, and in fact formally avoids some problems that occasionally crop up for "regular" density functional theory. These problems refer to what is known as v-representability, and are concerned with under what circumstances one can use the auxiliary noninteracting system to represent the density of the interacting one.

The most profound example of applications of EDFT comes in the fractional quantum Hall effect, about which we shall learn more in Chapter 10. These are strongly correlated systems (that is, the physics is determined entirely by the interactions), and there is no systematic perturbation theory that applies. Furthermore, it is known that the ground-state energy has cusps at the densities at which the fractional quantum Hall effect occurs. This implies that the LDA exchange-correlation potential has discontinuities, and so applications of DFT to fractional quantum Hall systems first have to deal with these discontinuities. If we assume that this can be done, and then naïvely apply the Kohn–Sham scheme we find that in general *all* Kohn–Sham eigenvalues are degenerate. This is apparently a situation where we have to take EDFT very seriously. A practical calculation scheme has been devised for these systems, and applications show that EDFT in the LDA gives very good results, and provides perhaps the only practical scheme with which to study inhomogeneous fractional quantum Hall systems.

Problems

5.1 A Wigner crystal forms when the potential energy gained by separating the electrons outweighs the cost of the extra kinetic energy needed to localize them. It is not easy to calculate the change in electrostatic energy per electron, but on dimensional grounds it must be of the form $-\alpha_1 e^2/a_0 r_s$, while the localization energy must be $\alpha_2 \hbar^2/m a_0^2 r_s^2$, with α_1

and α_2 constants. Estimate the value of r_s at which crystallization might occur.

5.2 Explicit calculation of even the kinetic energy functional $T_s[n]$ of a noninteracting electron gas is very difficult. Consider the effect of a weak external potential $V \cos gx$ for which $g \gg k_F$ on $n(\mathbf{r})$ and on the kinetic energy, and hence evaluate the function h_{22} in the expansion of $T_s[n]$ in powers of ∇n that is analogous to the $g_{22}(n)$ defined in Eq. (5.5.1).

5.3 Now consider the calculation of the kinetic energy functional $T_s[n]$ of a noninteracting electron gas in the opposite limit from Problem 5.2. Consider the effect of a weak external potential $V \cos gx$ for which $g \ll k_F$ as causing a periodic variation in the radius of the Fermi surface. The electron density $n(\mathbf{r})$ is again perturbed from its original uniform value $n_0(\mathbf{r})$, and this time the kinetic energy functional can be written as

$$T_s[n] = T_s[n_0]\left(1 + \alpha(\langle n^2 \rangle - n_0^2)/n_0^2\right).$$

Evaluate α.

5.4 In a collinear spin-polarized electron gas, the total electron density n can be separated into components $n = n_\uparrow + n_\downarrow$, where $n_\uparrow$ and $n_\downarrow$ are the densities of electrons with $\langle s_z \rangle = +\hbar/2$ and $\langle s_z \rangle = -\hbar/2$, respectively. In this case, the Hohenberg–Kohn theorem can be formulated in terms of $n_\uparrow$ and $n_\downarrow$, or, equivalently, in terms of n and the polarization $\xi = (n_\uparrow - n_\downarrow)/n$. In particular, the exchange energy can then be considered a functional $\mathcal{E}_x[n_\uparrow, n_\downarrow]$ of $n_\uparrow$ and $n_\downarrow$. Show that this functional must satisfy the relation

$$\mathcal{E}_x[n_\uparrow, n_\downarrow] = \tfrac{1}{2}\mathcal{E}_x[n_\uparrow, n_\uparrow] + \tfrac{1}{2}\mathcal{E}_x[n_\downarrow, n_\downarrow],$$

where $\mathcal{E}_x[n_\sigma, n_\sigma]$ means that the functional is evaluated with both arguments set equal to n_σ. You may use the fact that the exchange interaction acts only between electrons with the same value of $\langle s_z \rangle$. Since the exchange energy in the local density approximation is a simple function of electron density, we then have in this approximation in d dimensions

$$\mathcal{E}_x[n_\uparrow, n_\downarrow] \propto \int d^d r \left(n_\uparrow^\alpha + n_\downarrow^\alpha\right).$$

Find the value of α by using dimensional analysis. Then, by changing variables to n and ξ from $n_\uparrow$ and $n_\downarrow$, find the form of the interpolating function $f(\xi)$ for a two-dimensional electron gas [cf. Eq. (5.3.6)].

5.5 An old interpolation formula, but one that is simply expressed in closed form, for the correlation energy of a uniform unpolarized electron gas of N electrons is due to Nozières and Pines, who wrote

$$\mathcal{E}_c/N = 0.0311 \ln r_s - 0.115.$$

The unit of energy here is the hartree, which is equal to about 27.2 eV. Use this form of the correlation energy to estimate the ground-state energy of the helium atom. Use a simple singlet Ansatz wavefunction for the two electrons, the spatial part of which is

$$\Psi(\mathbf{r}_1, \mathbf{r}_2) = \frac{8}{\pi a_0^3} \exp[-2(r_1 + r_2)/a_0],$$

where $\mathbf{r}_1$ and $\mathbf{r}_2$ are the coordinates of the two electrons and a_0 is the Bohr radius. First, evaluate the expectation value of the kinetic and potential energies. To evaluate the Hartree and correlation energy, you need to derive an expression for the electron density using the Ansatz wavefunction. [You do not need to consider the exchange energy, since there is only one electron of each spin orientation, and it cannot exchange with itself!] Compare your result to the experimental value of -98.975 eV. How large is the correlation correction? Then calculate the electron interaction energy in first-order perturbation theory by forming the quantity $\langle \Psi | V(\mathbf{r}_1 - \mathbf{r}_2) | \Psi \rangle$. Discuss the difference you find from the previous result.

5.6 In Problem 2.3 you estimated the density at which the Hartree–Fock approximation predicts that the uniform electron gas has a paramagnetic to ferromagnetic transition. This critical density will change when the correlation energy is included. Estimate this revised critical density by using the scaling relation (5.3.3) and the Nozières–Pines expression for the correlation energy for an unpolarized electron gas given in Problem 5.5. You will need to estimate the ground-state energy (kinetic, exchange, and correlation) of both the unpolarized and of a fully polarized ($\xi = 1$) electron gas at a fixed density, and find when they become equal.

Chapter 6
Electron–phonon interactions

6.1 The Fröhlich Hamiltonian

In this chapter we shall consider some of the consequences of the interaction of phonons with electrons, and in particular with the electrons in a simple metal. The subject is a complicated and difficult one, in that we need to call on most of the knowledge that we have of the behavior of the electron gas and of lattice vibrations. A complete calculation should really start with the Hamiltonian of a lattice of bare ions whose mutual interaction would include the long-range Coulomb potential. One would then add the electron gas, which would shield the potential due to the ions in the manner indicated in Section 1.7. It is, however, possible to explore many of the consequences of the electron–phonon interaction by use of a simpler model. In this model we take for granted the concept of screening, and assume that the ions interact with each other and with the electrons only through a short-range screened potential, and we treat the electrons themselves as independent fermions. For a Bravais lattice our unperturbed Hamiltonian is then simply

$$\mathcal{H}_0 = \sum_{\mathbf{k}} \mathcal{E}_{\mathbf{k}} c_{\mathbf{k}}^\dagger c_{\mathbf{k}} + \sum_{\mathbf{q},s} \hbar \omega_{\mathbf{q}s} a_{\mathbf{q}s}^\dagger a_{\mathbf{q}s},$$

the phonon frequencies $\omega_{\mathbf{q}s}$ being proportional to q as $q \to 0$. To this we add the interaction, $\mathcal{H}_1$, of the electrons with the screened ions. We assume that at any point the potential due to a particular ion depends only on the distance from the center of the ion – an assumption sometimes known as the *rigid-ion approximation* – so that in second-quantized notation

$$\mathcal{H}_1 = \sum_{\mathbf{k},\mathbf{k}',l} \langle \mathbf{k} | V(\mathbf{r} - \mathbf{l} - \mathbf{y}_l) | \mathbf{k}' \rangle c_{\mathbf{k}}^\dagger c_{\mathbf{k}'}$$

$$= \sum_{\mathbf{k},\mathbf{k}',l} e^{i(\mathbf{k}' - \mathbf{k}) \cdot (\mathbf{l} + \mathbf{y}_l)} V_{\mathbf{k} - \mathbf{k}'} c_{\mathbf{k}}^\dagger c_{\mathbf{k}'}.$$

6.1 The Fröhlich Hamiltonian

Here $V(\mathbf{r})$ is the potential due to a single ion at the origin, and $V_{\mathbf{k}-\mathbf{k}'}$ its Fourier transform. With the assumption that the displacement $\mathbf{y}_l$ of the ion whose equilibrium position is l is sufficiently small that $(\mathbf{k}' - \mathbf{k}) \cdot \mathbf{y}_l \ll 1$ we can write

$$e^{i(\mathbf{k}'-\mathbf{k})\cdot\mathbf{y}_l} \simeq 1 + i(\mathbf{k}' - \mathbf{k}) \cdot \mathbf{y}_l$$
$$= 1 + iN^{-1/2}(\mathbf{k}' - \mathbf{k}) \cdot \sum_{\mathbf{q}} e^{i\mathbf{q}\cdot l}\mathbf{y}_{\mathbf{q}}.$$

Then $\mathcal{H}_1$ can be split into two parts,

$$\mathcal{H}_1 = \mathcal{H}_{\text{Bloch}} + \mathcal{H}_{e-p},$$

the first term, $\mathcal{H}_{\text{Bloch}}$, being independent of the lattice displacements. We have

$$\mathcal{H}_{\text{Bloch}} = \sum_{\mathbf{k},\mathbf{k}',l} e^{i(\mathbf{k}'-\mathbf{k})\cdot l} V_{\mathbf{k}-\mathbf{k}'} c_{\mathbf{k}}^{\dagger} c_{\mathbf{k}'}$$
$$= N \sum_{\mathbf{k},\mathbf{g}} V_{-\mathbf{g}} c_{\mathbf{k}-\mathbf{g}}^{\dagger} c_{\mathbf{k}},$$

where the $\mathbf{g}$ are reciprocal lattice vectors, and

$$\mathcal{H}_{e-p} = iN^{-1/2} \sum_{\mathbf{k},\mathbf{k}',l,\mathbf{q}} e^{i(\mathbf{k}'-\mathbf{k}+\mathbf{q})\cdot l}(\mathbf{k}' - \mathbf{k}) \cdot \mathbf{y}_{\mathbf{q}} V_{\mathbf{k}-\mathbf{k}'} c_{\mathbf{k}}^{\dagger} c_{\mathbf{k}'}$$
$$= iN^{1/2} \sum_{\mathbf{k},\mathbf{k}'} (\mathbf{k}' - \mathbf{k}) \cdot \mathbf{y}_{\mathbf{k}-\mathbf{k}'} V_{\mathbf{k}-\mathbf{k}'} c_{\mathbf{k}}^{\dagger} c_{\mathbf{k}'}.$$

In terms of the annihilation and creation operators defined in Eq. (3.6.4) this becomes

$$\mathcal{H}_{e-p} = i \sum_{\mathbf{k},\mathbf{k}',\mathbf{s}} \left(\frac{N\hbar}{2M\omega_{\mathbf{k}-\mathbf{k}',\mathbf{s}}}\right)^{1/2} (\mathbf{k}' - \mathbf{k}) \cdot \mathbf{s} V_{\mathbf{k}-\mathbf{k}'} (a_{\mathbf{k}'-\mathbf{k},\mathbf{s}}^{\dagger} + a_{\mathbf{k}-\mathbf{k}',\mathbf{s}}) c_{\mathbf{k}}^{\dagger} c_{\mathbf{k}'},$$

where the summation now also includes the three polarization vectors, $\mathbf{s}$, of the phonons. (Because $\mathbf{s}$ and $-\mathbf{s}$ represent the same phonon mode, care is necessary in using this expression. With a different convention for the direction of $\mathbf{s}$, the plus sign in this expression would become a minus sign.) For simplicity we shall assume the phonon spectrum to be isotropic, so that the phonons will be either longitudinally or transversely polarized. Only the longitudinal modes, for which $\mathbf{s}$ is parallel to $\mathbf{k}' - \mathbf{k}$, then enter $\mathcal{H}_{e-p}$ – a fact in accord with the semiclassical viewpoint of Section 1.7. We shall also neglect the effects of $\mathcal{H}_{\text{Bloch}}$, the periodic potential of the stationary lattice. With these simplifications we are left with the *Fröhlich Hamiltonian*,

$$\mathcal{H} = \sum_{\mathbf{k}} \mathcal{E}_{\mathbf{k}} c_{\mathbf{k}}^{\dagger} c_{\mathbf{k}} + \sum_{\mathbf{q}} \hbar\omega_{\mathbf{q}} a_{\mathbf{q}}^{\dagger} a_{\mathbf{q}} + \sum_{\mathbf{k},\mathbf{k}'} M_{\mathbf{k}\mathbf{k}'} (a_{-\mathbf{q}}^{\dagger} + a_{\mathbf{q}}) c_{\mathbf{k}}^{\dagger} c_{\mathbf{k}'}, \qquad (6.1.1)$$

where the electron–phonon matrix element is defined by

$$M_{\mathbf{k}\mathbf{k}'} = i \sqrt{\frac{N\hbar}{2M\omega_{\mathbf{q}}}} |\mathbf{k}' - \mathbf{k}| V_{\mathbf{k}-\mathbf{k}'}, \qquad (6.1.2)$$

with the phonon wavenumber $\mathbf{q}$ equal to $\mathbf{k} - \mathbf{k}'$, reduced to the first Brillouin zone if necessary.

The interaction $\mathcal{H}_{e-p}$ can be considered as being composed of two parts – terms involving $a_{-\mathbf{q}}^{\dagger} c_{\mathbf{k}}^{\dagger} c_{\mathbf{k}'}$ and terms involving $a_{\mathbf{q}} c_{\mathbf{k}}^{\dagger} c_{\mathbf{k}'}$. These may be represented by the diagrams shown in Figs. 6.1.1(a) and 6.1.1(b), respectively. In the first diagram an electron is scattered from $\mathbf{k}'$ to $\mathbf{k}$ with the emission of a phonon of wavenumber $\mathbf{k}' - \mathbf{k}$. The total wavenumber is then conserved, as is always the case in a periodic system, unless the vector $\mathbf{k}' - \mathbf{k}$ lies outside the first Brillouin zone, so that $\mathbf{q} = \mathbf{k}' - \mathbf{k} + \mathbf{g}$ for some nonzero $\mathbf{g}$. Such electron–phonon Umklapp processes do not conserve wavenumber, and are important in contributing to the electrical resistivity of metals.

In the remainder of this chapter we shall study some of the consequences of the electron–phonon interaction for the equilibrium properties of materials. We shall see that not only are the electron and phonon excitation spectra modified, but that there can also appear an effective attractive interaction between electrons caused by the electron–phonon interaction.

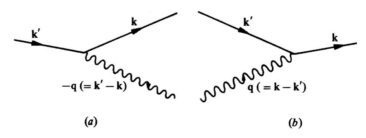

Figure 6.1.1. The Fröhlich Hamiltonian includes an interaction term in which an electron is scattered from $\mathbf{k}'$ to $\mathbf{k}$ with either emission (a) or absorption (b) of a phonon. In each case the total wavenumber is conserved.

6.2 Phonon frequencies and the Kohn anomaly

The effect of the electron–phonon interaction on the phonon spectrum may be seen by using perturbation theory to calculate the total energy of the system described by the Fröhlich Hamiltonian (6.1.1) to second order in $\mathcal{H}_{e-p}$. We have

$$\mathcal{E} = \mathcal{E}_0 + \langle \Phi | \mathcal{H}_{e-p} | \Phi \rangle + \langle \Phi | \mathcal{H}_{e-p} (\mathcal{E}_0 - \mathcal{H}_0)^{-1} \mathcal{H}_{e-p} | \Phi \rangle,$$

with $\mathcal{E}_0$ the unperturbed energy of the state Φ having $n_\mathbf{q}$ phonons in the longitudinally polarized mode $\mathbf{q}$ and $n_\mathbf{k}$ electrons in state $\mathbf{k}$. The first-order term vanishes from this expression, since the components of $\mathcal{H}_{e-p}$ act on Φ either to destroy or create one phonon, and the resulting wavefunction must be orthogonal to Φ. In second order there is a set of nonvanishing terms, as the phonon destroyed by the first factor of $\mathcal{H}_{e-p}$ to act on Φ can be replaced by the second factor of $\mathcal{H}_{e-p}$, and vice versa. We then find the contribution, $\mathcal{E}_2$, of the second-order terms to be

$$\begin{aligned}\mathcal{E}_2 &= \langle \Phi | \sum_{\mathbf{k},\mathbf{k}'} M_{\mathbf{k}\mathbf{k}'}(a^\dagger_{-\mathbf{q}} + a_\mathbf{q}) c^\dagger_\mathbf{k} c_{\mathbf{k}'} (\mathcal{E}_0 - \mathcal{H}_0)^{-1} \\ &\quad \times \sum_{\mathbf{k}'',\mathbf{k}'''} M_{\mathbf{k}''\mathbf{k}'''}(a^\dagger_{-\mathbf{q}'} + a_{\mathbf{q}'}) c^\dagger_{\mathbf{k}''} c_{\mathbf{k}'''} | \Phi \rangle \\ &= \langle \Phi | \sum_{\mathbf{k},\mathbf{k}'} |M_{\mathbf{k}\mathbf{k}'}|^2 [a^\dagger_{-\mathbf{q}} c^\dagger_\mathbf{k} c_{\mathbf{k}'} (\mathcal{E}_0 - \mathcal{H}_0)^{-1} a_{-\mathbf{q}} c^\dagger_{\mathbf{k}'} c_\mathbf{k} \\ &\quad + a_\mathbf{q} c^\dagger_\mathbf{k} c_{\mathbf{k}'} (\mathcal{E}_0 - \mathcal{H}_0)^{-1} a^\dagger_\mathbf{q} c^\dagger_{\mathbf{k}'} c_\mathbf{k}] | \Phi \rangle,\end{aligned} \quad (6.2.1)$$

all other terms having zero matrix element. The first term in the brackets in (6.2.1) can be represented as in Fig. 6.2.1(a). An electron is first scattered from $\mathbf{k}$ to $\mathbf{k}'$ with the absorption of a phonon of wavenumber $-\mathbf{q} = \mathbf{k}' - \mathbf{k}$. The factor $(\mathcal{E}_0 - \mathcal{H}_0)^{-1}$ then measures the amount of time the electron is allowed by the Uncertainty Principle to stay in the intermediate state $\mathbf{k}'$. In this case the energy difference between the initial and intermediate states is $\mathcal{E}_\mathbf{k} + \hbar\omega_{-\mathbf{q}} - \mathcal{E}_{\mathbf{k}'}$, and so a factor of $(\mathcal{E}_\mathbf{k} + \hbar\omega_{-\mathbf{q}} - \mathcal{E}_{\mathbf{k}'})^{-1}$ is contributed. The electron is then scattered back into its original state with the re-emission of the phonon. We can represent the second term in Eq. (6.2.1) by Fig. 6.2.1(b), and there find an energy denominator of $\mathcal{E}_\mathbf{k} - \hbar\omega_\mathbf{q} - \mathcal{E}_{\mathbf{k}'}$. A rearrangement of the a's and c's into the form of number

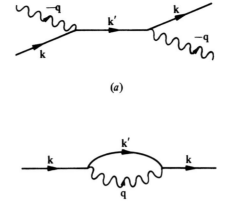

(a)

(b)

Figure 6.2.1. These two processes contribute to the energy of the electron–phonon system in second-order perturbation theory.

operators then gives

$$\mathcal{E}_2 = \sum_{k,k'} |M_{kk'}|^2 \langle n_k(1 - n_{k'})\rangle \left(\frac{\langle n_{-q}\rangle}{\mathcal{E}_k - \mathcal{E}_{k'} + \hbar\omega_{-q}} + \frac{\langle n_q + 1\rangle}{\mathcal{E}_k - \mathcal{E}_{k'} - \hbar\omega_q} \right). \quad (6.2.2)$$

Here $\langle n_k \rangle$ and $\langle n_{k'} \rangle$ are electron occupation numbers while $\langle n_q \rangle$ and $\langle n_{-q} \rangle$ refer to phonon states. It may be assumed that $\omega_q = \omega_{-q}$, and hence that in equilibrium $\langle n_q \rangle = \langle n_{-q} \rangle$. One may then rearrange Eq. (6.2.2), to find

$$\mathcal{E} = \mathcal{E}_0 + \sum_{k,k'} |M_{kk'}|^2 \langle n_k \rangle \left[\frac{2(\mathcal{E}_k - \mathcal{E}_{k'})\langle n_q \rangle}{(\mathcal{E}_k - \mathcal{E}_{k'})^2 - (\hbar\omega_q)^2} + \frac{1 - \langle n_{k'} \rangle}{\mathcal{E}_k - \mathcal{E}_{k'} - \hbar\omega_q} \right], \quad (6.2.3)$$

the term in $\langle n_k n_{k'} n_q \rangle$ cancelling by symmetry.

The effect of the electron–phonon interaction on the phonon spectrum is contained in the term proportional to $\langle n_q \rangle$ in Eq. (6.2.3). We again identify the perturbed phonon energy, $\hbar\omega_q^{(p)}$, with the energy required to increase $\langle n_q \rangle$ by unity, and so find

$$\hbar\omega_q^{(p)} = \frac{\partial \mathcal{E}}{\partial \langle n_q \rangle}$$

$$= \hbar\omega_q + \sum_k |M_{kk'}|^2 \frac{2\langle n_k \rangle(\mathcal{E}_k - \mathcal{E}_{k'})}{(\mathcal{E}_k - \mathcal{E}_{k'})^2 - (\hbar\omega_q)^2}.$$

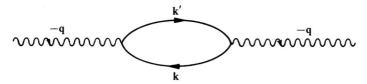

Figure 6.2.2. This alternative way of considering the process of Fig. 6.2.1(a) suggests that a phonon spends part of its time as a virtual electron–hole pair.

If we neglect the phonon energy in the denominator in comparison with the electron energies we have

$$\hbar\omega_{\mathbf{q}}^{(p)} = \hbar\omega_{\mathbf{q}} - \sum_{\mathbf{k}} 2|M_{\mathbf{k}\mathbf{k}'}|^2 \langle n_{\mathbf{k}}\rangle(\mathcal{E}_{\mathbf{k}'} - \mathcal{E}_{\mathbf{k}})^{-1}, \qquad (6.2.4)$$

where, as before, $\mathbf{k}' = \mathbf{k} - \mathbf{q}$. One may picture the origin of this change in phonon frequency by redrawing Fig. 6.2.1(a) in the form of Fig. 6.2.2, in which the first interaction is represented, not as the scattering of an electron, but as the creation of an electron–hole pair. One can then say that it is the fact that the phonon spends part of its time in the form of an electron–hole pair that modifies its energy.

One interesting consequence of Eq. (6.2.4) occurs in metals when $\mathbf{q}$ has a value close to the diameter, $2k_F$, of the Fermi surface. Let us suppose $\mathbf{q}$ to be in the x-direction and of magnitude $2k_F$, and evaluate $\hbar\partial\omega_{\mathbf{q}}^{(p)}/\partial q_x$. If we neglect the variation of $M_{\mathbf{k}\mathbf{k}'}$ with $\mathbf{q}$ the electron–phonon interaction contributes an amount

$$2\sum_{\mathbf{k}} |M_{\mathbf{k}\mathbf{k}'}|^2 \langle n_{\mathbf{k}}\rangle(\mathcal{E}_{\mathbf{k}-\mathbf{q}} - \mathcal{E}_{\mathbf{k}})^{-2} \frac{\partial \mathcal{E}_{\mathbf{k}-\mathbf{q}}}{\partial q_x}.$$

On substituting for $\mathcal{E}_{\mathbf{k}-\mathbf{q}}$ one finds the summation to contain the factors $\langle n_{\mathbf{k}}\rangle(k_x - k_F)^{-2}$. These cause a logarithmic divergence when the summation is performed, and thus indicate that the phonon spectrum has the form indicated qualitatively in Fig. 6.2.3. The kink in the spectrum when $q = 2k_F$ reflects the infinite group velocity of the phonons at that point, and constitutes the *Kohn anomaly*. Its importance lies in the fact that even for very complex metals there should always be such an image of the Fermi surface in the phonon spectrum. One should then in principle be able to gain information about the electronic structure of metals by studies of the phonon spectrum alone. This is important in that neutron diffraction experiments can be performed to determine $\omega_{\mathbf{q}}$ at high temperatures and in impure samples

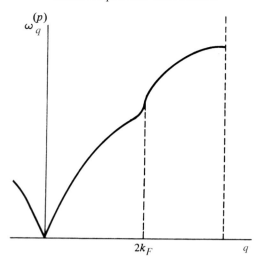

Figure 6.2.3. The kink that appears in the phonon dispersion curve when the phonon wavenumber q is equal to the diameter of the Fermi surface is known as the Kohn anomaly.

where many other techniques are not useful. Such effects have been seen most clearly in lead, in which the electron–phonon interaction is very strong.

6.3 The Peierls transition

When we calculated the change in phonon energy caused by the electron–phonon interaction, the answer we found contained a sum over terms of the form $(\mathcal{E}_{\mathbf{k}-\mathbf{q}} - \mathcal{E}_{\mathbf{k}})^{-1}$. Even though these could become infinite for certain values of $\mathbf{k}$, the sum in Eq. (6.2.4) fortunately remained finite. It is only the derivative $\partial \omega_{\mathbf{q}}^{(p)}/\partial \mathbf{q}$ that diverges, giving rise to the Kohn anomaly. We are saved from finding an infinite perturbation to the phonon frequency by the fact that we perform the sum in three-dimensional $\mathbf{k}$-space. The integration proceeds over both the magnitude of $\mathbf{k}$ and its direction, and it is the directional integration that rescues us from the embarrassment of finding an infinite negative perturbation to the phonon frequency.

However, before congratulating ourselves on this good fortune we might ponder the fact that there are a number of physical systems that approximate one-dimensional conductors. These include various organic materials, and in particular certain families of polymers. A polymer is a long chain molecule of covalently connected units (monomers), and a polymeric crystal can be thought of as an assembly of parallel chains. Most everyday polymers, like the polyethylene shown in Fig. 6.3.1(a), for example, have completely filled electronic bands, and are thus electrical insulators.

Polyacetylene, however, differs from polyethylene in having only one hydrogen atom on each carbon of the chain backbone as in Fig. 6.3.1(b). The double bond between every second pair of carbon atoms is the chemist's representation of the extra pair of electrons not used in bonding a hydrogen atom. There is then effectively only one conduction electron per unit cell if we think of the polymer as a one-dimensional lattice of spacing a.

The band structure for this Bravais lattice has the qualitative form shown in Fig. 6.3.2(a). The wavenumber at the Fermi level (in one dimension the Fermi surface has become a pair of Fermi points!) is $\pm\pi/2a$, as the band is

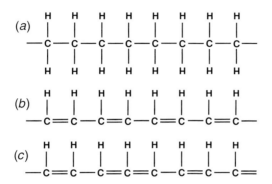

Figure 6.3.1. The chemical structure of polyethylene (a) makes it an insulator, but polyacetylene (b) can be a conductor through the motion of the defects (c) known as charged solitons.

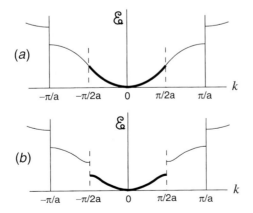

Figure 6.3.2. The half-filled band (a) of polyacetylene would make it a metallic conductor, but dimerization opens up an energy gap.

half-filled. This causes the phonon wavenumber q that connects the two Fermi points to be π/a, which is just half a reciprocal lattice vector. This phonon, of wavelength $2a$, is thus the mode of highest unperturbed frequency, and vibrates in such a way that adjacent monomers move in antiphase with each other. To estimate the perturbation to this frequency, we put $q = (\pi/a)(1+\delta)$ and perform the sum in the one-dimensional version of Eq. (6.2.4) to find the result

$$\hbar\omega_q^{(p)} - \hbar\omega_q \sim \ln \delta. \tag{6.3.1}$$

As expected, this tends to $-\infty$ as δ tends to zero.

The resolution of this difficulty comes when we realize that, on its way to negative infinity, $\hbar\omega_q^{(p)}$ must pass through zero, and a zero frequency phonon is simply a static distortion of the lattice. The lattice has thus *dimerized*, so that now the effective unit-cell dimension is $2a$. The Fermi "points" and the first Brillouin zone thus coincide. Bragg scattering of the electron states at the Fermi points thus causes an electronic band gap to open at $k = \pm\pi/2a$, as in Fig. 6.3.2(b). If we now repeat the sum in Eq. (6.2.4), but with the new form for $\mathcal{E}_\mathbf{k}$, the infinity is removed. This process of dimerization is known as the *Peierls transition*.

We could have reached a similar conclusion without the use of many-body theory by examining the total electronic energy as a function of the dimerization distortion. Displacing every second monomer by a distance u opens up a band gap $2V$ that is proportional to u. As only the lower band is occupied, the electronic energy of a k-state is lowered by an amount equal to $\mathcal{E}_\mathbf{k}^{(\text{new})} - \mathcal{E}_\mathbf{k}^{(\text{old})}$. From Eq. (4.3.2) we can evaluate the sum of these energies. The result is that the total electronic energy is lowered by an amount $\sim V^2 \ln V^2$, which is always negative for small V, and hence for small dimerization parameter u. Adding an elastic energy of distortion, which will be proportional to u^2, and hence to V^2, does not change this fact, and so the Peierls transition is always favored at low temperatures.

Because the energy of the distorted system is an even function of u, the ground state is degenerate. This corresponds to the chemical picture in which the double bond can form either between monomers $2n$ and $2n+1$, or between monomers $2n$ and $2n-1$. If the symmetry of the unperturbed state is broken differently in two different parts of the chain, then a defect occurs at the boundary of these regions, as shown in Fig. 6.3.1(c). This defect is mobile, since it just requires the displacement of one monomer to move the defect a distance $2a$ along the chain, and may also be charged, as it can represent a region with added electron concentration. The name *soliton* is used to describe this type of excitation. It bears some qualitative similarity to the solitary waves that we encountered in Section 1.3 when we looked at wave propagation in the Toda chain.

6.4 Polarons and mass enhancement

Just as the energy of the phonons in a crystal is altered by interaction with the electrons, so also does the converse process occur. We examine Eq. (6.2.3) in the limit of low temperatures, when $\langle n_q \rangle$ vanishes for all $\mathbf{q}$. The perturbed energy of an electron – once again the energy needed to fill an initially empty unperturbed state – is given by

$$\frac{\partial \mathcal{E}}{\partial \langle n_k \rangle} = \mathcal{E}_k + \sum_{k'} |M_{kk'}|^2 \left[\frac{1 - \langle n_{k'} \rangle}{\mathcal{E}_k - \mathcal{E}_{k'} - \hbar\omega_q} - \frac{\langle n_{k'} \rangle}{\mathcal{E}_{k'} - \mathcal{E}_k - \hbar\omega_q} \right]$$

$$= \mathcal{E}_k + \sum_{k'} |M_{kk'}|^2 \left[\frac{1}{\mathcal{E}_k - \mathcal{E}_{k'} - \hbar\omega_q} - \frac{2\hbar\omega_q \langle n_{k'} \rangle}{(\mathcal{E}_k - \mathcal{E}_{k'})^2 - (\hbar\omega_q)^2} \right]. \quad (6.4.1)$$

The first term in the brackets is independent of $n_{k'}$, and is thus a correction to the electron energy that would be present for a single electron in an insulating crystal. Indeed, in an ionic crystal the effect of this term may be so great as to change markedly the effective mass of an electron at the bottom of the conduction band. It then becomes reasonable to use the term *polaron* to describe the composite particle shown in Fig. 6.2.1(b) that is the electron with its attendant cloud of virtual phonons. The name arises because one considers the positive ions to be attracted towards the electron, and thus to polarize the lattice. If this polarization is too great then second-order perturbation theory is inadequate, and different methods must be used.

The second term in the brackets in Eq. (6.4.1) expresses the dependence of the electron energy on the occupancy of the other k-states. In a metal it has the effect of causing a kink in the graph of energy against wavenumber for the perturbed electron states, as shown in Fig. 6.4.1. This kink occurs at the Fermi wavenumber k_F, and leads to a change in the group velocity $\mathbf{v}_k$ of the electron. We find an expression for $\hbar \mathbf{v}_k$ by differentiating Eq. (6.4.1) with respect to $\mathbf{k}$. Instead of the free-electron expression we find

$$\hbar \mathbf{v}_k = \frac{\partial \mathcal{E}_k}{\partial \mathbf{k}} \left[1 - \frac{d}{d\mathcal{E}_k} \sum_{k'} |M_{kk'}|^2 \frac{2\hbar\omega_q \langle n_{k'} \rangle}{(\mathcal{E}_k - \mathcal{E}_{k'})^2 - (\hbar\omega_q)^2} \right].$$

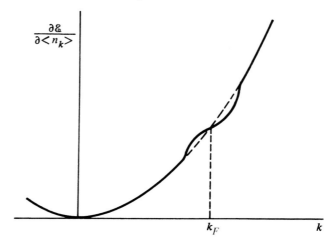

Figure 6.4.1. The electron–phonon interaction changes the effective electron energy in such a way that the velocity is lowered in the vicinity of the Fermi surface. This gives rise to an increase in the observed electronic specific heat.

We then argue that the major contribution to the derivative of the summation comes from the rapid variation of $\langle n_{\mathbf{k}'} \rangle$ at the Fermi surface. We change the summation over $\mathbf{k}'$ to an integral over $\mathcal{E}_{\mathbf{k}'}$ and make the approximations of replacing $M_{\mathbf{k}\mathbf{k}'}$ and $\omega_{\mathbf{q}}$ by their average values $\bar{M}$ and $\bar{\omega}$, and $D(\mathcal{E}_{\mathbf{k}'})$ by its value at the Fermi energy μ. With $\mathcal{E}_{\mathbf{k}'} - \mathcal{E}_{\mathbf{k}}$ written as η we then have

$$\hbar v_{\mathbf{k}} \simeq \frac{\partial \mathcal{E}_{\mathbf{k}}}{\partial \mathbf{k}} \left[1 - 2\hbar\bar{\omega}D(\mu)|\bar{M}|^2 \frac{d}{d\mathcal{E}_{\mathbf{k}}} \int \frac{\langle n(\mathcal{E}_{\mathbf{k}} + \eta) \rangle}{\eta^2 - (\hbar\bar{\omega})^2} \, d\eta \right].$$

Since

$$\frac{d\langle n(\mathcal{E}_{\mathbf{k}} + \eta) \rangle}{d\mathcal{E}_{\mathbf{k}}} = -\delta(\mathcal{E}_{\mathbf{k}} + \eta - \mu)$$

we find

$$\hbar v_{\mathbf{k}} \simeq \frac{\partial \mathcal{E}_{\mathbf{k}}}{\partial \mathbf{k}} \left[1 + \frac{2\hbar\bar{\omega}D(\mu)|\bar{M}|^2}{(\mathcal{E}_{\mathbf{k}} - \mu)^2 - (\hbar\bar{\omega})^2} \right].$$

The infinities that this expression predicts when $\mathcal{E}_{\mathbf{k}} = \mu \pm \hbar\bar{\omega}$ are a spurious consequence of our averaging procedure. The value predicted when $\mathbf{k}$ lies in the Fermi surface, where $\mathcal{E}_{\mathbf{k}} = \mu$, is more plausible, and gives us the result

$$\mathbf{v}_{\mathbf{k}} \simeq \mathbf{v}_{\mathbf{k}}^0 (1 - \alpha)$$

where $\mathbf{v}_\mathbf{k}^0$ is the unperturbed velocity and

$$\alpha = \frac{2|\bar{M}|^2 D(\mu)}{\hbar\bar{\omega}}.$$

This decrease in the electron velocity is equivalent to an increase in the density of states by the factor $(1-\alpha)^{-1}$. Because $\mathbf{v}_\mathbf{k}^0$ is inversely proportional to the electron mass, m, it is common to discuss the increase in the density of states in terms of an increase in the effective mass of the electron. One refers to $(1-\alpha)^{-1}$ as the *mass enhancement* factor due to electron–phonon interactions.

It is important to note that the enhanced density of states must only be used in interpreting experiments in which the Fermi energy is not altered. This is because the kink in the electron spectrum depicted in Fig. 6.4.1 is in fact tied to the Fermi energy; if one were to increase the density of electrons in the metal the kink in the curve would ride up with μ. We thus predict that it is the enhanced density of states that determines the electronic specific heat, since the effect of raising the temperature above zero is just to excite a few electrons from states below the Fermi energy to states above the Fermi energy. The semiclassical dynamics of a conduction electron in a magnetic field considered in Section 4.6 will also involve the corrected electron velocity, as again the Fermi energy is unchanged. The cyclotron mass m^* is thus enhanced by the factor $(1-\alpha)^{-1}$.

A different situation is encountered in the theory of the Pauli spin susceptibility χ_p of the electron gas. We recall that a suitably oriented magnetic field increases the size of the spin-up Fermi surface while diminishing the spin-down one in the manner shown in Fig. 6.4.2, in which the densities

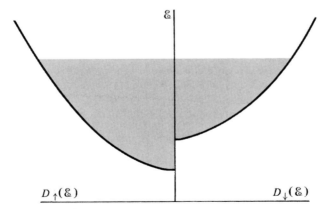

Figure 6.4.2. A magnetic field lowers the energy of the spin-up states while increasing the energy of those with spin down. The magnetization is proportional to the difference in area between the two halves of this diagram.

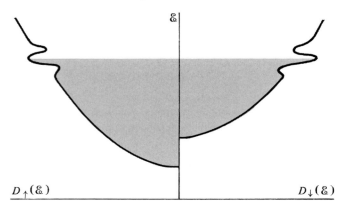

Figure 6.4.3. The electron–phonon interaction changes the shape of the densities of states of the spin-up and spin-down systems, but does not alter the difference in areas of the two halves of the diagram. The magnetization due to the electron spins thus remains unenhanced.

of states, $D_\uparrow$ and $D_\downarrow$, are plotted as functions of energy. The magnetization is proportional to $N_\uparrow - N_\downarrow$, and hence to the difference in the areas enclosed on the left and on the right; in a free-electron model χ_p is thus proportional to $D(\mathcal{E})$. If we now include the electron–phonon interaction in our calculation the spin-up and spin-down systems remain independent of each other, but two kinks are introduced in the densities of states, as shown in Fig. 6.4.3. The difference in areas enclosed, however, remains unaltered by this modification. We conclude that we should use the *unenhanced* density of states when discussing such properties as the Pauli paramagnetism or the change in Fermi surface dimensions on alloying, since in these experiments the position of the Fermi surface is altered.

6.5 The attractive interaction between electrons

The fact that the energy of the electron–phonon system (as expressed in Eq. (6.2.3)) contains terms proportional to $\langle n_\mathbf{k} n_{\mathbf{k}'} \rangle$ is an indication that there is an effective electron–electron interaction mediated by phonons. The nature of this interaction is illustrated by redrawing Fig. 6.2.1(b) in the form of Fig. 6.5.1, in which we deliberately fail to notice that the two intermediate electron states are the same. The passage of the phonon from one electron to the other contributes an energy $\mathcal{E}_c$ which from Eq. (6.2.3) may

6.5 The attractive interaction between electrons

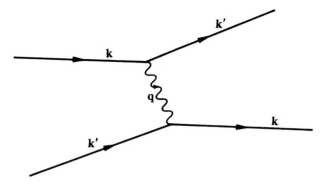

Figure 6.5.1. This way of redrawing Fig. 6.2.1(b) looks similar enough to Fig. 2.3.1 that we are led to believe there may be an effective electron–electron interaction caused by the electron–phonon interaction.

be written as

$$\mathcal{E}_c = -\sum_{\mathbf{k},\mathbf{k'}} |M_{\mathbf{k}\mathbf{k'}}|^2 \frac{\langle n_\mathbf{k} n_{\mathbf{k'}} \rangle}{\mathcal{E}_\mathbf{k} - \mathcal{E}_{\mathbf{k'}} - \hbar\omega_\mathbf{q}}$$

$$= \frac{1}{2} \sum_{\mathbf{k},\mathbf{k'}} |M_{\mathbf{k}\mathbf{k'}}|^2 \frac{2\hbar\omega_\mathbf{q}}{(\mathcal{E}_\mathbf{k} - \mathcal{E}_{\mathbf{k'}})^2 - (\hbar\omega_\mathbf{q})^2} \langle -n_\mathbf{k} n_{\mathbf{k'}} \rangle. \quad (6.5.1)$$

Comparison with Eq. (2.4.2) suggests that this may be considered as the exchange energy of the electron gas when interacting through a potential V whose matrix element $V_{\mathbf{k}\mathbf{k'}}$ is given by

$$V_{\mathbf{k}\mathbf{k'}} = |M_{\mathbf{k}\mathbf{k'}}|^2 \frac{2\hbar\omega_\mathbf{q}}{(\mathcal{E}_\mathbf{k} - \mathcal{E}_{\mathbf{k'}})^2 - (\hbar\omega_\mathbf{q})^2}. \quad (6.5.2)$$

This is significant in that although $\mathcal{E}_\mathbf{k}$ and $\mathcal{E}_{\mathbf{k'}}$ are much greater than $\hbar\omega_\mathbf{q}$, and thus usually give a positive $V_{\mathbf{k}\mathbf{k'}}$, there is always the possibility of $\mathcal{E}_\mathbf{k}$ and $\mathcal{E}_{\mathbf{k'}}$ being close enough that the matrix element could become negative. While it is not necessarily true that an interaction that is negative in k-space is negative in r-space, this form for $V_{\mathbf{k}\mathbf{k'}}$ does open the door to the possibility of an attractive interaction between electrons. Because the wavefunction of a system always modifies itself to maximize the effects of attractive interactions, while it tends to minimize those of repulsive interactions, even a weak attraction between electrons can lead to important consequences.

We can verify that such an attractive interaction does exist by performing a *canonical transformation* of the Fröhlich Hamiltonian (6.1.1). We write

$$\mathcal{H} = \mathcal{H}_0 + \mathcal{H}_{e-p}$$

and then look for a transformation of the form

$$\mathcal{H}' = e^{-s}\mathcal{H}e^{s} \qquad (6.5.3)$$

that will eliminate $\mathcal{H}_{e-p}$ to first order. On expansion of the exponentials in Eq. (6.5.3) we have

$$\begin{aligned}\mathcal{H}' &= (1 - s + \tfrac{1}{2}s^2 - \cdots)\mathcal{H}(1 + s + \tfrac{1}{2}s^2 + \cdots) \\ &= \mathcal{H}_0 + \mathcal{H}_{e-p} + [\mathcal{H}_0, s] + [\mathcal{H}_{e-p}, s] + \tfrac{1}{2}[[\mathcal{H}_0, s], s] + \cdots.\end{aligned}$$

We choose s in such a way that its commutator with $\mathcal{H}_0$ cancels the term $\mathcal{H}_{e-p}$. With this achieved we have

$$\mathcal{H}' = \mathcal{H}_0 + \tfrac{1}{2}[\mathcal{H}_{e-p}, s] + \cdots,$$

the omitted terms being of order s^3 or higher. Now since

$$\mathcal{H}_0 = \sum_{\mathbf{k}} \mathcal{E}_{\mathbf{k}} c^{\dagger}_{\mathbf{k}} c_{\mathbf{k}} + \sum_{\mathbf{q}} \hbar\omega_{\mathbf{q}} a^{\dagger}_{\mathbf{q}} a_{\mathbf{q}}$$

and

$$\mathcal{H}_{e-p} = \sum_{\mathbf{k},\mathbf{k}'} M_{\mathbf{k}\mathbf{k}'}(a^{\dagger}_{-\mathbf{q}} + a_{\mathbf{q}}) c^{\dagger}_{\mathbf{k}} c_{\mathbf{k}'}$$

we try

$$s = \sum_{\mathbf{k},\mathbf{k}'} (A a^{\dagger}_{-\mathbf{q}} + B a_{\mathbf{q}}) M_{\mathbf{k}\mathbf{k}'} c^{\dagger}_{\mathbf{k}} c_{\mathbf{k}'},$$

with A and B coefficients to be determined. We then find that in order to satisfy

$$\mathcal{H}_{e-p} + [\mathcal{H}_0, s] = 0$$

we must have

$$A = -(\mathcal{E}_{\mathbf{k}} - \mathcal{E}_{\mathbf{k}'} + \hbar\omega_{-\mathbf{q}})^{-1}; \quad B = -(\mathcal{E}_{\mathbf{k}} - \mathcal{E}_{\mathbf{k}'} - \hbar\omega_{\mathbf{q}})^{-1}.$$

6.5 The attractive interaction between electrons

Then

$$\mathcal{H}' = \mathcal{H}_0 - \tfrac{1}{2}\Bigg[\bigg\{\sum_{k,k'} M_{kk'}(a^\dagger_{-q} + a_q)c^\dagger_k c_{k'}\bigg\},$$

$$\bigg\{\sum_{k'',k'''} M_{k''k'''}\bigg(\frac{a^\dagger_{-q'}}{\mathcal{E}_{k''} - \mathcal{E}_{k'''} + \hbar\omega_{-q'}} + \frac{a_{q'}}{\mathcal{E}_{k''} - \mathcal{E}_{k'''} - \hbar\omega_{q'}}\bigg)c^\dagger_{k''} c_{k'''}\bigg\}\Bigg].$$

Of the many terms in the commutator, we examine particularly the set that arise from commuting the phonon operators. Using the fact that $M_{kk'}$ is a function only of $\mathbf{k} - \mathbf{k}' = \mathbf{q}$ we find

$$\mathcal{H}' = \mathcal{H}_0 + \sum_{k,k',q} |M_q|^2 \frac{\hbar\omega_q}{(\mathcal{E}_k - \mathcal{E}_{k-q})^2 - (\hbar\omega_q)^2} c^\dagger_{k'+q} c^\dagger_{k-q} c_k c_{k'}$$

$$+ \text{(terms involving only two electron operators).} \quad (6.5.4)$$

If we were now to take the expectation value of $\mathcal{H}'$ in an eigenstate of $\mathcal{H}_0$ we should regain expression (6.2.3). (Because of the transformation we performed, the energy of the system described by $\mathcal{H}'$ is the same in first-order perturbation theory as that of the system described by $\mathcal{H}$ in second order.) The terms that are not displayed explicitly in Eq. (6.5.4) contain one $c^\dagger$ and one c, and have diagonal elements equal to the terms in $\langle n_k \rangle$ and $\langle n_k n_q \rangle$ in Eq. (6.2.3). The diagonal elements of the terms involving four electron operators have the sum shown in Eq. (6.5.1), and represent the Hartree–Fock approximation applied to $\mathcal{H}'$. (We recall that the direct term is absent as there is no phonon for which $q = 0$ in a Bravais lattice.)

The power of the method of canonical transformations which we have just employed lies in the fact that we are not restricted to using perturbation theory to find the eigenstates of $\mathcal{H}'$. This is particularly important for the theory of superconductivity which, as we shall see in the next chapter, involves a phase transition in the electron gas brought about by attractive electron interactions and which cannot be accounted for by perturbation theory. Just such an attractive interaction is present in $\mathcal{H}'$ whenever $|\mathcal{E}_k - \mathcal{E}_{k-q}| < \hbar\omega_q$. This may allow pairs of electrons to form a bound state of lower energy than that of the two free electrons. The existence of *Cooper pairs*, in which two electrons of opposite wavenumber and spin form a bound state, provides the foundation for the BCS theory of superconductivity.

6.6 The Nakajima Hamiltonian

While the Fröhlich Hamiltonian (6.1.1) provides a useful model that exhibits many of the interesting properties of metals, it does have the failing that the phonon frequencies $\omega_\mathbf{q}$ and the electron–phonon matrix elements, $M_{\mathbf{kk}'}$, must be assumed known. The results of the previous sections cannot then be thought of as calculations "from first principles." A better theory would start with the Hamiltonian of a lattice of bare ions interacting through a Coulomb potential. To this would be added the electrons, which would interact with each other and with the ions.

The total Hamiltonian obtained in this way would differ from the Fröhlich Hamiltonian in a number of aspects. Firstly we should have to replace the phonon Hamiltonian

$$\mathcal{H}_p = \sum_\mathbf{q} \hbar\omega_\mathbf{q} a_\mathbf{q}^\dagger a_\mathbf{q}$$

by the phonon Hamiltonian, $\mathcal{H}_p^i$, for the lattice of bare ions, which, as we saw in Chapter 1, does not support longitudinally polarized acoustic phonons. We should have

$$\mathcal{H}_p^i = \sum_\mathbf{q} \hbar\Omega_\mathbf{q} A_\mathbf{q}^\dagger A_\mathbf{q}$$

with $A_\mathbf{q}^\dagger$ the operator creating a phonon whose frequency $\Omega_\mathbf{q}$ in a monovalent metal approaches Ω_p, the ion plasma frequency, $(4\pi Ne^2/M\Omega)^{1/2}$, for small q. By making use of the definitions (3.6.4) of both types of phonon operator we can write $\mathcal{H}_p^i$ in terms of the operators $a^\dagger$ and a. With neglect of the off-diagonal terms and some constants we have

$$\mathcal{H}_p^i - \mathcal{H}_p \simeq \sum_\mathbf{q} \hbar \frac{(\Omega_\mathbf{q}^2 - \omega_\mathbf{q}^2)}{4\omega_\mathbf{q}} (2n_\mathbf{q} + 1).$$

The next correction to the Fröhlich Hamiltonian is the replacement of the screened matrix element, $M_{\mathbf{kk}'}$, by the scattering matrix element, $M_{\mathbf{kk}'}^i$, of an ion. Because of the long-range Coulomb potential due to the ion we expect $M_{\mathbf{kk}'}^i$ to diverge as $q \to 0$. Finally we must add the mutual interaction of the electrons. Our total Hamiltonian, $\mathcal{H}_N$, can then be written as the Fröhlich Hamiltonian, $\mathcal{H}_F$, plus a set of correction terms.

We have

$$\mathcal{H}_N = \mathcal{H}_F + \sum_{k,k'} (M^i_{kk'} - M_{kk'})(a^\dagger_{-q} + a_q) c^\dagger_k c_{k'}$$
$$+ \frac{1}{2} \sum_{k,k',q} V_q c^\dagger_{k-q} c^\dagger_{k'+q} c_{k'} c_k + \sum_q \hbar \frac{(\Omega_q^2 - \omega_q^2)}{4\omega_q} (2n_q + 1). \quad (6.6.1)$$

This model was studied by Nakajima.

In writing the Fröhlich Hamiltonian we made the assumption that if ω_q and $M_{kk'}$ were correctly chosen, then it was permissible to ignore the effects of the correction terms in Eq. (6.6.1). If this is really to be the case then the canonical transformation (6.5.3) must not only eliminate the interaction terms in Eq. (6.1.1), but also the correction terms in (6.6.1). That is to say, we must demand that in some approximation

$$e^{-s}(\mathcal{H}_N - \mathcal{H}_F)e^s = 0,$$

the operator s being the same as before. Let us write Eq. (6.6.1) as

$$\mathcal{H}_N = \mathcal{H}_F + \delta\mathcal{H}_{e-p} + \mathcal{H}_{e-e} + \delta\mathcal{H}_p.$$

Then we impose the condition that all terms linear or bilinear in the phonon operators must vanish in the expression

$$\delta\mathcal{H}_{e-p} + \mathcal{H}_{e-e} + \delta\mathcal{H}_p + [\delta\mathcal{H}_{e-p}, s] + [\mathcal{H}_{e-e}, s] + [\delta\mathcal{H}_p, s] + \cdots. \quad (6.6.2)$$

If we neglect $[\delta\mathcal{H}_p, s]$ then the two sets of terms of first order in the phonon operators that arise in this expression are the correction to the electron–phonon interaction, $\delta\mathcal{H}_{e-p}$, and the commutator, $[\mathcal{H}_{e-e}, s]$, of the Coulomb interaction of the electrons with s. For the latter term we can build on the calculation of Eq. (2.7.1) in which we saw that

$$\left[\left(\tfrac{1}{2} \sum_{k,k',q'} V_{q'} c^\dagger_{k-q'} c^\dagger_{k'+q'} c_{k'} c_k \right), (c^\dagger_{p+q} c_p) \right]$$
$$= \tfrac{1}{2} \sum_{q'} V_{q'} [(c^\dagger_{p+q-q'} c_p - c^\dagger_{p+q} c_{p+q'}) \rho^\dagger_{q'} + \rho_{q'} (c^\dagger_{p+q'+q} c_p - c^\dagger_{p+q} c_{p-q'})] \Omega.$$

We then make the random phase approximation of retaining only those terms containing number operators, and so we must put $q' = \pm q$. We are left with just

$$V_\mathbf{q}\rho_\mathbf{q}^\dagger(n_\mathbf{p} - n_\mathbf{p+q}).$$

Since we can write

$$s = -\sum_{\mathbf{p,q}} M_{\mathbf{p,p+q}} c_{\mathbf{p+q}}^\dagger c_\mathbf{p} \left(\frac{a_{-\mathbf{q}}^\dagger}{\mathcal{E}_\mathbf{p} - \mathcal{E}_\mathbf{p+q} + \hbar\omega_{-\mathbf{q}}} + \frac{a_\mathbf{q}}{\mathcal{E}_\mathbf{p} - \mathcal{E}_\mathbf{p+q} - \hbar\omega_\mathbf{q}} \right),$$

then

$$\delta\mathcal{H}_{e-p} + [\mathcal{H}_{e-e}, s] = 0$$

provided

$$M_{\mathbf{kk'}}^i - M_{\mathbf{kk'}} = -M_{\mathbf{kk'}} V_\mathbf{q} \sum_\mathbf{p} \frac{\langle n_{\mathbf{p+q}} - n_\mathbf{p}\rangle}{\mathcal{E}_\mathbf{p+q} - \mathcal{E}_\mathbf{p} + \hbar\omega_\mathbf{q}} \qquad (6.6.3)$$

where $\mathbf{k} - \mathbf{k'} = \mathbf{q}$ as before. In terms of the dielectric constant $\epsilon(\mathbf{q}, \omega)$ defined in Eq. (2.7.6) and Problem 2.9 this simplifies to

$$M_{\mathbf{kk'}} = \frac{M_{\mathbf{kk'}}^i}{\epsilon(\mathbf{q}, \omega_\mathbf{q})}, \qquad (6.6.4)$$

which is a result we might intuitively have expected. We note that since $\epsilon(\mathbf{q}, \omega_\mathbf{q})$ varies as q^{-2} for small q, the screened matrix element $M_{\mathbf{kk'}}$ no longer diverges as $q \to 0$.

The phonon frequency $\omega_\mathbf{q}$ is fixed by the condition that the diagonal elements of $\delta\mathcal{H}_p$ are cancelled by the other terms in the transformed Hamiltonian of order $n_\mathbf{q}$. This leads to the condition

$$\delta\mathcal{H}_p + \tfrac{1}{2}[(\mathcal{H}_{e-p} + \delta\mathcal{H}_{e-p}), s] = 0,$$

the term in $\mathcal{H}_{e-p}$ occurring since this part of the Fröhlich Hamiltonian gives rise to terms in $n_\mathbf{q}$. [These were included in the unspecified part of Eq. (6.5.4).] We then find

$$\hbar \frac{\Omega_\mathbf{q}^2 - \omega_\mathbf{q}^2}{4\omega_\mathbf{q}} = -\frac{1}{2} M_{\mathbf{kk'}}^i M_{\mathbf{k'k}} \sum_\mathbf{p} \frac{\langle n_{\mathbf{p+q}} - n_\mathbf{p}\rangle}{\mathcal{E}_\mathbf{p+q} - \mathcal{E}_\mathbf{p} + \hbar\omega_\mathbf{q}}. \qquad (6.6.5)$$

If, following Eq. (6.1.2), we write

$$M^i_{\mathbf{kk'}} = i\sqrt{\frac{N\hbar}{2M\omega_\mathbf{q}}}|\mathbf{k'}-\mathbf{k}|V^i_{\mathbf{k}-\mathbf{k'}},$$

with $V^i_{\mathbf{k}-\mathbf{k'}}$ the Fourier transform of the Coulomb potential of a bare ion, we can rewrite (6.6.5) as

$$\Omega_\mathbf{q}^2 - \omega_\mathbf{q}^2 = -\frac{N}{M}(\mathbf{k'}-\mathbf{k})^2 V^i_{\mathbf{k}-\mathbf{k'}} V_{\mathbf{k}-\mathbf{k'}} \sum_\mathbf{p} \frac{\langle n_{\mathbf{p}+\mathbf{q}} - n_\mathbf{p}\rangle}{\mathcal{E}_{\mathbf{p}+\mathbf{q}} - \mathcal{E}_\mathbf{p} + \hbar\omega_\mathbf{q}}$$

$$= -\frac{N}{M}q^2 \frac{(V^i_{\mathbf{k}-\mathbf{k'}})^2}{V_\mathbf{q}}\left(\frac{1-\epsilon(\mathbf{q},\omega_\mathbf{q})}{\epsilon(\mathbf{q},\omega_\mathbf{q})}\right).$$

For our monovalent metal we can put

$$V^i_{\mathbf{k}-\mathbf{k'}} = -V_\mathbf{q} = -\frac{4\pi e^2}{q^2\Omega},$$

and then

$$\omega_\mathbf{q}^2 = \Omega_\mathbf{q}^2 - \frac{4\pi N e^2}{M\Omega}\left(1-\frac{1}{\epsilon}\right)$$

$$= \Omega_\mathbf{q}^2 - \Omega_p^2 + \frac{\Omega_p^2}{\epsilon(\mathbf{q},\omega_\mathbf{q})}. \tag{6.6.6}$$

Since for small $\mathbf{q}$ the phonon frequency $\Omega_\mathbf{q}$ of the Coulomb lattice tends to Ω_p this equation correctly predicts that $\omega_\mathbf{q}$ should vanish as $q \to 0$. An expansion of the right-hand side of Eq. (6.6.6) in even powers of q leads to a prediction for the velocity of sound. One can write

$$\Omega_p^2 - \Omega_\mathbf{q}^2 = \frac{\alpha\Omega_p^2 q^2}{k_F^2} + \cdots$$

and

$$\frac{\Omega_p^2}{\epsilon(\mathbf{q},\omega_\mathbf{q})} = \frac{\beta\Omega_p^2 q^2 v_F^2}{\omega_p^2}$$

with α and β constants of order unity and ω_p the plasma frequency. Thus

$$\omega_\mathbf{q} \simeq q\left(\frac{m}{M}\right)^{1/2} v_F\left[\beta - \alpha\left(\frac{\hbar\omega_p}{2\mathcal{E}_F}\right)^2\right]^{1/2} \tag{6.6.7}$$

in rough agreement with the semiclassical arguments of Section 1.7.

We note that the electron–electron interaction has not been removed from $\mathcal{H}_N$ by our canonical transformation, but only modified. Further transformations are necessary to exhibit explicitly such consequences of $\mathcal{H}_{e-e}$ as the plasma oscillations.

Problems

6.1 Show that the enhancement of the electron density of states in a metal by the electron–phonon interaction is independent of the isotopic mass of the ions.

6.2 Use the method of canonical transformations discussed in Section 6.5 to diagonalize the magnon–phonon system described in Problem 3.6.

6.3 Investigate to second order the T-matrix for scattering of a free electron by a single impurity fixed in a crystal lattice. Assume the impurity to participate in the motion of the lattice without perturbing the phonon spectrum. How is the probability of elastic scattering of the electron affected by the motion of the impurity?

6.4 What errors would one make if, instead of using Nakajima's approach, one used a Fröhlich Hamiltonian with matrix elements given by Eq. (6.6.4)?

6.5 Discuss whether the infinite group velocity of phonons predicted in the theory of the Kohn anomaly violates the principle of special relativity.

6.6 Estimate the effective mass of a single polaron in the model described by the Fröhlich Hamiltonian (6.1.1) with ω_q equal (for all $\mathbf{q}$) to the ion plasma frequency Ω_p defined in Eq. (1.5.1) and with $V_{\mathbf{k'k}}$ equal to $-4\pi e^2/\Omega(\mathbf{k}-\mathbf{k'})^2$. Why is this an unrealistic description of a polaron in an ionic crystal?

6.7 A one-dimensional system consists of electrons and phonons in interaction, and is described by a Hamiltonian of the form

$$\mathcal{H} = \sum_k \mathcal{E}_k c_k^\dagger c_k + \sum_q \hbar\omega_q a_q^\dagger a_q + M \sum_{k,k'} (a_{-q}^\dagger + a_q) c_k^\dagger c_{k'}.$$

Here $\hbar\omega_q = W \sin(qa/2)$ and in the electron–phonon interaction term, $q = k - k'$. The lattice, originally of spacing a, has become dimerized by the electron–phonon interaction, so that the spacing is now $2a$. A gap has consequently appeared in the electron energy-band structure at $k = \pi/2a$, and now $\mathcal{E}_k = -A - B\cos(2ka)$ in the filled valence band and $\mathcal{E}_k = A + B\cos(2ka)$ in the empty conduction band. The energies A and B are both positive, and $A - B \gg W > 0$. The matrix element M is a constant and $M^2 = G/N$, with N the number of atoms. Find a condition on the magnitude of G in terms of W, A, and B in order for the phonon energy at $q = \pi/a$ to be reduced to zero.

6.8 Why doesn't benzene dimerize?

(a) Consider a periodic polyacetylene chain of six monomers with lattice spacing a, and evaluate the perturbed frequency of the phonon of wavelength $2a$. Determine the condition for this frequency to vanish when the Hamiltonian is

$$\mathcal{H} = \sum_k \mathcal{E}_k c_k^\dagger c_k + \sum_q \hbar\omega_q a_q^\dagger a_q + M \sum_{k,k'} (a_{-q}^\dagger + a_q) c_k^\dagger c_{k'}$$

with $q = k - k'$, $M =$ constant, $\mathcal{E}_k = A(1 - \cos ka)$, and $\hbar\omega_q = W\sin(qa/2)$.

(b) In a completely different approach, calculate the total energy, $\mathcal{E}_{\text{elastic}} + \mathcal{E}_{\text{electronic}}$, of a static periodic polyacetylene chain of six monomers with periodic boundary conditions. The distance between adjacent monomers alternates between $a + u$ and $a - u$. The Hookian spring constant for a single monomer–monomer bond is K. The electronic energy $\mathcal{E}_k$ is given by Eq. (4.3.2) with $g = \pi/a$ and $V_g = \gamma u$. Find the condition for γ in terms of $K, \hbar, m$, and a for dimerization to be energetically favored.

6.9 A hydrogen atom may be absorbed into an interstitial position in palladium with evolution of energy $\mathcal{E}_s$. Make an order-of-magnitude estimate of the amount by which $\mathcal{E}_s$ is altered by the vibrational motion of the hydrogen. [Hint: Consider the atom as a three-dimensional harmonic oscillator of frequency ω_0 moving in a free-electron gas of Fermi energy $\mathcal{E}_F$ and carrying its potential $V_a(r)$ rigidly with it. Calculate to second order in the electron-vibration interaction in analogy with the treatment of the electron–phonon interaction of Section 6.1.]

6.10 Find a condition for the minimum value of the electron density in a metal by requiring that the expression (6.6.7) for the phonon frequency be real.

Chapter 7

Superconductivity

7.1 The superconducting state

The discovery, in Section 6.5, of an attractive interaction between electrons in a metal has mentally prepared us for the existence of a phase transition in the electron gas at low temperatures. It would, however, never have prepared us to expect a phenomenon as startling and varied as superconductivity if we were not already familiar with the experimental evidence. The ability to pass an electrical current without any measurable resistance has now been found in a wide range of types of material, including simple elements like mercury, metallic alloys, organic salts containing five- or six-membered rings of carbon and sulfur atoms, and ceramic oxides containing planes of copper and oxygen atoms.

In this chapter we shall concentrate mainly on the simplest type of superconductor, typified by elements such as tin, zinc, or aluminum. The organic superconductors and the ceramic oxides have properties that are so anisotropic that the theories developed to treat elemental materials are not applicable. Accordingly, with the exception of the final Section 7.11, the discussion that follows in this chapter applies only to the classic low-temperature superconductors.

Some of the properties of these materials are shown in Fig. 7.1.1, in which the resistivity, ρ, specific heat, C, and damping coefficient for phonons, α, are plotted as functions of temperature for a typical superconductor. At the transition temperature, T_c, a second-order phase transition occurs, the most spectacular consequence being the apparent total disappearance of resistance to weak steady electric currents. The contribution of the electrons to the specific heat is found no longer to be proportional to the absolute temperature, as it is in normal (i.e., nonsuperconducting) metals and in superconductors when $T > T_c$, but to vary at the lowest temperatures as

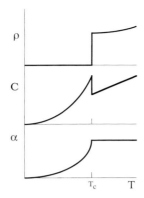

Figure 7.1.1. The resistivity ρ, the electronic specific heat C, and the coefficient of attenuation for sound waves all change sharply as the temperature is lowered through the transition temperature T_c.

$e^{-\Delta/kT}$, with Δ an energy of the order of kT_c. This leads one to suppose that there is an *energy gap* in the excitation spectrum – an idea that is confirmed by the absorption spectrum for electromagnetic radiation. Only when the energy $\hbar\omega$ of the incident photons is greater than about 2Δ does absorption occur, which suggests that the excitations that give the exponential specific heat are created in pairs.

A rod-shaped sample of superconductor held parallel to a weak applied magnetic field $\mathbf{H}_0$ has the property that the field can penetrate only a short distance λ into the sample. Beyond this distance, which is known as the *penetration depth* and is typically of the order of 10^{-5} cm, the field decays rapidly to zero. This is known as the *Meissner effect* and is sometimes thought of as "perfect diamagnetism." This rather misleading term refers to the fact that if the magnetic moment of the rod were not due to currents flowing in the surface layers (the true situation) but were instead the consequence of a uniform magnetization, then the magnetic susceptibility would have to be $-1/4\pi$, which is the most negative value thermodynamically permissible. If the strength of the applied field is increased the superconductivity is eventually destroyed, and this can happen in two ways. In a *type I superconductor* the whole rod becomes normal at an applied field $\mathbf{H}_c$, and then the magnetic field $\mathbf{B}$ in the interior of a large sample changes from zero to a value close to $\mathbf{H}_0$ (Fig. 7.1.2(a)). In a *type II superconductor*, on the other hand, although the magnetic field starts to penetrate the sample at an applied field, $\mathbf{H}_{c1}$, it is not until a greater field, $\mathbf{H}_{c2}$, is reached, that $\mathbf{B}$ approaches $\mathbf{H}_0$ within the rod, and a thin surface layer may remain superconducting up to a yet higher field, $\mathbf{H}_{c3}$ (Fig. 7.1.2(b)). For applied fields between $\mathbf{H}_{c1}$ and $\mathbf{H}_{c2}$

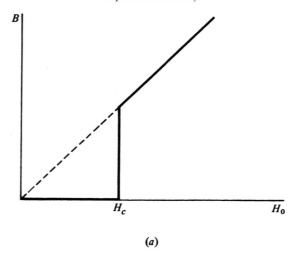

(a)

Figure 7.1.2(a). A magnetic field $\mathbf{H}_0$ applied parallel to a large rod-shaped sample of a type I superconductor is completely excluded from the interior of the specimen when $\mathbf{H}_0 < \mathbf{H}_c$, the critical field, and completely penetrates the sample when $\mathbf{H}_0 > \mathbf{H}_c$.

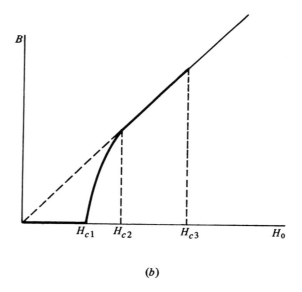

(b)

Figure 7.1.2(b). In a type II superconductor there is a partial penetration of the magnetic field into the sample when $\mathbf{H}_0$ lies between the field values $\mathbf{H}_{c1}$ and $\mathbf{H}_{c2}$. Small surface supercurrents may still flow up to an applied field $\mathbf{H}_{c3}$.

the sample is in a *mixed state* consisting of superconductor penetrated by threads of the material in its normal phase.

These filaments form a regular two-dimensional array in the plane perpendicular to H_0. In many cases it is found possible to predict whether a superconductor will be of type I or II from measurements of Δ and λ. One defines a *coherence length* ξ_0 equal to $\hbar v_F/\pi\Delta$ with v_F the Fermi velocity. (This length is of the order of magnitude of $\mathcal{E}_F/\Delta$ times a lattice spacing.) It is those superconductors for which $\lambda \gg \xi_0$ that tend to exhibit properties of type II.

7.2 The BCS Hamiltonian

The existence of such an obvious phase transition as that involved in superconductivity led to a long search for a mechanism that would lead to an attractive interaction between electrons. Convincing evidence that the electron–phonon interaction was indeed the mechanism responsible was provided with the discovery of the *isotope effect*, when it was found that for some metals the transition temperature T_c was dependent on the mass A of the nucleus. For the elements first measured it appeared that T_c was proportional to $A^{-1/2}$, and hence to the Debye temperature Θ. More recent measurements have shown a wider variety of power laws, varying from an almost complete absence of an isotope effect in osmium to a dependence of the approximate form $T_c \propto A^2$ in α-uranium. It is accordingly reasonable to turn to the Fröhlich electron–phonon Hamiltonian discussed in the preceding chapter as a simple model of a system that might exhibit superconductivity. The canonical transformation of Section 6.5 allowed us to write the Hamiltonian in the form of Eq. (6.5.4), which states that

$$\mathcal{H}' = \mathcal{H}_0 + \sum_{\mathbf{k},s,\mathbf{k}',s',\mathbf{q}} W_{\mathbf{k}\mathbf{k}'\mathbf{q}} c^\dagger_{\mathbf{k}'+\mathbf{q},s'} c^\dagger_{\mathbf{k}-\mathbf{q},s} c_{\mathbf{k},s} c_{\mathbf{k}',s'}. \qquad (7.2.1)$$

Here $\mathcal{H}_0$ was the Hamiltonian of the noninteracting system of electrons and phonons, and $W_{\mathbf{k}\mathbf{k}'\mathbf{q}}$ was a matrix element of the form

$$W_{\mathbf{k}\mathbf{k}'\mathbf{q}} = \frac{|M_\mathbf{q}|^2 \hbar\omega_\mathbf{q}}{(\mathcal{E}_\mathbf{k} - \mathcal{E}_{\mathbf{k}-\mathbf{q}})^2 - (\hbar\omega_\mathbf{q})^2}. \qquad (7.2.2)$$

The spin s of the electrons has also been explicitly included in this transformed Hamiltonian.

Because this interaction represents an attractive force between electrons, we do not expect perturbation theory to be useful in finding the eigenstates of a Hamiltonian like Eq. (7.2.1). In fact, an infinitesimal attraction can change the entire character of the ground state in a way not accessible to perturbation theory. We could now turn to a variational approach, in which we make a brilliant guess at the form the correct wavefunction $|\Psi\rangle$ will take, and then pull and push at it until the expectation value $\langle\Psi|\mathcal{H}'|\Psi\rangle$ is minimized. This is the approach that Bardeen, Cooper, and Schrieffer originally took, and is described in their classic paper of 1957. We shall take a slightly different route to the same result, and turn for inspiration to the only problem that we have yet attempted without using perturbation theory – the Bogoliubov theory of helium discussed in Section 3.4.

The starting point of the Bogoliubov theory was the assumption of the existence of a *condensate* of particles having zero momentum. This led to the approximate Hamiltonian (3.4.1) in which the interactions that took place involved the scattering of *pairs* of particles of equal but opposite momentum. Now we can consider the superconducting electron system as being in some sort of condensed phase, and it thus becomes reasonable to make the hypothesis that in the superconductivity problem the scattering of pairs of electrons having equal but opposite momentum will be similarly important. We refer to these as Cooper pairs, and accordingly retain from the interaction in Eq. (7.2.1) only those terms for which $\mathbf{k} = -\mathbf{k}'$; in this way we find a reduced Hamiltonian

$$\mathcal{H}_{\text{BCS}} = \sum_{\mathbf{k}s}\mathcal{E}_{\mathbf{k}}c^{\dagger}_{\mathbf{k}s}c_{\mathbf{k}s} - \tfrac{1}{2}\sum_{\mathbf{k}\mathbf{k}'s}V_{\mathbf{k}\mathbf{k}'}c^{\dagger}_{\mathbf{k}'s'}c^{\dagger}_{-\mathbf{k}'s}c_{-\mathbf{k}s}c_{\mathbf{k}s'} \qquad (7.2.3)$$

where in the notation of Eq. (7.2.2)

$$V_{\mathbf{k}\mathbf{k}'} = -2W_{-\mathbf{k},\mathbf{k},\mathbf{k}'-\mathbf{k}} - U_{\mathbf{k}\mathbf{k}'}$$

with $U_{\mathbf{k}\mathbf{k}'}$ a screened Coulomb repulsion term which we add to the Fröhlich Hamiltonian. A positive value of $V_{\mathbf{k}\mathbf{k}'}$ thus corresponds to a net attractive interaction between electrons.

One other question that must be answered before we attempt to diagonalize the Hamiltonian $\mathcal{H}_{\text{BCS}}$ concerns the spins of the electrons: do we pair electrons of like spin (so that in Eq. (7.2.3) we put $s = s'$) or of opposite spin? The answer is that to minimize the ground-state energy it appears that we must pair electrons of *opposite* spin. We shall assume this to be the case, and leave it as a challenge to the sceptical reader to find a wavefunction that leads to a lower expectation value of Eq. (7.2.3) with $s = s'$ than we shall find when s and s' represent spins in opposite directions. With this assumption we can perform the sum over s in Eq. (7.2.3). Since

$$\sum_s c^\dagger_{\mathbf{k}'s'} c^\dagger_{-\mathbf{k}'s} c_{-\mathbf{k}s} c_{\mathbf{k}s'} = c^\dagger_{\mathbf{k}'\downarrow} c^\dagger_{-\mathbf{k}'\uparrow} c_{-\mathbf{k}\uparrow} c_{\mathbf{k}\downarrow} + c^\dagger_{\mathbf{k}'\uparrow} c^\dagger_{-\mathbf{k}'\downarrow} c_{-\mathbf{k}\downarrow} c_{\mathbf{k}\uparrow}$$

$$= c^\dagger_{\mathbf{k}'\downarrow} c^\dagger_{-\mathbf{k}'\uparrow} c_{-\mathbf{k}\uparrow} c_{\mathbf{k}\downarrow} + c^\dagger_{-\mathbf{k}'\downarrow} c^\dagger_{\mathbf{k}'\uparrow} c_{\mathbf{k}\uparrow} c_{-\mathbf{k}\downarrow}$$

and $V_{\mathbf{k}\mathbf{k}'} = V_{-\mathbf{k},-\mathbf{k}'}$ the summation over s is equivalent to a factor of 2. This allows us to abbreviate further the notation of Eq. (7.2.3) by adopting the convention that an operator written with an explicit minus sign in the subscript refers to a spin-down state while an operator without a minus sign refers to a spin-up state. Thus

$$c^\dagger_{\mathbf{k}'} \equiv c^\dagger_{\mathbf{k}'\uparrow}; \quad c^\dagger_{-\mathbf{k}'} \equiv c^\dagger_{(-\mathbf{k}')\downarrow}, \text{ etc.}$$

We then have

$$\mathcal{H}_{\text{BCS}} = \sum_{\mathbf{k}} \mathcal{E}_{\mathbf{k}}(c^\dagger_{\mathbf{k}} c_{\mathbf{k}} + c^\dagger_{-\mathbf{k}} c_{-\mathbf{k}}) - \sum_{\mathbf{k},\mathbf{k}'} V_{\mathbf{k}\mathbf{k}'} c^\dagger_{\mathbf{k}'} c^\dagger_{-\mathbf{k}'} c_{-\mathbf{k}} c_{\mathbf{k}}. \quad (7.2.4)$$

This is the model Hamiltonian of Bardeen, Cooper, and Schrieffer, of which the eigenstates and eigenvalues must now be explored.

7.3 The Bogoliubov–Valatin transformation

In the Bogoliubov theory of helium described in Section 3.4 it was found to be possible to diagonalize a Hamiltonian that contained scattering terms like $a^\dagger_\mathbf{k} a^\dagger_{-\mathbf{k}}$ by means of a transformation to new operators

$$\alpha_\mathbf{k} = (\cosh \theta_\mathbf{k}) a_\mathbf{k} - (\sinh \theta_\mathbf{k}) a^\dagger_{-\mathbf{k}}.$$

The $\alpha_\mathbf{k}$ and their conjugates $\alpha^\dagger_\mathbf{k}$ were found to have the commutation relations of boson operators, and allowed an exact solution of the model Hamiltonian (3.4.1). This suggests that we try a similar transformation for the fermion problem posed by $\mathcal{H}_{\text{BCS}}$, and so we define two new operators

$$\gamma_\mathbf{k} = u_\mathbf{k} c_\mathbf{k} - v_\mathbf{k} c^\dagger_{-\mathbf{k}}; \quad \gamma_{-\mathbf{k}} = u_\mathbf{k} c_{-\mathbf{k}} + v_\mathbf{k} c^\dagger_\mathbf{k} \quad (7.3.1)$$

with conjugates

$$\gamma^\dagger_\mathbf{k} = u_\mathbf{k} c^\dagger_\mathbf{k} - v_\mathbf{k} c_{-\mathbf{k}}; \quad \gamma^\dagger_{-\mathbf{k}} = u_\mathbf{k} c^\dagger_{-\mathbf{k}} + v_\mathbf{k} c_\mathbf{k}. \quad (7.3.2)$$

The constants u_k and v_k are chosen to be real and positive and to obey the condition

$$u_k^2 + v_k^2 = 1$$

in order that the new operators have the fermion anticommutation relations

$$\{\gamma_k, \gamma_{k'}\} = \{\gamma_k, \gamma_{-k'}\} = \{\gamma_k^\dagger, \gamma_{-k'}^\dagger\} = 0$$

$$\{\gamma_k^\dagger, \gamma_{k'}\} = \{\gamma_{-k}^\dagger, \gamma_{-k'}\} = \delta_{kk'},$$

as was verified in Problem 2.4. Equations (7.3.1) and (7.3.2) comprise the *Bogoliubov–Valatin transformation*, which allows us to write the BCS Hamiltonian in terms of new operators. We do not expect to be able to diagonalize $\mathcal{H}_{BCS}$ completely, as this Hamiltonian contains terms involving products of four electron operators, and is intrinsically more difficult than Eq. (3.4.1); we do, however, hope that a suitable choice of u_k and v_k will allow the elimination of the most troublesome off-diagonal terms.

We rewrite the BCS Hamiltonian by first forming the inverse transformations to Eqs. (7.3.1) and (7.3.2). These are

$$c_k = u_k \gamma_k + v_k \gamma_{-k}^\dagger; \quad c_{-k} = u_k \gamma_{-k} - v_k \gamma_k^\dagger \tag{7.3.3}$$

$$c_k^\dagger = u_k \gamma_k^\dagger + v_k \gamma_{-k}; \quad c_{-k}^\dagger = u_k \gamma_{-k}^\dagger - v_k \gamma_k. \tag{7.3.4}$$

The first part of Eq. (7.2.4) represents the kinetic energy $\mathcal{H}_T$, and on substitution from Eqs. (7.3.3) and (7.3.4) is given by

$$\mathcal{H}_T = \sum_k \varepsilon_k [u_k^2 \gamma_k^\dagger \gamma_k + v_k^2 \gamma_{-k} \gamma_{-k}^\dagger + u_k v_k (\gamma_k^\dagger \gamma_{-k}^\dagger + \gamma_{-k} \gamma_k)$$
$$+ v_k^2 \gamma_k \gamma_k^\dagger + u_k^2 \gamma_{-k}^\dagger \gamma_{-k} - u_k v_k (\gamma_{-k}^\dagger \gamma_k^\dagger + \gamma_k \gamma_{-k})].$$

The diagonal parts of this expression can be simplified by making use of the anticommutation relations of the γ's and defining a new pair of number operators

$$m_k = \gamma_k^\dagger \gamma_k; \quad m_{-k} = \gamma_{-k}^\dagger \gamma_{-k}.$$

Then

$$\mathcal{H}_T = \sum_k \varepsilon_k [2v_k^2 + (u_k^2 - v_k^2)(m_k + m_{-k}) + 2u_k v_k (\gamma_k^\dagger \gamma_{-k}^\dagger + \gamma_{-k} \gamma_k)]. \tag{7.3.5}$$

7.3 The Bogoliubov–Valatin transformation

We note here the presence of three types of term – a constant, a term containing the number operators $m_{\mathbf{k}}$ and $m_{-\mathbf{k}}$, and off-diagonal terms containing the product $\gamma_{\mathbf{k}}^{\dagger}\gamma_{-\mathbf{k}}^{\dagger}$ or $\gamma_{-\mathbf{k}}\gamma_{\mathbf{k}}$. The potential energy $\mathcal{H}_V$ is given by the second part of $\mathcal{H}_{\mathrm{BCS}}$ and leads to a more complicated expression. We find

$$\mathcal{H}_V = -\sum_{\mathbf{k},\mathbf{k}'} V_{\mathbf{k}\mathbf{k}'}(u_{\mathbf{k}'}\gamma_{\mathbf{k}'}^{\dagger} + v_{\mathbf{k}'}\gamma_{-\mathbf{k}'})(u_{\mathbf{k}'}\gamma_{-\mathbf{k}'}^{\dagger} - v_{\mathbf{k}'}\gamma_{\mathbf{k}'})$$

$$\times (u_{\mathbf{k}}\gamma_{-\mathbf{k}} - v_{\mathbf{k}}\gamma_{\mathbf{k}}^{\dagger})(u_{\mathbf{k}}\gamma_{\mathbf{k}} + v_{\mathbf{k}}\gamma_{-\mathbf{k}}^{\dagger})$$

$$= -\sum_{\mathbf{k},\mathbf{k}'} V_{\mathbf{k}\mathbf{k}'}[u_{\mathbf{k}'}v_{\mathbf{k}'}u_{\mathbf{k}}v_{\mathbf{k}}(1 - m_{\mathbf{k}'} - m_{-\mathbf{k}'})(1 - m_{\mathbf{k}} - m_{-\mathbf{k}})$$

$$+ u_{\mathbf{k}'}v_{\mathbf{k}'}(1 - m_{\mathbf{k}'} - m_{-\mathbf{k}'})(u_{\mathbf{k}}^2 - v_{\mathbf{k}}^2)(\gamma_{-\mathbf{k}}\gamma_{\mathbf{k}} + \gamma_{\mathbf{k}}^{\dagger}\gamma_{-\mathbf{k}}^{\dagger})]$$

$$+ \text{(fourth-order off-diagonal terms)}. \quad (7.3.6)$$

We now argue that if we can eliminate the off-diagonal terms in $\mathcal{H}_{\mathrm{BCS}}$ by ensuring that those in Eq. (7.3.5) are cancelled by those in Eq. (7.3.6), then we shall be left with the Hamiltonian of a system of independent fermions. We first assume that the state of this system of lowest energy has all the occupation numbers $m_{\mathbf{k}}$ and $m_{-\mathbf{k}}$ equal to zero; this assumption may be verified at a later stage of the calculation. To find the form of the Bogoliubov–Valatin transformation that is appropriate to a superconductor in its ground state we then let all the $m_{\mathbf{k}}$ and $m_{-\mathbf{k}}$ vanish in Eqs. (7.3.5) and (7.3.6) and stipulate that the sum of off-diagonal terms also vanish. We find

$$\sum_{\mathbf{k}} 2\mathcal{E}_{\mathbf{k}}u_{\mathbf{k}}v_{\mathbf{k}}(\gamma_{\mathbf{k}}^{\dagger}\gamma_{-\mathbf{k}}^{\dagger} + \gamma_{-\mathbf{k}}\gamma_{\mathbf{k}}) - \sum_{\mathbf{k},\mathbf{k}'} V_{\mathbf{k}\mathbf{k}'}u_{\mathbf{k}'}v_{\mathbf{k}'}(u_{\mathbf{k}}^2 - v_{\mathbf{k}}^2)(\gamma_{\mathbf{k}}^{\dagger}\gamma_{-\mathbf{k}}^{\dagger} + \gamma_{-\mathbf{k}}\gamma_{\mathbf{k}})$$

$$+ \text{(fourth-order terms)} = 0. \quad (7.3.7)$$

If we make the approximation that the fourth-order terms can be neglected (Problem 7.6) then this reduces to

$$2\mathcal{E}_{\mathbf{k}}u_{\mathbf{k}}v_{\mathbf{k}} - (u_{\mathbf{k}}^2 - v_{\mathbf{k}}^2)\sum_{\mathbf{k}'} V_{\mathbf{k}\mathbf{k}'}u_{\mathbf{k}'}v_{\mathbf{k}'} = 0. \quad (7.3.8)$$

Because $u_{\mathbf{k}}$ and $v_{\mathbf{k}}$, being related by the condition that $u_{\mathbf{k}}^2 + v_{\mathbf{k}}^2 = 1$, are not independent it becomes convenient to express them in terms of a single variable $x_{\mathbf{k}}$, defined by

$$u_{\mathbf{k}} = (\tfrac{1}{2} - x_{\mathbf{k}})^{1/2}; \quad v_{\mathbf{k}} = (\tfrac{1}{2} + x_{\mathbf{k}})^{1/2}. \quad (7.3.9)$$

Then Eq. (7.3.8) becomes

$$2\mathcal{E}_{\mathbf{k}}(\tfrac{1}{4} - x_{\mathbf{k}}^2)^{1/2} + 2x_{\mathbf{k}} \sum_{\mathbf{k}'} V_{\mathbf{k}\mathbf{k}'}(\tfrac{1}{4} - x_{\mathbf{k}'}^2)^{1/2} = 0. \qquad (7.3.10)$$

If we define a new quantity $\Delta_{\mathbf{k}}$ by writing

$$\Delta_{\mathbf{k}} = \sum_{\mathbf{k}'} V_{\mathbf{k}\mathbf{k}'}(\tfrac{1}{4} - x_{\mathbf{k}'}^2)^{1/2} \qquad (7.3.11)$$

then Eq. (7.3.10) leads to the result

$$x_{\mathbf{k}} = \pm \frac{\mathcal{E}_{\mathbf{k}}}{2(\mathcal{E}_{\mathbf{k}}^2 + \Delta_{\mathbf{k}}^2)^{1/2}}. \qquad (7.3.12)$$

Substitution of this expression in Eq. (7.3.11) gives an integral equation for $\Delta_{\mathbf{k}}$ of the form

$$\Delta_{\mathbf{k}} = \frac{1}{2} \sum_{\mathbf{k}'} V_{\mathbf{k}\mathbf{k}'} \frac{\Delta_{\mathbf{k}'}}{(\mathcal{E}_{\mathbf{k}'}^2 + \Delta_{\mathbf{k}'}^2)^{1/2}}. \qquad (7.3.13)$$

If $V_{\mathbf{k}\mathbf{k}'}$ is known then this equation can in principle be solved and re-substituted in Eq. (7.3.12) to give $x_{\mathbf{k}}$. In doing so we once more note that the zero of energy of the electrons must be chosen to be the chemical potential μ if the total number of electrons is to be kept constant. To see this we note that

$$N = \sum_{\mathbf{k}} (c_{\mathbf{k}}^\dagger c_{\mathbf{k}} + c_{-\mathbf{k}}^\dagger c_{-\mathbf{k}})$$

$$= \sum_{\mathbf{k}} [2v_{\mathbf{k}}^2 + (u_{\mathbf{k}}^2 - v_{\mathbf{k}}^2)(m_{\mathbf{k}} + m_{-\mathbf{k}}) + 2u_{\mathbf{k}}v_{\mathbf{k}}(\gamma_{\mathbf{k}}^\dagger \gamma_{-\mathbf{k}}^\dagger + \gamma_{-\mathbf{k}}\gamma_{\mathbf{k}})],$$

and so the expectation value of N in the ground state of the system is just

$$\langle N \rangle = \sum_{\mathbf{k}} 2v_{\mathbf{k}}^2$$

$$= \sum_{\mathbf{k}} (1 + 2x_{\mathbf{k}}). \qquad (7.3.14)$$

In the absence of interactions

$$\langle N \rangle = \sum_{k < k_F} 2,$$

and so we deduce that $x_{\mathbf{k}} = \tfrac{1}{2}$ if $\mathcal{E}_{\mathbf{k}} \leq \mu$ and $x_{\mathbf{k}} = -\tfrac{1}{2}$ if $\mathcal{E}_{\mathbf{k}} > \mu$, as illustrated

7.3 The Bogoliubov–Valatin transformation

in Fig. 7.3.1(a). This shows that x_k is an odd function of $\mathcal{E}_k - \mu$ in the noninteracting case, and that if we make sure x_k remains an odd function of $\mathcal{E}_k - \mu$ in the presence of interactions, then Eq. (7.3.14) tells us that $\langle N \rangle$ will be unchanged (the energy dependence of the density of states being neglected). We also want the form of x_k given by Eq. (7.3.12) to reduce to the free-electron case when $V_{kk'}$ vanishes, and so we choose the negative square root and obtain a form for v_k and x_k like that shown in Fig. 7.3.1(b). To remind ourselves that $\mathcal{E}_k$ is measured relative to μ we use the symbol $\hat{\mathcal{E}}_k = \mathcal{E}_k - \mu$ to rewrite Eq. (7.3.12) as

$$x_k = -\frac{\hat{\mathcal{E}}_k}{2(\hat{\mathcal{E}}_k^2 + \Delta_k^2)^{1/2}}.$$

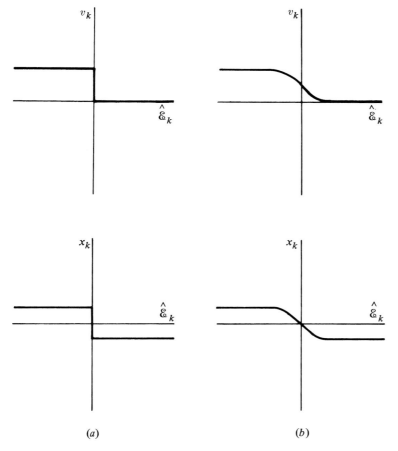

Figure 7.3.1. In this diagram we compare the form that the functions v_k and x_k take in a normal metal at zero temperature (a) and in a BCS superconductor (b).

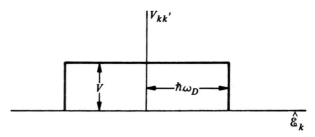

Figure 7.3.2. This simple model for the matrix element of the attractive interaction between electrons was used in the original calculations of Bardeen, Cooper, and Schrieffer.

To make these ideas more explicit we next consider a simple model that allows us to solve the integral equation for $\Delta_{\mathbf{k}}$ exactly. The matrix element $V_{\mathbf{kk'}}$ has its origin in the electron–phonon interaction, and, as Eq. (7.2.2) indicates, is only attractive when $|\hat{\mathcal{E}}_{\mathbf{k}} - \hat{\mathcal{E}}_{\mathbf{k'}}|$ is less than the energy $\hbar\omega_q$ of the phonon involved. In the simple model first chosen by BCS the matrix element was assumed to be of the form shown in Fig. 7.3.2, in which

$$V_{\mathbf{kk'}} \begin{cases} = V & \text{if } |\hat{\mathcal{E}}_{\mathbf{k}}| < \hbar\omega_D \\ = 0 & \text{otherwise,} \end{cases} \quad (7.3.15)$$

with V a constant and $\hbar\omega_D$ the Debye energy. It then follows that $\Delta_{\mathbf{k}}$ is also a constant, since Eq. (7.3.13) reduces to

$$\Delta_{\mathbf{k}} = \frac{1}{2} \int_{-\infty}^{\infty} D(\mathcal{E}_{\mathbf{k'}}) \, d\mathcal{E}_{\mathbf{k'}} \frac{\Delta_{\mathbf{k'}}}{(\hat{\mathcal{E}}_{\mathbf{k'}}^2 + \Delta_{\mathbf{k'}}^2)^{1/2}} V_{\mathbf{kk'}}$$

$$= \frac{1}{2} V D(\mu) \int_{-\hbar\omega_D}^{\hbar\omega_D} \frac{\Delta}{(\hat{\mathcal{E}}^2 + \Delta^2)^{1/2}} \, d\hat{\mathcal{E}},$$

the energy density of states $D(\mathcal{E})$ here referring to states of one spin only and again being taken as constant. This has the solution

$$\Delta = \frac{\hbar\omega_D}{\sinh[1/VD(\mu)]}. \quad (7.3.16)$$

The magnitude of the product $VD(\mu)$ can be estimated by noting that from Eq. (7.2.2)

$$V \sim \frac{|M_{\mathbf{q}}|^2}{\hbar\omega_D}$$

and from Eq. (6.1.2)

$$|M_\mathbf{q}|^2 \sim \frac{N\hbar k_F^2}{M\omega_D}|V_\mathbf{k}|^2$$

$$\sim \frac{m}{M}\frac{\mu}{\hbar\omega_D}N|V_\mathbf{k}|^2$$

with $V_\mathbf{k}$ the Fourier transform of a screened ion potential. Since $D(\mu) \sim N/\mu$ we find that

$$VD(\mu) \sim \frac{m}{M}\left(\frac{NV_\mathbf{k}}{\hbar\omega_D}\right)^2.$$

While $\hbar\omega_D$ might typically be 0.03 eV, the factor $NV_\mathbf{k}$ is something like the average of the screened ion potential over the unit cell containing the ion, and might have a value of a few electron volts, making $(NV_\mathbf{k}/\hbar\omega_D)^2$ of the order of 10^4. The ratio of electron mass to ion mass, m/M, however, is only of the order of 10^{-5}, and so it is in most cases reasonable to make the approximation of *weak coupling*, and replace Eq. (7.3.16) by

$$\Delta = 2\hbar\omega_D e^{-1/VD(\zeta)}, \qquad (7.3.17)$$

the difference between $2\sinh(10)$ and e^{10} being negligible. In *strong-coupling superconductors* such as mercury or lead, however, the electron–phonon interaction is too strong for such a simplification to be valid; for these metals the rapid damping of the quasiparticle states must also be taken into account.

The important fact that this rough calculation tells us is that Δ is a very small quantity indeed, being generally about one percent of the Debye energy, and hence corresponding to thermal energies at temperatures of the order of 1 K. The parameter $x_\mathbf{k}$ thus only differs from $\pm\frac{1}{2}$ within this short distance of the Fermi energy and our new operators $\gamma_\mathbf{k}$ and $\gamma_{-\mathbf{k}}$ reduce to simple electron annihilation or creation operators everywhere except within this thin shell of states containing the Fermi surface.

7.4 The ground-state wavefunction and the energy gap

Our first application of the Bogoliubov–Valatin formalism must be an evaluation of the ground-state energy $\mathcal{E}_S$ of the superconducting system. We hope to find a result that is lower than $\mathcal{E}_N$, the energy of the normal system, by some amount which we shall call the condensation energy $\mathcal{E}_c$. The ground-state energy of the BCS state is given by the sum of the expectation values of Eqs. (7.3.5) and (7.3.6) under the conditions that $m_\mathbf{k} = m_{-\mathbf{k}} = 0$. As we have already eliminated the off-diagonal terms we are only left with the constant terms, and find

$$\mathcal{E}_S = \sum_\mathbf{k} 2\hat{\mathcal{E}}_\mathbf{k} v_\mathbf{k}^2 - \sum_{\mathbf{k},\mathbf{k}'} V_{\mathbf{k}\mathbf{k}'} u_{\mathbf{k}'} v_{\mathbf{k}'} u_\mathbf{k} v_\mathbf{k}$$

$$= \sum_\mathbf{k} \hat{\mathcal{E}}_\mathbf{k}(1 + 2x_\mathbf{k}) - \sum_{\mathbf{k},\mathbf{k}'} V_{\mathbf{k}\mathbf{k}'}[(\tfrac{1}{4} - x_{\mathbf{k}'}^2)(\tfrac{1}{4} - x_\mathbf{k}^2)]^{1/2}. \quad (7.4.1)$$

It is interesting to pause at this point and note that we could have considered the BCS Hamiltonian from a variational point of view. Instead of eliminating the off-diagonal elements of $\mathcal{H}_{BCS}$ we could have decided to choose the $x_\mathbf{k}$ in such a way as to minimize $\mathcal{E}_S$. It is reassuring to see that this approach leads to the same solution as before; if we differentiate Eq. (7.4.1) with respect to $x_\mathbf{k}$ and equate the result to zero we obtain an equation that is identical with Eq. (7.3.10). We write expression (7.4.1) as

$$\mathcal{E}_S = \sum_\mathbf{k}[\hat{\mathcal{E}}_\mathbf{k}(1 + 2x_\mathbf{k}) - (\tfrac{1}{4} - x_\mathbf{k}^2)^{1/2}\Delta]$$

with $x_\mathbf{k}$ and Δ defined as before in Eqs. (7.3.12) and (7.3.11). In the normal system $x_\mathbf{k}^2 = \tfrac{1}{4}$ for all $\mathbf{k}$ and so the condensation energy, defined as $\mathcal{E}_S - \mathcal{E}_N$, is

$$\mathcal{E}_c = \sum_{k<k_F} \hat{\mathcal{E}}_\mathbf{k}(2x_\mathbf{k} - 1) + \sum_{k>k_F} \hat{\mathcal{E}}_\mathbf{k}(2x_\mathbf{k} + 1) - \sum_\mathbf{k}(\tfrac{1}{4} - x_\mathbf{k}^2)^{1/2}\Delta$$

$$= 2D(\mu)\int_0^{\hbar\omega_D}\left[\hat{\mathcal{E}}_\mathbf{k} - \frac{2\hat{\mathcal{E}}_\mathbf{k}^2 + \Delta^2}{2(\hat{\mathcal{E}}_\mathbf{k}^2 + \Delta^2)^{1/2}}\right]d\hat{\mathcal{E}}_\mathbf{k}$$

$$= D(\mu)\{(\hbar\omega_D)^2 - \hbar\omega_D[(\hbar\omega_D)^2 + \Delta^2]^{1/2}\}$$

$$= (\hbar\omega_D)^2 D(\mu)\left[1 - \coth\left(\frac{1}{VD(\mu)}\right)\right].$$

In the weak-coupling case this becomes

$$\mathcal{E}_c \simeq -2(\hbar\omega_D)^2 D(\mu) e^{-2/VD(\mu)}$$

$$= -\tfrac{1}{2}D(\mu)\Delta^2. \quad (7.4.2)$$

This condensation energy is surprisingly small, being of the order of only 10^{-7} eV per electron, which is the equivalent of a thermal energy of about a millidegree Kelvin. This is a consequence of the fact that only the electrons

7.4 The ground-state wavefunction and the energy gap

with energies in the range $\mu - \Delta$ to $\mu + \Delta$ are affected by the attractive interaction, and these are only a small fraction of the order of Δ/μ of the whole. We note that we cannot expand $\mathcal{E}_c$ in a power series in the interaction strength V since the function $\exp[-2/VD(\mu)]$ has an essential singularity at $V = 0$, which means that while the function and all its derivatives vanish as $V \to +0$, they all become infinite as $V \to -0$. This shows the qualitative difference between the effects of an attractive and a repulsive interaction, and tells us that we could never have been successful in calculating $\mathcal{E}_c$ by using perturbation theory.

The wavefunction Ψ_0 of the superconducting system in its ground state may be found by recalling that it must be the eigenfunction of the diagonalized BCS Hamiltonian that has $m_\mathbf{k} = m_{-\mathbf{k}} = 0$ for all $\mathbf{k}$, so that

$$\gamma_\mathbf{k}|\Psi_0\rangle = \gamma_{-\mathbf{k}}|\Psi_0\rangle = 0. \quad (7.4.3)$$

Now since $\gamma_\mathbf{k}\gamma_\mathbf{k} = \gamma_{-\mathbf{k}}\gamma_{-\mathbf{k}} = 0$ we can form the wavefunction that satisfies Eq. (7.4.3) simply by operating on the vacuum state with all the $\gamma_\mathbf{k}$ and all the $\gamma_{-\mathbf{k}}$. From Eq. (7.3.1) we have

$$\left(\prod_\mathbf{k} \gamma_\mathbf{k}\gamma_{-\mathbf{k}}\right)|0\rangle = \left[\prod_\mathbf{k} (u_\mathbf{k} c_\mathbf{k} - v_\mathbf{k} c^\dagger_{-\mathbf{k}})(u_\mathbf{k} c_{-\mathbf{k}} + v_\mathbf{k} c^\dagger_\mathbf{k})\right]|0\rangle$$

$$= \left[\prod_\mathbf{k} (u_\mathbf{k} v_\mathbf{k} + v_\mathbf{k}^2 c^\dagger_\mathbf{k} c^\dagger_{-\mathbf{k}})\right]|0\rangle.$$

To normalize this we divide by the product of all the $v_\mathbf{k}$ to obtain

$$|\Psi_0\rangle = \left[\prod_\mathbf{k} (u_\mathbf{k} + v_\mathbf{k} c^\dagger_\mathbf{k} c^\dagger_{-\mathbf{k}})\right]|0\rangle. \quad (7.4.4)$$

This wavefunction is a linear combination of simpler wavefunctions containing different numbers of particles, which means that it is not an eigenstate of the total number operator N. Our familiarity with the concept of the chemical potential μ teaches us not to be too concerned about this fact, however, as long as we make sure that the average value is kept constant.

The quasiparticle excitations of the system are created by the operators $\gamma^\dagger_\mathbf{k}$ and $\gamma^\dagger_{-\mathbf{k}}$ acting on Ψ_0. By adding Eqs. (7.3.5) and (7.3.6) one can write the Hamiltonian in the form

$$\mathcal{H}_{BCS} = \mathcal{E}_S + \sum_\mathbf{k} (m_\mathbf{k} + m_{-\mathbf{k}})\left[(u_\mathbf{k}^2 - v_\mathbf{k}^2)\hat{\mathcal{E}}_\mathbf{k} + 2u_\mathbf{k} v_\mathbf{k} \sum_{\mathbf{k}'} V_{\mathbf{k}\mathbf{k}'} u_{\mathbf{k}'} v_{\mathbf{k}'}\right]$$

$$+ \text{higher-order terms,}$$

which on substitution of our solution for $u_\mathbf{k}$ and $v_\mathbf{k}$ becomes

$$\mathcal{H}_{\text{BCS}} = \mathcal{E}_S + \sum_\mathbf{k} (\hat{\mathcal{E}}_\mathbf{k}^2 + \Delta^2)^{1/2}(m_\mathbf{k} + m_{-\mathbf{k}}) + \cdots .$$

The energies $E_\mathbf{k}$ of these elementary excitations are thus given by

$$E_\mathbf{k} = (\hat{\mathcal{E}}_\mathbf{k}^2 + \Delta^2)^{1/2}. \tag{7.4.5}$$

These excitations cannot be created singly, for that would mean operating on Ψ_0 with a single $\gamma^\dagger$, which is a sum containing just one c and one $c^\dagger$. Now any physical perturbation that we apply to Ψ_0 will contain at least two electron operators, since such perturbations as electric and magnetic fields act to scatter rather than to create or destroy electrons. For instance

$$c_\mathbf{k}^\dagger c_{\mathbf{k}'} |\Psi_0\rangle = (u_\mathbf{k} \gamma_\mathbf{k}^\dagger + v_\mathbf{k} \gamma_{-\mathbf{k}})(u_{\mathbf{k}'} \gamma_{\mathbf{k}'} + v_{\mathbf{k}'} \gamma_{-\mathbf{k}'}^\dagger)|\Psi_0\rangle$$

$$= u_\mathbf{k} v_{\mathbf{k}'} \gamma_\mathbf{k}^\dagger \gamma_{-\mathbf{k}'}^\dagger |\Psi_0\rangle,$$

the other terms vanishing. We thus conclude that *only pairs of quasiparticles can be excited*, and that from Eq. (7.4.5) the minimum energy necessary to create such a pair of excitations is 2Δ. This explains the exponential form of the electronic specific heat at low temperatures and also the absorption edge for electromagnetic radiation at $\hbar\omega = 2\Delta$.

It is interesting to compare these quasiparticle excitations with the particle–hole excitations of a normal Fermi system. In the noninteracting electron gas at zero temperature the operator $c_\mathbf{k}^\dagger c_{\mathbf{k}'}$ creates a hole at $\mathbf{k}'$ and an electron at $\mathbf{k}$ provided $\hat{\mathcal{E}}_{\mathbf{k}'} < 0$ and $\hat{\mathcal{E}}_\mathbf{k} > 0$. The energy of this excitation is $\hat{\mathcal{E}}_\mathbf{k} - \hat{\mathcal{E}}_{\mathbf{k}'}$, which can be written as $|\hat{\mathcal{E}}_\mathbf{k}| + |\hat{\mathcal{E}}_{\mathbf{k}'}|$, and is thus equal to the sum of the lengths of the arrows in Fig. 7.4.1(a). In the superconducting system the operator $c_\mathbf{k}^\dagger c_{\mathbf{k}'}$ has a component $\gamma_\mathbf{k}^\dagger \gamma_{-\mathbf{k}'}^\dagger$, which creates an excitation of total energy $E_\mathbf{k} + E_{\mathbf{k}'}$ equal to the sum of lengths of the arrows in Fig. 7.4.1(b). The density of states is inversely proportional to the slope of $\hat{\mathcal{E}}(k)$ in the normal metal, and this leads us to think of an effective density of states in the superconductor inversely proportional to $dE/d|\mathbf{k}|$. As $E_\mathbf{k}$ and $\hat{\mathcal{E}}_\mathbf{k}$ are related by Eq. (7.4.5) we find this effective density of states to be equal to

$$D(E) = 2D(\mathcal{E}) \frac{d\hat{\mathcal{E}}}{dE}$$

$$\simeq \begin{cases} 2D(\mu) \dfrac{|E|}{(E^2 - \Delta^2)^{1/2}} & \text{if } |E| > \Delta \\ 0 & \text{if } |E| \leq \Delta. \end{cases} \tag{7.4.6}$$

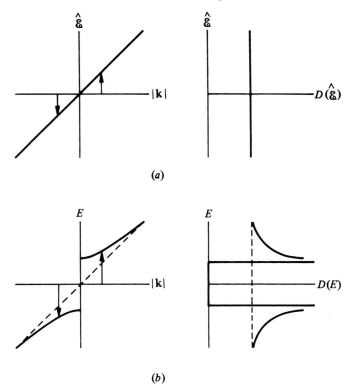

Figure 7.4.1. In a normal metal (a) an electron–hole pair has an excitation energy equal to the sum of the lengths of the two arrows in the left-hand diagram; the density of states is inversely proportional to the slope of $\hat{\mathcal{E}}(\mathbf{k})$ and is thus roughly constant. In a superconductor (b) the energy of the excitations is the sum of the lengths of the arrows when $E_\mathbf{k}$ is plotted; an effective density of states can again be drawn which is inversely proportional to the slope of $E_\mathbf{k}$, as shown on the right.

The factor of 2 enters when $D(\mu)$ is the normal density of states for one spin direction because the states $\gamma_\mathbf{k}^\dagger|\Psi_0\rangle$ and $\gamma_{-\mathbf{k}}^\dagger|\Psi_0\rangle$ are degenerate.

7.5 The transition temperature

Our method of diagonalizing $\mathcal{H}_{\mathrm{BCS}}$ in Section 7.3 was to fix $u_\mathbf{k}$ and $v_\mathbf{k}$ so that the sum of off-diagonal terms from Eqs. (7.3.5) and (7.3.6) vanished. The resulting equation was of the form

$$\sum_\mathbf{k} \left[2\hat{\mathcal{E}}_\mathbf{k} u_\mathbf{k} v_\mathbf{k} - \sum_{\mathbf{k}'} V_{\mathbf{k}\mathbf{k}'} u_{\mathbf{k}'} v_{\mathbf{k}'} (1 - m_{\mathbf{k}'} - m_{-\mathbf{k}'})(u_\mathbf{k}^2 - v_\mathbf{k}^2) \right]$$
$$\times (\gamma_\mathbf{k}^\dagger \gamma_{-\mathbf{k}}^\dagger + \gamma_{-\mathbf{k}} \gamma_\mathbf{k}) = 0, \qquad (7.5.1)$$

and to solve this we first put $m_{\mathbf{k}'} = m_{-\mathbf{k}'} = 0$. While this approach remains valid in the presence of a few quasiparticle excitations it clearly needs modification whenever the proportion of excited states becomes comparable to unity, for then the terms in $m_{\mathbf{k}'}$ and $m_{-\mathbf{k}'}$ will contribute significantly to the summation over $\mathbf{k}'$. This will be the case if the temperature is such that kT is not much less than Δ.

We resolve this difficulty by first of all assuming that it *is* possible to eliminate these off-diagonal terms. We are then left with a Hamiltonian that is the sum of the diagonal parts of Eqs. (7.3.5) and (7.3.6), so that

$$\mathcal{H}_{\text{BCS}} = \sum_{\mathbf{k}} 2\hat{\mathcal{E}}_{\mathbf{k}} v_{\mathbf{k}}^2 + \sum_{\mathbf{k}} (u_{\mathbf{k}}^2 - v_{\mathbf{k}}^2)\hat{\mathcal{E}}_{\mathbf{k}}(m_{\mathbf{k}} + m_{-\mathbf{k}})$$
$$- \sum_{\mathbf{k},\mathbf{k}'} V_{\mathbf{k}\mathbf{k}'} u_{\mathbf{k}'} v_{\mathbf{k}'} u_{\mathbf{k}} v_{\mathbf{k}} (1 - m_{\mathbf{k}'} - m_{-\mathbf{k}'})(1 - m_{\mathbf{k}} - m_{-\mathbf{k}}). \quad (7.5.2)$$

The energy $E_{\mathbf{k}}$ necessary to create a quasiparticle excitation will be

$$E_{\mathbf{k}} = \frac{\partial \langle \mathcal{H}_{\text{BCS}} \rangle}{\partial \langle m_{\mathbf{k}} \rangle}$$
$$= \hat{\mathcal{E}}_{\mathbf{k}}(u_{\mathbf{k}}^2 - v_{\mathbf{k}}^2) + 2u_{\mathbf{k}} v_{\mathbf{k}} \sum_{\mathbf{k}'} V_{\mathbf{k}\mathbf{k}'} u_{\mathbf{k}'} v_{\mathbf{k}'} (1 - \langle m_{\mathbf{k}'} \rangle - \langle m_{-\mathbf{k}'} \rangle). \quad (7.5.3)$$

Now if we had a system of *independent* fermions at temperature T then we should know from Fermi–Dirac statistics just what the average occupancy of each state would be; from Eq. (3.3.3) we could immediately write

$$\bar{m}_{\mathbf{k}} = \bar{m}_{-\mathbf{k}} = \frac{1}{\exp(E_{\mathbf{k}}/kT) + 1}. \quad (7.5.4)$$

(No chemical potential appears in this function because the total number of quasiparticles is not conserved.) Although the terms in $m_{\mathbf{k}} m_{\mathbf{k}'}$ in Eq. (7.5.2) represent interactions among the fermions, the definition of the quasiparticle energy adopted in Eq. (7.5.3) allows us to treat them as independent. We can then obtain an approximate solution of Eq. (7.5.1) by replacing $m_{\mathbf{k}'}$ and $m_{-\mathbf{k}'}$ by their thermal averages. If we abbreviate the Fermi function of Eq. (7.5.4) by $f(E_{\mathbf{k}})$ then we find that to satisfy Eq. (7.5.1) we must put

$$2\hat{\mathcal{E}}_{\mathbf{k}} u_{\mathbf{k}} v_{\mathbf{k}} - (u_{\mathbf{k}}^2 - v_{\mathbf{k}}^2) \sum_{\mathbf{k}'} V_{\mathbf{k}\mathbf{k}'} u_{\mathbf{k}'} v_{\mathbf{k}'} [1 - 2f(E_{\mathbf{k}'})] = 0.$$

7.5 The transition temperature

The only difference between this equation and Eq. (7.3.8) lies in the extra factor of $1 - 2f(E_{\mathbf{k}'})$ that multiplies the matrix element $V_{\mathbf{kk}'}$. Consequently if we again make the substitution (7.3.9), but this time replace the definition of $\Delta_{\mathbf{k}}$ given in Eq. (7.3.11) by the definition

$$\Delta_{\mathbf{k}}(T) = \sum_{\mathbf{k}'} V_{\mathbf{kk}'}(\tfrac{1}{4} - x_{\mathbf{k}'}^2)^{1/2}[1 - 2f(E_{\mathbf{k}'})] \tag{7.5.5}$$

we regain Eq. (7.3.12). The temperature-dependent gap parameter $\Delta(T)$ is then found by resubstituting Eq. (7.3.12) in Eq. (7.5.5). One has

$$\Delta_{\mathbf{k}}(T) = \frac{1}{2}\sum_{\mathbf{k}'} V_{\mathbf{kk}'} \frac{\Delta_{\mathbf{k}'}(T)}{[\hat{\mathcal{E}}_{\mathbf{k}'}^2 + \Delta_{\mathbf{k}'}^2(T)]^{1/2}}[1 - 2f(E_{\mathbf{k}'})]. \tag{7.5.6}$$

This equation still contains the excitation energy $E_{\mathbf{k}}$ which we evaluate by taking the thermal average of Eq. (7.5.3). We find

$$E_{\mathbf{k}} = \hat{\mathcal{E}}_{\mathbf{k}}(u_{\mathbf{k}}^2 - v_{\mathbf{k}}^2) + 2u_{\mathbf{k}}v_{\mathbf{k}}\sum_{\mathbf{k}'} V_{\mathbf{kk}'} u_{\mathbf{k}'}v_{\mathbf{k}'}[1 - 2f(E_{\mathbf{k}'})]$$

$$= [\hat{\mathcal{E}}_{\mathbf{k}}^2 + \Delta_{\mathbf{k}}^2(T)]^{1/2}, \tag{7.5.7}$$

which is identical with our previous expression (7.4.5) except that Δ is now a function of temperature. On substituting for $E_{\mathbf{k}'}$ and $f(E_{\mathbf{k}'})$ in Eq. (7.5.6) we find

$$\Delta_{\mathbf{k}} = \frac{1}{2}\sum_{\mathbf{k}'} V_{\mathbf{kk}'} \frac{\Delta_{\mathbf{k}'}}{(\hat{\mathcal{E}}_{\mathbf{k}'}^2 + \Delta_{\mathbf{k}'}^2)^{1/2}} \tanh\left[\frac{(\hat{\mathcal{E}}_{\mathbf{k}'}^2 + \Delta_{\mathbf{k}'}^2)^{1/2}}{2kT}\right]. \tag{7.5.8}$$

In the simple model defined by Eq. (7.3.15) this reduces to

$$VD(\mu)\int_0^{\hbar\omega_D} \frac{\tanh[(\hat{\mathcal{E}}^2 + \Delta^2)^{1/2}/2kT]}{(\hat{\mathcal{E}}^2 + \Delta^2)^{1/2}}\, d\hat{\mathcal{E}} = 1. \tag{7.5.9}$$

At zero temperature this equation for Δ reduces to our previous solution (7.3.16). When the temperature is raised above zero the numerator of the integrand is reduced, and so in order for Eq. (7.5.9) to be satisfied the denominator must also decrease. This implies that Δ is a monotonically decreasing function of T; in fact it has the form shown qualitatively in Fig. 7.5.1. The initial decrease is exponentially slow until kT becomes of the order of $\Delta(0)$ and the quasiparticle excitations become plentiful; $\Delta(T)$ then

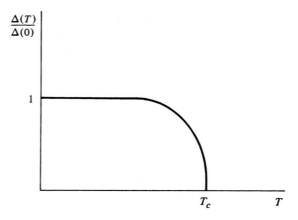

Figure 7.5.1. The energy gap parameter Δ decreases as the temperature is raised from zero, and vanishes at the transition temperature T_c.

begins to drop more rapidly until at the transition temperature T_c it vanishes. The magnitude of T_c in the BCS model is found from Eq. (7.5.9) by putting $\Delta(T_c) = 0$. We then have

$$VD(\mu) \int_0^{\hbar\omega_D/2kT_c} x^{-1} \tanh x \, dx = 1$$

or

$$[\ln x \tanh x]_0^{\hbar\omega_D/2kT_c} - \int_0^{\hbar\omega_D/2kT_c} \text{sech}^2 x \ln x \, dx = \frac{1}{VD(\mu)}.$$

For weak-coupling superconductors we can replace $\tanh(\hbar\omega_D/2kT_c)$ by unity and extend the upper limit of the integral to infinity to find

$$\ln\left(\frac{\hbar\omega_D}{2kT_c}\right) - \int_0^\infty \text{sech}^2 x \ln x \, dx = \frac{1}{VD(\mu)}.$$

The integral is more easily looked up than evaluated, but either way is equal to $\ln 0.44$, from which

$$kT_c = 1.14\hbar\omega_D e^{-1/VD(\mu)}. \tag{7.5.10}$$

Comparison with Eq. (7.3.17) shows that in this model

$$\frac{2\Delta(0)}{kT_c} = 3.50,$$

a result in adequate agreement with the experimentally observed values of this parameter, which for most elements lie between two and five.

The existence of an isotope effect is an obvious consequence of Eq. (7.5.10). The simplest form of isotope effect occurs when $VD(\mu)$ is independent of the ionic mass A, as then T_c depends on A only through the Debye energy $\hbar\omega_D$, which is proportional to $A^{-1/2}$. The fact that the electron–phonon enhancement of $D(\mu)$ is independent of A has already been considered in Problem 6.1, and the demonstration that V should also have this property follows similar lines. One should, however, note that T_c is very sensitive to changes in the density of states $D(\mu)$ as a consequence of the fact that $VD(\mu) \ll 1$. Thus if $VD(\mu) = \frac{1}{8}$ then from Eq. (7.5.10) one sees that a one percent decrease in $D(\mu)$ will cause an eight percent decrease in T_c. This makes it not very surprising that inclusion of the Coulomb repulsion in V or the use of other more complicated models can lead to different kinds of isotope effect.

The electronic specific heat C may now also be calculated for the BCS model. The energy $\mathcal{E}(T)$ of the superconductor at temperature T will be the average expectation value of the Hamiltonian (7.5.2). We find

$$\mathcal{E}(T) = \sum_{\mathbf{k}} [2\hat{\mathcal{E}}_{\mathbf{k}} v_{\mathbf{k}}^2 + (u_{\mathbf{k}}^2 - v_{\mathbf{k}}^2) 2 f_{\mathbf{k}} \hat{\mathcal{E}}_{\mathbf{k}} - (1 - 2f_{\mathbf{k}}) u_{\mathbf{k}} v_{\mathbf{k}} \Delta]$$

$$= \sum_{\mathbf{k}} \left[\hat{\mathcal{E}}_{\mathbf{k}} - \left(E_{\mathbf{k}} - \frac{\Delta^2}{2 E_{\mathbf{k}}} \right)(1 - 2f_{\mathbf{k}}) \right]$$

from which

$$C = \frac{d\mathcal{E}(T)}{dT}$$

$$= \sum_{\mathbf{k}} \left[2\left(E - \frac{\Delta^2}{2E}\right) \frac{df}{dT} + \frac{1}{2}(1 - 2f)\Delta^2 \frac{d(1/E)}{dT} \right]$$

$$= \sum_{\mathbf{k}} \left[2E \frac{df}{dT} + \Delta^2 \frac{d}{dT}\left(\frac{1 - 2f}{2E}\right) \right]$$

$$= \sum_{\mathbf{k}} \left(\frac{d\Delta^2}{dT} - \frac{2E^2}{T} \right) \frac{\partial f}{\partial E}. \tag{7.5.11}$$

Reasonably good agreement with experiment is usually obtained with this formula. The observed discontinuity in C at the transition temperature arises from the term $d\Delta^2/dT$, which is zero for $T > T_c$ but finite for $T < T_c$. The prediction of Eq. (7.5.11) is that C increases by a factor of 2.43 as the sample is cooled through T_c; observed increases are within a factor of four of this value.

7.6 Ultrasonic attenuation

In Fig. 7.1.1 the damping coefficient α for low-frequency sound waves in a superconductor was shown as a function of temperature. The rapid decrease of α as the sample is cooled below T_c suggests that the attenuation of an ultrasonic wave depends on the presence of quasiparticle excitations in the material. We can see why this should be so when we recall that the frequencies used in such ultrasonic experiments are typically less than 100 MHz, so that each phonon has an energy of less than 10^{-6} eV. Unless T is very close to T_c this phonon energy will be much less than 2Δ, which is the minimum energy required to create a pair of quasiparticles. The phonons can then only be absorbed by scattering already existing quasiparticles from one k-state to another.

The canonical transformation that we adopted in Section 6.5 allowed us to ignore the phonon system in calculating the BCS ground state; we had removed the first-order effects of the electron–phonon interaction and we assumed the transformed phonon system just to form a passive background that did not affect our calculations. In the theory of ultrasonic attenuation, however, the occupation number $n_\mathbf{q}$ of the applied sound wave must be allowed to change as the wave is damped, and so we must exclude this particular phonon mode $\mathbf{q}$ from the canonical transformation that leads to Eq. (6.5.4). We are consequently left with a Hamiltonian that still contains the set of terms $\mathcal{H}_\mathbf{q}$ and $\mathcal{H}_{-\mathbf{q}}$ where

$$\mathcal{H}_\mathbf{q} = \sum_{\mathbf{k},s} M_\mathbf{q} c^\dagger_{\mathbf{k}+\mathbf{q},s} c_{\mathbf{k},s} (a^\dagger_{-\mathbf{q}} + a_\mathbf{q}).$$

The annihilation operator $a_\mathbf{q}$ will reduce the occupation number of the phonon mode $\mathbf{q}$ by unity, and thus effect the damping of the applied sound wave. The effect of this process on the electron system is seen by transforming from the electron operators c to the quasiparticle operators γ by means of Eqs. (7.3.3) and (7.3.4). In the notation in which $\mathbf{k} \equiv \mathbf{k}\uparrow$ and $-\mathbf{k} \equiv -\mathbf{k}\downarrow$ this gives us

$$\mathcal{H}_\mathbf{q} = M_\mathbf{q}(a^\dagger_{-\mathbf{q}} + a_\mathbf{q}) \sum_\mathbf{k} (c^\dagger_{\mathbf{k}+\mathbf{q}} c_\mathbf{k} + c^\dagger_{-\mathbf{k}} c_{-(\mathbf{k}+\mathbf{q})})$$

$$= M_\mathbf{q}(a^\dagger_{-\mathbf{q}} + a_\mathbf{q}) \sum_\mathbf{k} [(u_{\mathbf{k}+\mathbf{q}} \gamma^\dagger_{\mathbf{k}+\mathbf{q}} + v_{\mathbf{k}+\mathbf{q}} \gamma_{-(\mathbf{k}+\mathbf{q})})(u_\mathbf{k} \gamma_\mathbf{k} + v_\mathbf{k} \gamma^\dagger_{-\mathbf{k}})$$

$$+ (u_\mathbf{k} \gamma^\dagger_{-\mathbf{k}} - v_\mathbf{k} \gamma_\mathbf{k})(u_{\mathbf{k}+\mathbf{q}} \gamma_{-(\mathbf{k}+\mathbf{q})} - v_{\mathbf{k}+\mathbf{q}} \gamma^\dagger_{\mathbf{k}+\mathbf{q}})]$$

$$= M_\mathbf{q}(a^\dagger_{-\mathbf{q}} + a_\mathbf{q}) \sum_\mathbf{k} [(u_{\mathbf{k}+\mathbf{q}} u_\mathbf{k} - v_{\mathbf{k}+\mathbf{q}} v_\mathbf{k})(\gamma^\dagger_{\mathbf{k}+\mathbf{q}} \gamma_\mathbf{k} + \gamma^\dagger_{-\mathbf{k}} \gamma_{-(\mathbf{k}+\mathbf{q})})$$

$$+ (u_{\mathbf{k}+\mathbf{q}} v_\mathbf{k} + u_\mathbf{k} v_{\mathbf{k}+\mathbf{q}})(\gamma^\dagger_{\mathbf{k}+\mathbf{q}} \gamma^\dagger_{-\mathbf{k}} + \gamma_{-(\mathbf{k}+\mathbf{q})} \gamma_\mathbf{k})]. \quad (7.6.1)$$

7.6 Ultrasonic attenuation

The terms in $\gamma^\dagger_{k+q}\gamma^\dagger_{-k}$ and $\gamma_{-(k+q)}\gamma_k$ lead to the simultaneous creation or destruction of two quasiparticles and, as we have seen, cannot represent energy-conserving processes when $\hbar\omega_q$ is negligible. We are thus left with the terms $\gamma^\dagger_{k+q}\gamma_k$ and $\gamma^\dagger_{-k}\gamma_{-(k+q)}$, which can lead to energy-conserving processes if either $E_{k+q} = E_k + \hbar\omega_q$ or $E_{-k} = E_{-(k+q)} + \hbar\omega_q$. For such processes to be allowed, the quasiparticle states $\mathbf{k}$ and $-(\mathbf{k}+\mathbf{q})$ must initially be filled and the final states $(\mathbf{k}+\mathbf{q})$ and $-\mathbf{k}$ must be empty. We then find the total probability P_a of a phonon of wavenumber $\mathbf{q}$ being absorbed to be given by

$$P_a \propto \sum_k (u_{k+q}u_k - v_{k+q}v_k)^2 [f_k(1-f_{k+q})\delta(E_{k+q} - E_k - \hbar\omega_q)$$
$$+ f_{-(k+q)}(1 - f_{-k})\delta(E_{-k} - E_{-(k+q)} - \hbar\omega_q)].$$

The set of terms $\mathcal{H}_{-q}$ is similarly calculated to give a certain probability P_e of a phonon of wavenumber $\mathbf{q}$ being emitted when a quasiparticle is scattered from $\mathbf{k}$ to $\mathbf{k} - \mathbf{q}$. We assume this phonon mode to be macroscopically occupied, so that we can neglect the difference between n_q and $n_q + 1$. We then find the net probability P that a phonon is absorbed to be

$$P = P_a - P_e$$
$$\propto \sum_k (u_k^2 - v_k^2)^2 [(f_k - f_{k+q})\delta(E_{k+q} - E_k - \hbar\omega_q)$$
$$+ (f_{-(k+q)} - f_{-k})\delta(E_{-k} - E_{-(k+q)} - \hbar\omega_q)], \quad (7.6.2)$$

where $(u_{k+q}u_k - v_{k+q}v_k)^2$ has been replaced by $(u_k^2 - v_k^2)^2$, the justification being that q is small while in the present context u_k and v_k are slowly varying functions. We similarly make the approximation

$$f_k - f_{k+q} = -\mathbf{q} \cdot \frac{\partial f}{\partial \mathbf{k}}$$
$$= -\mathbf{q} \cdot \frac{d\mathcal{E}}{d\mathbf{k}}\frac{dE}{d\mathcal{E}}\frac{\partial f}{\partial E}$$
$$= -q\hbar v_k \cos\theta \, \frac{\hat{\mathcal{E}}}{E}\frac{\partial f}{\partial E},$$

with θ the angle between $\mathbf{q}$ and $\mathbf{k}$. On changing the sum in Eq. (7.6.2) to an integral over energy and solid angle and substituting for $(u_k^2 - v_k^2)^2$

we find

$$P \propto D(\mu) \int \left(\frac{\hat{\mathcal{E}}}{E}\right)^2 q\hbar v_{\mathbf{k}} \cos\theta \, \frac{\hat{\mathcal{E}}}{E} \frac{\partial f}{\partial E} \delta\left(q\hbar v_{\mathbf{k}} \cos\theta \, \frac{\hat{\mathcal{E}}}{E} - \hbar\omega_{\mathbf{q}}\right) d(\cos\theta) \, d\hat{\mathcal{E}}$$

$$= D(\mu) \int \left(\frac{\hat{\mathcal{E}}}{E}\right)^2 |\cos\theta| \frac{\partial f}{\partial E} \delta\left(\cos\theta - \frac{\hbar\omega_{\mathbf{q}} E}{q\hbar v_{\mathbf{k}} \hat{\mathcal{E}}}\right) d(\cos\theta) \, d\hat{\mathcal{E}}$$

$$\simeq -2D(\mu) \int_{\Delta}^{\infty} \frac{\omega_{\mathbf{q}}}{q v_{\mathbf{k}}} \frac{\partial f}{\partial E} \, dE$$

$$\propto f(\Delta).$$

The ratio of the damping α_s in the superconducting state to that in the normal state is thus approximately

$$\frac{\alpha_s}{\alpha_n} = \frac{f(\Delta)}{f(0)}$$

$$= \frac{2}{\exp(\Delta/kT) + 1},$$

in good agreement with experiment.

7.7 The Meissner effect

In the Meissner effect the vanishing of the magnetic field inside a bulk superconductor must be attributed to the existence of an electric current flowing in the surface of the sample; the magnetic field due to this current must exactly cancel the applied field $\mathbf{H}_0$. We investigate this phenomenon within the BCS model by applying a weak magnetic field to a superconductor and calculating the resulting current density $\mathbf{j}$ to first order in the applied field. We avoid the difficulties of handling surface effects by applying a spatially varying magnetic field defined by a vector potential $\mathbf{A} = \mathbf{A}_{\mathbf{q}} e^{i\mathbf{q}\cdot\mathbf{r}}$ and examining the response $\mathbf{j} = \mathbf{j}_{\mathbf{q}} e^{i\mathbf{q}\cdot\mathbf{r}}$ in the limit that $\mathbf{q}$ becomes very small.

It had been realized for many years before the BCS theory that any first-order response of $\mathbf{j}$ to the vector potential $\mathbf{A}$ would lead to a Meissner effect. If we define the constant of proportionality between $\mathbf{j}$ and $\mathbf{A}$ so that

$$\mathbf{j} = -\left(\frac{c}{4\pi\lambda^2}\right)\mathbf{A}, \tag{7.7.1}$$

7.7 The Meissner effect

with c the speed of light, and then take the curl of this relation we obtain the *London equation*,

$$\nabla \times \mathbf{j} = -\left(\frac{c}{4\pi\lambda^2}\right)\mathbf{B}.$$

The use of the Maxwell equations

$$\nabla \times \mathbf{H} = \left(\frac{4\pi}{c}\right)\mathbf{j}; \quad \nabla \cdot \mathbf{B} = 0$$

and the fact that $\mathbf{B} = \mathbf{H}$ when all magnetization is attributed to the current $\mathbf{j}$ gives us the equation

$$\nabla^2 \mathbf{B} = \lambda^{-2}\mathbf{B}.$$

The solutions of this equation show a magnetic field that decays exponentially with a characteristic length λ, which can thus be identified with the penetration depth discussed in Section 7.1. In a normal metal λ would be infinite, and so from Eq. (7.7.1) we should expect to find no first-order term in an expansion of $\mathbf{j}$ in powers of $\mathbf{A}$.

We now proceed to perform the calculation of $\mathbf{j_q}$ in the BCS model. As we saw in Section 3.10, a magnetic field enters the Hamiltonian of a single electron as a perturbation term

$$\mathcal{H}_1 = \frac{e}{2mc}\left(\frac{e}{c}\mathbf{A}^2 - \mathbf{p}\cdot\mathbf{A} - \mathbf{A}\cdot\mathbf{p}\right),$$

relativistic effects and the spin of the electron being neglected. We drop the term in $\mathbf{A}^2$ and write $\mathcal{H}_1$ in the notation of second quantization as

$$\mathcal{H}_1 = \frac{e}{2mc}\sum_{\mathbf{k},\mathbf{k}',s}\langle\mathbf{k}|-\mathbf{p}\cdot\mathbf{A}-\mathbf{A}\cdot\mathbf{p}|\mathbf{k}'\rangle c^\dagger_{\mathbf{k},s}c_{\mathbf{k}',s}$$

$$= \frac{e\hbar}{2mc}\sum_{\mathbf{k},\mathbf{k}',s}\langle\mathbf{k}|-\mathbf{A}\cdot(\mathbf{k}+\mathbf{k}')|\mathbf{k}'\rangle c^\dagger_{\mathbf{k},s}c_{\mathbf{k}',s}.$$

When the vector potential is of the form $\mathbf{A_q}e^{i\mathbf{q}\cdot\mathbf{r}}$ we have

$$\mathcal{H}_1 = -\mu_B \mathbf{A_q} \cdot \sum_{\mathbf{k},s}(2\mathbf{k}-\mathbf{q})c^\dagger_{\mathbf{k},s}c_{\mathbf{k}-\mathbf{q},s} \qquad (7.7.2)$$

with $\mu_B = e\hbar/2mc$. The current density $e\mathbf{v}/\Omega$ is found from the continuity equation

$$\Omega^{-1}\nabla\cdot\mathbf{v} = -\frac{\partial\rho}{\partial t},$$

of which the Fourier transform is

$$i\Omega^{-1}\mathbf{q}\cdot\mathbf{v_q} = -\frac{\partial\rho_\mathbf{q}}{\partial t} = \frac{1}{i\hbar}[\mathcal{H},\rho_\mathbf{q}].$$

We thus find

$$\mathbf{q}\cdot\mathbf{j_q} = -\frac{e}{\hbar}[(\mathcal{H}_0+\mathcal{H}_1),\rho_\mathbf{q}].$$

The commutator of the zero-field Hamiltonian $\mathcal{H}_0$ with $\rho_\mathbf{q}$ was evaluated in Section 2.7, and $[\mathcal{H}_1,\rho_\mathbf{q}]$ can be similarly calculated. The result is that

$$\mathbf{j_q} = \sum_{\mathbf{k},s}\left[\frac{e\hbar}{2m\Omega}(2\mathbf{k}-\mathbf{q})c^\dagger_{\mathbf{k}-\mathbf{q},s}c_{\mathbf{k},s} - \frac{e^2\mathbf{A_q}}{mc\Omega}c^\dagger_{\mathbf{k},s}c_{\mathbf{k},s}\right]. \tag{7.7.3}$$

The effect of the magnetic field is to perturb the wavefunction of the superconductor from its initial state $|\Psi_0\rangle$ to the new state $|\Psi\rangle$ that is given to first order in $\mathcal{H}_1$ by the usual prescription

$$|\Psi\rangle = |\Psi_0\rangle + \frac{1}{\mathcal{E}_0-\mathcal{H}_0}\mathcal{H}_1|\Psi_0\rangle.$$

The current in the presence of the applied magnetic field will then be

$$\langle\mathbf{j_q}\rangle = \langle\Psi_0|\mathbf{j_q}|\Psi_0\rangle + \langle\Psi_0|\mathbf{j_q}\frac{1}{\mathcal{E}_0-\mathcal{H}_0}\mathcal{H}_1|\Psi_0\rangle$$
$$+ \langle\Psi_0|\mathcal{H}_1\frac{1}{\mathcal{E}_0-\mathcal{H}_0}\mathbf{j_q}|\Psi_0\rangle.$$

We substitute in this expression for $\mathbf{j_q}$ and $\mathcal{H}_1$ from Eqs. (7.7.2) and (7.7.3)

7.7 The Meissner effect

and drop terms of order A_q^2 to find

$$\langle j_q \rangle = -\frac{Ne^2 A_q}{mc\Omega} - \langle \Psi_0 | \sum_{k,s,k',s'} \frac{e\hbar\mu_B A_q}{2m\Omega} \cdot (2k' - q)$$

$$\times (2k - q) \left(c^\dagger_{k-q,s} c_{k,s} \frac{1}{\mathcal{E}_0 - \mathcal{H}_0} c^\dagger_{k',s'} c_{k'-q,s'} + c^\dagger_{k',s'} c_{k'-q,s'} \right.$$

$$\left. \times \frac{1}{\mathcal{E}_0 - \mathcal{H}_0} c^\dagger_{k-q,s} c_{k,s} \right) |\Psi_0\rangle. \tag{7.7.4}$$

In this expression the operator $\mathcal{H}_0$ is the BCS Hamiltonian in the absence of the magnetic field. To evaluate $\langle j_q \rangle$ it is then necessary to express the electron operators in terms of the quasiparticle operators that change $|\Psi_0\rangle$ into the other eigenstates of $\mathcal{H}_0$. We accordingly use Eqs. (7.3.3) and (7.3.4) to write

$$c^\dagger_{k-q\uparrow} c_{k\uparrow} = (u_{k-q} \gamma^\dagger_{k-q} + v_{k-q} \gamma_{-(k-q)})(u_k \gamma_k + v_k \gamma^\dagger_{-k})$$

and other similar expressions. There then follows a straightforward but lengthy set of manipulations in which the first step is the argument that whatever quasiparticles are created or destroyed when $c^\dagger_{k'\uparrow} c_{k'-q\uparrow}$ acts on Ψ_0 must be replaced when $c^\dagger_{k-q\uparrow} c_{k\uparrow}$ acts if the matrix element is not to vanish. Similar reasoning is applied to all such combinations of operators so that the summation over k and k' is reduced to a single sum. When q is very small u_{k-q} and v_{k-q} are replaced by u_k and v_k and a large amount of cancellation occurs. Eventually one reduces Eq. (7.7.4) to

$$\langle j_q \rangle = -\frac{2Ne\mu_B}{\hbar\Omega} A_q - \frac{e\hbar\mu_B}{2m\Omega}$$

$$\times A_q \cdot \sum_k 2k(u_k^2 + v_k^2) 4k \langle \Psi_0 | \left(\frac{m_k - m_{k+q}}{E_k - E_{k+q}} \right) |\Psi_0\rangle.$$

The summation over k is then replaced by an integral over energy multiplied by $D(\mu)/3$, the factor of $\frac{1}{3}$ arising from the angular integration. The expectation of the quasiparticle number operators m_k is replaced by their averages f_k, while for small q the ratio of the differences $m_k - m_{k+q}$ and $E_k - E_{k+q}$ becomes the ratio of their derivatives so that one has

$$\langle j_q \rangle = -\frac{Ne^2}{mc\Omega} A_q - \frac{e^2\hbar^2}{4m^2 c\Omega} A_q \frac{1}{3} D(\mu) 8k_F^2 \int \frac{\partial f_k}{\partial E_k} d\mathcal{E}.$$

Since $D(\mu)$ may be written as $3Nm/2\hbar^2 k_F^2$ this becomes

$$\langle \mathbf{j_q} \rangle = -\frac{Ne^2}{mc\Omega} \mathbf{A_q} \left(1 + \int \frac{\partial f}{\partial E} d\mathcal{E}\right)$$

$$= -\frac{Ne^2}{mc\Omega} \mathbf{A_q} \left(1 + 2 \int_\Delta^\infty \frac{E}{(E^2 - \Delta^2)^{1/2}} \frac{df}{dE} dE\right) \quad (7.7.5)$$

with $f(E)$ the Fermi–Dirac function.

That this expression is of the expected form can be verified by examining the cases where $\Delta = 0$ and where $T = 0$. At temperatures greater than T_c the gap parameter Δ vanishes and the integration may be performed immediately: the two terms in parentheses cancel exactly and no Meissner effect is predicted. At $T = 0$, on the other hand, the integral over E itself vanishes and Eq. (7.7.5) may be written as

$$\langle \mathbf{j_q} \rangle = -\frac{c}{4\pi\lambda^2} \mathbf{A_q}$$

if λ is chosen such that

$$\lambda = \sqrt{\frac{mc^2\Omega}{4\pi Ne^2}}$$

$$= \frac{c}{\omega_p}$$

with ω_p the plasma frequency. A Meissner effect is thus predicted with a penetration depth λ which is of the correct order of magnitude and which depends only on the electron density. At intermediate temperatures Eq. (7.7.5) predicts a penetration depth that increases monotonically with T and becomes infinite at T_c, in accord with experiment.

7.8 Tunneling experiments

Shortly after the development of the BCS theory it was discovered that a great deal of information about superconductors could be obtained by studying the current–voltage characteristics of devices composed of two pieces of metal separated by a thin layer of insulator. A typical device of this kind might consist of a layer of magnesium that had been exposed to the atmosphere to allow an insulating layer of MgO to form on its surface. After this layer had reached a thickness of about 20 Å a layer of lead might be deposited on top of the oxide, and the differential conductance, dI/dV, measured at some temperature low enough for the lead to be superconducting. The result would be of the form shown in Fig. 7.8.1; a sharp peak in dI/dV is observed when the potential difference V between the magnesium and the lead is such that the gap parameter Δ of the lead is equal to eV.

To discuss effects such as these we choose a simplified model in which the two halves of a box are separated by a thin potential barrier so that the variation of potential in the z-direction is as shown in Fig. 7.8.2. The

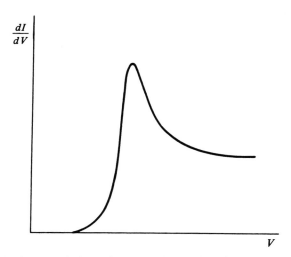

Figure 7.8.1. A device consisting of a normal metal and a superconductor separated by a thin layer of insulator has a differential conductance that exhibits a sharp peak at a voltage V such that eV is equal to the gap parameter Δ.

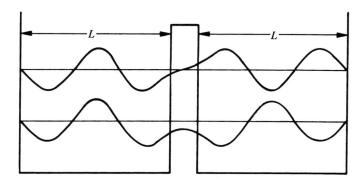

Figure 7.8.2. In the simplest model of a tunneling junction the two halves of a box are separated by a narrow potential barrier.

wavefunction of a single electron in this system will then be of the form

$$\psi = e^{ik_x x + ik_y y}\phi(z),$$

where, from the symmetry of the potential, $\phi(z)$ will be either an odd function ϕ_a of z or an even function ϕ_s. For each state ϕ_s, there will exist a corresponding ϕ_a that has just one more node (which will be located at the center of the barrier). The difference in energy of these states will be governed by the fact that the phase of the wavefunction at the edge of the barrier will be different in the two cases by some amount $\delta\theta$; if in the left-hand side of the box ϕ_s is of the form $\sin k_z z$ then ϕ_a will be of the form $\sin[(k_z + \delta\theta/L)z]$. The energy difference $2T$ will then be given by

$$2T = \frac{\hbar^2}{2m}\left[\left(k_z + \frac{\delta\theta}{L}\right)^2 - k_z^2\right], \tag{7.8.1}$$

and will be proportional to v_z, the component of the electron velocity perpendicular to the barrier, when $\delta\theta$ is small.

In the discussion of tunneling experiments it is more useful to work in terms of the wavefunctions $\phi_s \pm \phi_a$ rather than ϕ_s and ϕ_a themselves, as the sum or difference is largely localized on one side or the other of the barrier. We thus do not write the Hamiltonian of the system as

$$\mathcal{H} = \mathcal{E}_s c_s^\dagger c_s + \mathcal{E}_a c_a^\dagger c_a, \tag{7.8.2}$$

with

$$c_s^\dagger|0\rangle = |\phi_s\rangle, \quad c_a^\dagger|0\rangle = |\phi_a\rangle;$$

instead we form new fermion operators

$$c^\dagger = \frac{c_a^\dagger - c_s^\dagger}{\sqrt{2}}, \quad d^\dagger = \frac{c_a^\dagger + c_s^\dagger}{\sqrt{2}}.$$

In terms of these the Hamiltonian (7.8.2) becomes

$$\mathcal{H} = \mathcal{E}(c^\dagger c + d^\dagger d) + T(c^\dagger d + d^\dagger c),$$

where

$$\mathcal{E} = \tfrac{1}{2}(\mathcal{E}_a + \mathcal{E}_s).$$

7.8 Tunneling experiments

We now have a picture of two independent systems connected by a perturbation term. The Hamiltonian

$$\mathcal{H}_l = \sum_{\mathbf{k},s} \mathcal{E}_\mathbf{k} c^\dagger_{\mathbf{k}s} c_{\mathbf{k}s}$$

describes electrons in the left-hand side of the box, while

$$\mathcal{H}_r = \sum_{\mathbf{k},s} \mathcal{E}_\mathbf{k} d^\dagger_{\mathbf{k}s} d_{\mathbf{k}s}$$

describes those on the right. The perturbation

$$\mathcal{H}_t = \sum_{\mathbf{k},\mathbf{k}',s} T_{\mathbf{k}\mathbf{k}'}(c^\dagger_{\mathbf{k}s} d_{\mathbf{k}'s} + d^\dagger_{\mathbf{k}'s} c_{\mathbf{k}s}) \qquad (7.8.3)$$

acts to transfer electrons through the insulating barrier from one side to the other, the transition probability in lowest order depending on the square of the modulus of the matrix element T.

When a voltage V is applied across this device the energies of those states on the left of the barrier will be raised by an amount eV. Such an electron of kinetic energy $\mathcal{E}_\mathbf{k}$ can then only be elastically scattered by $\mathcal{H}_t$ to a state $\mathbf{k}'$ on the right of the barrier of kinetic energy $\mathcal{E}_\mathbf{k} + eV$. The net flow of electrons from left to right will then be proportional to

$$\sum_{\mathbf{k},\mathbf{k}',s} |T|^2 [f_\mathbf{k}(1-f_{\mathbf{k}'}) - f_{\mathbf{k}'}(1-f_\mathbf{k})] \delta(\mathcal{E}_{\mathbf{k}'} - \mathcal{E}_\mathbf{k} - eV),$$

from which the current I can be considered as governed by the relation

$$I \propto \int T^2 [f(\mathcal{E}_\mathbf{k}) - f(\mathcal{E}_\mathbf{k} + eV)] D(\mathcal{E}_\mathbf{k}) D(\mathcal{E}_\mathbf{k} + eV) \, d\mathcal{E}_\mathbf{k}$$

with T^2 the square of some average matrix element. The fact that T is, as we have seen in Eq. (7.8.1), proportional to $\partial \mathcal{E}/\partial k_z$, while

$$D(\mathcal{E}_\mathbf{k}) = \frac{(dN/dk)}{(d\mathcal{E}/dk)}$$

leads to a cancellation in the energy dependence of the integrand when $eV \ll \mu$. One is left with a relation of the form

$$I \propto \int [f(\mathcal{E}_k) - f(\mathcal{E}_k + eV)] d\mathcal{E}_k$$
$$= \int \{f(\mathcal{E}_k) - [f(\mathcal{E}_k) + eVf'(\mathcal{E}_k) + \tfrac{1}{2}(eV)^2 f''(\mathcal{E}_k) + \cdots]\} d\mathcal{E}_k$$
$$\simeq eV$$

and the device is predicted to obey Ohm's law. We might picture the calculation of this kind of tunneling as in Fig. 7.8.3, where the densities of states of the two halves of the device are plotted horizontally and energy is measured vertically. The density of *occupied* states per unit energy is the product $f(\mathcal{E}_k)D(\mathcal{E}_k)$, and is represented by the shaded areas. The tunneling current is then proportional to the difference in the shaded areas on the two sides.

We now consider how this calculation should be modified if the metal on the right of the barrier becomes superconducting, so that our system becomes a model of the Mg–MgO–Pb device whose conductance was shown in Fig. 7.8.1. Our first step must be to replace the electron operators $d_{ks}^\dagger$ and d_{ks} by the quasiparticle operators that act on the BCS state. We accordingly rewrite

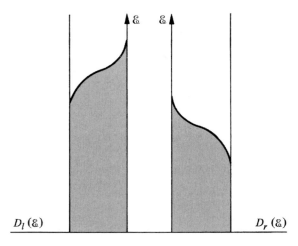

Figure 7.8.3. In this diagram the densities of states in the two halves of a tunneling junction are plotted to left and right. The densities of *occupied* states are found by multiplying these by the Fermi–Dirac distribution function to give the shaded areas. A voltage difference raises the energy of one side relative to the other and leads to a current flow proportional to the difference in areas.

the tunneling perturbation (7.8.3) as

$$\mathcal{H}_t = \sum_{\mathbf{k},\mathbf{k'}} T[c_{\mathbf{k}}^\dagger (u_{\mathbf{k'}}\gamma_{\mathbf{k'}} + v_{\mathbf{k'}}\gamma_{-\mathbf{k'}}^\dagger) + c_{-\mathbf{k}}^\dagger (u_{\mathbf{k'}}\gamma_{-\mathbf{k'}} - v_{\mathbf{k'}}\gamma_{\mathbf{k'}}^\dagger)$$
$$+ (u_{\mathbf{k'}}\gamma_{\mathbf{k'}}^\dagger + v_{\mathbf{k'}}\gamma_{-\mathbf{k'}})c_{\mathbf{k}} + (u_{\mathbf{k'}}\gamma_{-\mathbf{k'}}^\dagger - v_{\mathbf{k'}}\gamma_{\mathbf{k'}})c_{-\mathbf{k}}]. \tag{7.8.4}$$

This perturbation no longer simply takes an electron from one side of the barrier and replaces it on the other; the first term, for instance, creates an electron on the left of the barrier and either creates or destroys a quasiparticle excitation on the right. If the normal metal is at a voltage V relative to the superconductor then energy conservation demands that for the term $c_{\mathbf{k}}^\dagger \gamma_{\mathbf{k'}}$ to cause a real scattering process we must have $\hat{\mathcal{E}}_{\mathbf{k}} = E_{\mathbf{k'}} - eV$ (all energies now being measured relative to the chemical potential μ). For the term $c_{\mathbf{k}}^\dagger \gamma_{-\mathbf{k'}}^\dagger$ to cause scattering, on the other hand, we must have $\hat{\mathcal{E}}_{\mathbf{k}} = -E_{\mathbf{k'}} - eV$. Applying these arguments to all the terms in expression (7.8.4) we find that

$$I \propto \sum_{\mathbf{k},\mathbf{k'}} |T|^2 (u_{\mathbf{k'}}^2 \{f(\hat{\mathcal{E}}_{\mathbf{k}})[1 - f(E_{\mathbf{k'}})] - f(E_{\mathbf{k'}})[1 - f(\hat{\mathcal{E}}_{\mathbf{k}})]\} \delta(E_{\mathbf{k'}} - \hat{\mathcal{E}}_{\mathbf{k}} - eV)$$
$$+ v_{\mathbf{k'}}^2 \{f(\hat{\mathcal{E}}_{\mathbf{k}})f(E_{\mathbf{k'}}) - [1 - f(\hat{\mathcal{E}}_{\mathbf{k}})][1 - f(E_{\mathbf{k'}})]\} \delta(\hat{\mathcal{E}}_{\mathbf{k}} + E_{\mathbf{k'}} + eV))$$
$$= \sum_{\mathbf{k},\mathbf{k'}} |T|^2 \{u_{\mathbf{k'}}^2 [f(\hat{\mathcal{E}}_{\mathbf{k}}) - f(E_{\mathbf{k'}})] \delta(\hat{\mathcal{E}}_{\mathbf{k}} - E_{\mathbf{k'}} + eV)$$
$$+ v_{\mathbf{k'}}^2 [f(\hat{\mathcal{E}}_{\mathbf{k}}) - 1 + f(E_{\mathbf{k'}})] \delta(\hat{\mathcal{E}}_{\mathbf{k}} + E_{\mathbf{k'}} + eV)\}.$$

We change the sums over $\mathbf{k}$ and $\mathbf{k'}$ to integrals over $\hat{\mathcal{E}}_{\mathbf{k}}$ and $E_{\mathbf{k'}}$ by multiplying by $D_l(\hat{\mathcal{E}}_{\mathbf{k}})D_r(E_{\mathbf{k'}})$, with $D_r(E_{\mathbf{k'}})$ the effective density of states defined in Eq. (7.4.6). The first δ-function vanishes unless $E_{\mathbf{k'}} = \hat{\mathcal{E}}_{\mathbf{k}} + eV$, which from Eq. (7.5.7) means that $\mathbf{k'}$ must satisfy the condition

$$\hat{\mathcal{E}}_{\mathbf{k'}} = \pm[(\hat{\mathcal{E}}_{\mathbf{k}} + eV)^2 - \Delta^2]^{1/2}.$$

But since from Eqs. (7.3.9) and (7.3.12)

$$u^2(\hat{\mathcal{E}}_{\mathbf{k'}}) + u^2(-\hat{\mathcal{E}}_{\mathbf{k'}}) = 1,$$

the coefficients $u_{\mathbf{k'}}^2$ will vanish from the expression for the current when we perform the integration. Similar arguments applied to the second δ-function and the coefficients $v_{\mathbf{k}'}^2$ give us

$$I \propto \int T^2 D_l(\hat{\mathcal{E}}_{\mathbf{k}}) D_r(E = \hat{\mathcal{E}}_{\mathbf{k}} + eV)[f(\hat{\mathcal{E}}_{\mathbf{k}}) - f(\hat{\mathcal{E}}_{\mathbf{k}} + eV)] d\hat{\mathcal{E}}_{\mathbf{k}}$$

where use has been made of the fact that

$$f(-E) = 1 - f(E).$$

Because of the form of $D_r(E)$, which was shown in Fig. 7.4.1(b), the integral is no longer proportional to V, but leads to a current of the form shown in Figs. 7.8.4 and 7.8.1. For small V little current will flow, since $f(\hat{\mathcal{E}}_{\mathbf{k}} + eV)$ will differ appreciably from $f(\hat{\mathcal{E}}_{\mathbf{k}})$ only when $|\hat{\mathcal{E}}_{\mathbf{k}}| < kT$, and in this region the density of quasiparticle states, $D_r(E)$, vanishes. This is illustrated in Fig. 7.8.5, where the occupied states are again represented by shaded areas. It is only when $eV > \Delta$ that a large current flows, giving rise to the observed peak in dI/dV at this voltage.

One effect that does not emerge from this elementary calculation concerns the need to distinguish between the bare electron created by the operator $c_{\mathbf{k}}^{\dagger}$ and the electron in interaction with the phonon system. The canonical transformation of Section 6.5 is applied only to the electrons in the superconducting half of the device, and this must be allowed for in any more careful calculation of the tunneling current. One finds that the observed characteristics of the junction are greatly affected by the shape of the phonon

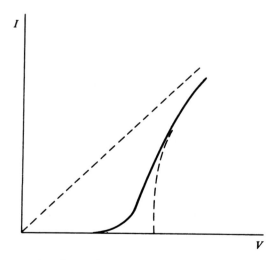

Figure 7.8.4. This current–voltage characteristic is the same as that shown in Fig. 7.8.1, and is typical of a superconductor–insulator–normal metal junction.

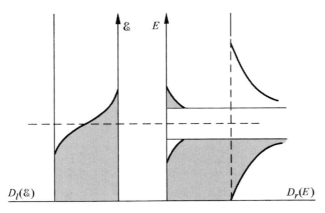

Figure 7.8.5. In this generalization of Fig. 7.8.3 the density of states on the right has been replaced by the effective density of states $D_r(E)$ of the superconductor.

density of states in strong-coupling superconductors, and may even give useful information about phonon modes in alloys that is not easily obtained by other means.

7.9 Flux quantization and the Josephson effect

The third possible type of tunnel junction is that in which the metal on both sides of the insulating barrier is superconducting. The calculation of the characteristics of this device is more difficult than the previous examples in that the total number of electrons is now not well defined on either side of the barrier; special operators must be defined that add pairs of electrons to one side or the other of the device. One result of such a calculation is that a current is predicted to flow that varies with applied voltage in the way shown in Fig. 7.9.1. and which is in accord with the simple interpretation of Fig. 7.9.2 for a device composed of two dissimilar superconductors. There is, however, also another type of current that may flow in such a device – a current associated with the tunneling through the barrier of *bound pairs* of electrons.

We can gain some insight into the nature of these currents by returning to a consideration of the effect of magnetic fields on the current carried by an electron. We have seen (in Eq. (3.10.8), for example) that the Hamiltonian of a free particle of mass m^* and charge e^* in a magnetic field $\mathbf{H} = \nabla \times \mathbf{A}$ is of the form

$$\mathcal{H} = \frac{(\mathbf{p} - e^*\mathbf{A}/c)^2}{2m^*}.$$

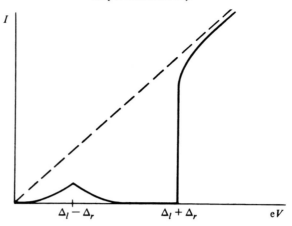

Figure 7.9.1. The current–voltage characteristic of a tunnel junction of two superconductors shows a discontinuity at a voltage V such that eV equals the sum of the gap parameters of the two materials.

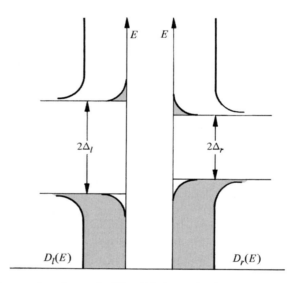

Figure 7.9.2. The results shown in Fig. 7.9.1 can be interpreted with the aid of this diagram of the effective densities of states.

If the vector potential were of the form

$$\mathbf{A} = (A, 0, 0) \tag{7.9.1}$$

with A a constant, then $\nabla \times \mathbf{A}$ would vanish and there would be no magnetic field. The eigenstates of $\mathcal{H}$ would be

$$\psi = \exp i(k_x x + k_y y + k_z z) \tag{7.9.2}$$

7.9 Flux quantization and the Josephson effect

with energies

$$\mathcal{E} = \frac{\hbar^2}{2m^*}\left[\left(k_x - \frac{e^*A}{\hbar c}\right)^2 + k_y^2 + k_z^2\right].$$

In the absence of applied fields we imposed periodic boundary conditions by stipulating that ψ should be equal at the points (x, y, z), $(x+L, y, z)$, $(x, y+L, z)$, and $(x, y, z+L)$. As long as no external electric or magnetic fields were acting, this was a permissible step whose only effect was to make the counting of states a little easier. If applied fields are present, however, this joining of the wavefunction on opposite faces of a cubical box must be treated more carefully. We cannot, for instance, apply a uniform electric field to the system and impose periodic boundary conditions, as such a procedure would result in the particle being continuously accelerated in one direction! In the particular case where only a vector potential $(A, 0, 0)$ acts on the particle we can commit the topological sin of imposing periodic boundary conditions if we remain awake to the physical implications. Because **A** is equal at the points (x, y, z) and $(x+L, y, z)$ the Hamiltonian itself is periodic in the x-direction. Periodicity of ψ in this direction then demands that $k_x = 2\pi n/L$ with n an integer. The contribution of the motion in the x-direction to the energy is thus

$$\mathcal{E}_x = \frac{\hbar^2}{2m^*}\left(\frac{2\pi n}{L} - \frac{e^*A}{\hbar c}\right)^2. \tag{7.9.3}$$

The motion in the y- and z-directions is unaffected by **A** and so we can also retain periodicity in these directions. The physical picture of this situation is that we have a closed loop of superconducting material, as shown in Fig. 7.9.3. Since **A** acts in the x-direction we physically join these two opposite faces of the material. Because the electrons occupy pair states we interpret the charge e^* of the current carriers as $2e$.

While we assume that the Meissner effect obliges the magnetic field to vanish within the superconductor (we assume dimensions large compared with the penetration depth) the magnetic flux Φ threading the ring does not necessarily vanish, since

$$\Phi = \int \nabla \times \mathbf{A} \cdot d\mathbf{S}$$
$$= \oint \mathbf{A} \cdot d\mathbf{l}$$
$$= AL, \tag{7.9.4}$$

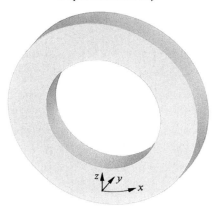

Figure 7.9.3. The free energy of a superconducting ring like this is a minimum when the magnetic flux threading it is quantized.

the line integral being taken around a closed path within the ring. From Eq. (7.9.3) the additional energy of the electron pair due to the vector potential will be

$$\Delta\mathcal{E}_x = \frac{\hbar^2}{2m^*}\left[-\frac{4\pi n}{L}\frac{2eA}{\hbar c} + \left(\frac{2eA}{\hbar c}\right)^2\right]. \tag{7.9.5}$$

If for every state of positive n the corresponding state of negative n is also occupied then there will be no contribution to the total energy of the electron gas from the term linear in A, and the total energy change will be

$$\Delta\mathcal{E}_{\text{total}} = \frac{2N\hbar^2}{m^*L^2}\left(\frac{e\Phi}{\hbar c}\right)^2.$$

The presence of a finite magnetic flux threading the ring thus increases the energy of the system.

If the flux is greater than $\pi\hbar c/2e$ it becomes energetically favorable for an electron pair with $k_x = -2\pi n/L$ to make a transition to a state for which $k_x = 2\pi(n+1)/L$. The total energy change due to A is then

$$\Delta\mathcal{E}_{\text{total}} = \frac{2N\hbar^2}{m^*L^2}\left(\pi - \frac{e\Phi}{\hbar c}\right)^2,$$

which takes on its minimum value of zero when $\Phi = \pi\hbar c/e$. These arguments can be extended to show that the minimum possible energy of the system is in

7.9 Flux quantization and the Josephson effect

fact a *periodic function* of Φ of the form shown in Fig. 7.9.4. There is thus a tendency for the magnetic flux Φ threading the ring to be *quantized* in units of the flux quantum ϕ_0, equal to $\pi\hbar c/e$. This is verified experimentally in a number of delicate measurements in which minute hollow cylinders of superconductor have been cooled down through T_c in the presence of applied magnetic fields. Subsequent measurements of the flux trapped in this way show values that unmistakably cluster around integral multiples of ϕ_0.

The current carried by a single pair of electrons will be proportional to $\hbar k_x - 2eA/c$, which from Eqs. (7.9.4) and (7.9.5) is in turn proportional to $\partial(\Delta\mathcal{E}_x)/\partial\Phi$. The total current I flowing in the ring is thus proportional to $d\mathcal{E}_{\text{total}}/d\Phi$; one differentiates the curve of Fig. 7.9.4 to find the sawtooth graph of Fig. 7.9.5. This graph leads to two results of great importance in the theory of superconductivity when we realize that this theory will apply not only to a closed ring of superconductor but also to a device such as that shown in Fig. 7.9.6, in which the ring is broken and a thin insulating layer inserted. Provided the gap is narrow enough that an appreciable number of electron pairs can tunnel through and that no flux quanta are contained in the gap itself, then the current will still be a periodic function of Φ with period ϕ_0.

The first experiment we consider is a measurement of the current I when the flux Φ is held constant. Since the electromotive force in the ring is equal

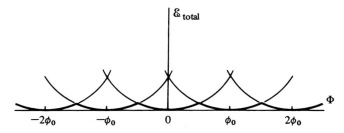

Figure 7.9.4. This construction shows the energy of a ring of superconductor at zero temperature to be a periodic function of the magnetic flux Φ, and to have minima at integral multiples of the flux quantum ϕ_0.

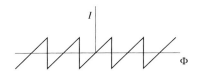

Figure 7.9.5. This sawtooth curve is the derivative of that shown in Fig. 7.9.4, and represents the current I.

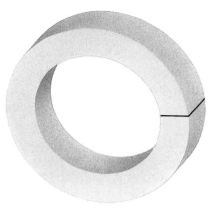

Figure 7.9.6. In this simple version of a Josephson junction a superconducting ring has been cut through at one place and has had a thin insulating layer inserted at the break.

to $-c^{-1}\,d\Phi/dt$ we know that any current flowing must do so in the absence of an electric field. If Φ is maintained at a value different from $n\phi_0$ it is thus energetically favorable for a small supercurrent to flow, and the I–V characteristic of Fig. 7.9.1 should exhibit a delta-function peak in the current at $V = 0$. This phenomenon is known as the *dc Josephson effect* (dc standing for "direct current").

The second experiment consists of causing the flux to increase uniformly with time. This constant value of $-c^{-1}\,d\Phi/dt$ represents a constant electromotive force V acting in the circuit. From Fig. 7.9.5 we then see that I will in fact alternate in sign as Φ is increased, the frequency of this alternating current being equal to the number of flux quanta introduced per unit time. The angular frequency ω of the current in this *ac Josephson effect* is thus

$$\omega = \frac{2eV}{\hbar},$$

which is of the order of magnitude of $\tfrac{1}{2}$ GHz per microvolt.

It is interesting to note that this is just the frequency difference that we would associate with the wavefunction of a single pair of electrons placed in this device; the time-dependent Schrödinger equation tells us that a particle of energy $\mathcal{E}$ has a wavefunction of the form

$$\Psi(\mathbf{r}, t) = \psi(\mathbf{r})e^{-i\mathcal{E}t/\hbar}$$

while the wavefunction for a particle of energy $\mathcal{E} + 2eV$ varies with time as $e^{-i(\mathcal{E}+2eV)t/\hbar}$. For a single particle such as this we can never expect to measure

7.10 The Ginzburg–Landau equations

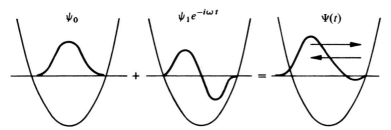

Figure 7.9.7. A wavefunction consisting of a linear combination of different harmonic-oscillator states describes a system in which both the number of particles and the phase of the wavefunction can be partially specified.

the phase of the wavefunction, as physically measurable quantities like the particle density always involve $|\psi|^2$. This may be thought of as another form of the Uncertainty Principle, and states that if the *number* of particles is known then the *phase* is unknown. We can, however, form a wavefunction of known phase if we form a wave packet of different numbers of particles. Adding equal amounts of the $n = 0$ ground state and the $n = 1$ state of a harmonic oscillator, for example, gives a wavefunction that oscillates back and forth with the classical oscillator frequency (Fig. 7.9.7). We thus look at the Josephson junction as a device in which the uncertainty in the number of pairs on each side of the barrier allows us to measure the relative phase of the wavefunction of the two parts of the system. This concept can be extended to the theory of superfluid liquid helium, where effects analogous to the ac Josephson effect have been detected when a pressure difference is maintained between two parts of a container separated by a small hole.

7.10 The Ginzburg–Landau equations

The destruction of superconductivity by a strong enough magnetic field may be understood by considering the energy associated with the Meissner effect. The expulsion of all magnetic flux from the interior of a long sample held parallel to an applied field $\mathbf{H}_0$ gives it an effective magnetization per unit volume of $-\mathbf{H}_0/4\pi$. Since the magnetic moment operator is given by $-\partial \mathcal{H}/\partial \mathbf{H}$ one finds the energy of the sample due to its magnetization to be

$$\mathcal{E}_M = \Omega \int_0^{\mathbf{H}_0} \frac{\mathbf{H}}{4\pi} \cdot d\mathbf{H}$$

$$= \frac{\Omega \mathbf{H}_0^2}{8\pi}.$$

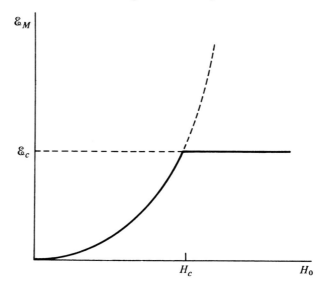

Figure 7.10.1. The magnetic energy of a superconductor varies as the square of the applied magnetic field H_0. For fields greater than H_c it is energetically favorable for the sample to make a transition to the normal state.

When this energy becomes greater than the condensation energy $\mathcal{E}_c$ then, as illustrated in Fig. 7.10.1, it is no longer energetically favorable for the sample to remain in the superconducting state. In a weak-coupling superconductor at zero temperature, for instance, the critical field H_c above which the metal could not remain uniformly superconducting would be given by

$$\Omega \frac{H_c^2}{8\pi} = \tfrac{1}{2} D(\mu)\Delta^2(0).$$

At finite temperatures, $H_c(T)$ can be found by identifying the condensation energy with the difference in Helmholtz energies of the normal and superconducting phases.

If the sample geometry is changed to that shown in Fig. 7.10.2 the magnetic energy is greatly increased, for there is now a large region outside the sample from which the applied field is partially excluded. Since the condensation energy remains constant the specimen starts to become normal at an applied field well below H_c. The sample does not become entirely nonsuperconducting, but enters what is known as the *intermediate state*, in which a large number of normal and superconducting regions exist side by side (Fig. 7.10.3). In this way the magnetic energy is greatly reduced while a

7.10 *The Ginzburg–Landau equations* 273

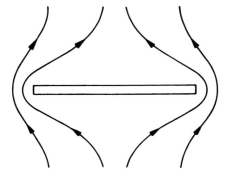

Figure 7.10.2. The magnetic field at the rim of this disc-shaped sample of superconductor is greater than the applied field; the superconductivity thus starts to be destroyed at weaker applied fields than is the case for a rod-shaped sample.

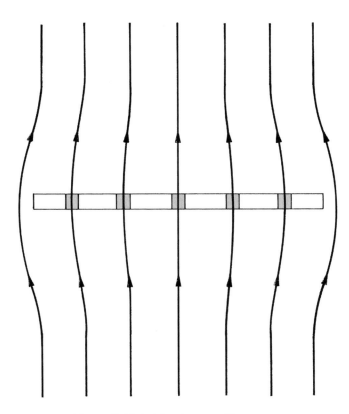

Figure 7.10.3. The intermediate state.

large part of the condensation energy is retained. This behavior is similar to that described at the beginning of this chapter, where a type II superconductor was defined. The difference is that while the type I superconductor only forms a mixture of normal and superconducting regions when the magnetic energy is magnified by geometric factors, the type II superconductor will enter the mixed state even when in the form of a long sample held parallel to a strong enough applied field. In the mixed state the distance between normal regions is typically 0.3 µm, which may be compared with the coarser structure of the intermediate state, which is characterized by distances typically of the order of 100 µm.

The analysis of these phenomena in terms of the BCS theory is very complicated. Because the magnetic field will be a rapidly varying function of position within the sample we must return to the arguments of Section 7.7 and ask for the response to the vector potential $\mathbf{A_q}e^{i\mathbf{q}\cdot\mathbf{r}}$ when $\mathbf{q}$ is no longer vanishingly small. There we saw in Eq. (7.7.5) that the current $\langle \mathbf{j_q} \rangle$ due to the vector potential could be expressed as the sum of two parts – a negative (or diamagnetic) part and a positive (or paramagnetic) part. In a superconductor at zero temperature the paramagnetic part vanished for $\mathbf{q} \to 0$, while in a normal metal it exactly cancelled the diamagnetic part and left no Meissner effect. A more careful investigation shows that as $\mathbf{q}$ is increased from zero the paramagnetic response of a superconductor also increases, until at large enough $\mathbf{q}$ it approximates the response of a normal metal. If one expresses this result in the form

$$\langle \mathbf{j_q} \rangle = -L_\mathbf{q} \mathbf{A_q} \qquad (7.10.1)$$

then one finds that $L_\mathbf{q}$ first becomes appreciably lower than its zero-q value when the approximation

$$E_{\mathbf{k+q}} - E_\mathbf{k} \simeq \mathbf{q} \cdot \frac{\partial E_\mathbf{k}}{\partial \mathbf{k}}$$

becomes invalid. This occurs when

$$\mathbf{q} \cdot \frac{\partial \mathcal{E}_\mathbf{k}}{\partial \mathbf{k}} \sim \Delta$$

which is equivalent to the condition

$$q\xi_0 \sim 1$$

with $\xi_0 = \hbar v_F/\pi\Delta$, the coherence length discussed in Section 7.1. The Fourier

7.10 The Ginzburg–Landau equations

transform of Eq. (7.10.1) is an equation of the form

$$\langle \mathbf{j}(\mathbf{r}) \rangle = -\int L(\mathbf{r}')\mathbf{A}(\mathbf{r}-\mathbf{r}')\,d\mathbf{r}' \qquad (7.10.2)$$

with $L(\mathbf{r}')$ a function that can be reasonably well approximated by $L_0 e^{-r'/\xi_0}$. An equation of this kind had been suggested by Pippard on macroscopic grounds before the development of the BCS theory.

When the coherence length ξ_0 is very much smaller than the penetration depth λ then $\mathbf{A}(\mathbf{r}-\mathbf{r}')$ will not vary appreciably within the range of $L(\mathbf{r}')$. Equation (7.10.2) then tells us that the current density at a point will be approximately proportional to the vector potential in the gauge that we have chosen, and the London equation (7.7.1) will be valid. Under these circumstances it will be energetically favorable for type II superconductivity to occur, as the magnetic field can penetrate the superconductor and reduce the magnetic energy without reducing the condensation energy. If, on the other hand, $\lambda \ll \xi_0$ then Eq. (7.10.2) predicts a nonlocal relation between the magnetic field and the current density. The wavefunction of the superconductor may then be modified up to a distance ξ_0 from the surface of the specimen, with a consequent reduction in the condensation energy. Because the magnetic field only penetrates a short distance λ, little magnetic energy is gained, and the sample will be a type I superconductor. We can illustrate this situation if we make the generalization that in a spatially varying magnetic field the gap parameter Δ should be considered as a function of position. The variation of $B(\mathbf{r})$ and $\Delta(\mathbf{r})$ at the boundary separating normal and superconducting regions of a metal can then be depicted as in Figs. 7.10.4(a) and 7.10.4(b) for type I and II superconductors respectively, where Δ_∞ is the value of Δ deep inside the superconductor.

A prediction of the geometry of the mixed state in a type II superconductor can be obtained by considering the Helmholtz energy $\mathcal{F}_S$ in a superconductor in which the gap parameter Δ varies with position. We saw in Eq. (7.4.2) that the condensation energy is proportional to $-\Delta^2$ in a homogeneous superconductor at zero temperature, and so it is natural to expect the dominant term in the Helmholtz energy to vary as $-\Delta^2(\mathbf{r})$ in the more general case. The fact that Eq. (7.10.2) shows the current (and hence the wavefunction) at a point $\mathbf{r}$ to depend on the conditions at points distant ξ_0 from $\mathbf{r}$ suggests that a term proportional to $[\xi_0 \nabla \Delta(\mathbf{r})]^2$ should be included in $\mathcal{F}_S$. In the presence of a magnetic field this contribution must be modified to preserve gauge invariance to $[\xi_0(\nabla - ie^*\mathbf{A}(\mathbf{r})/\hbar c)\Delta(\mathbf{r})]^2$, with e^* again chosen equal to $2e$; a magnetic energy density of $\mathbf{H}^2/8\pi$ must also be added.

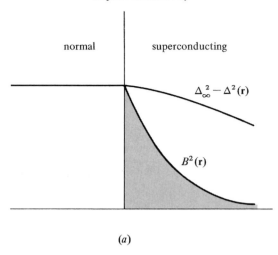

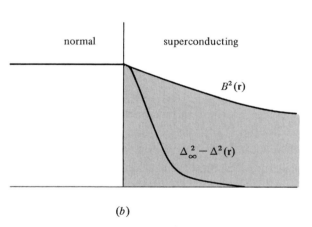

Figure 7.10.4. The area under the curve of $B^2(\mathbf{r})$ represents the magnetic energy gained in forming the normal–superconducting interface, while the area under the curve of $\Delta_\infty^2 - \Delta^2(\mathbf{r})$ is related to the condensation energy lost. In a type I material (a) the net energy is positive but in a type II superconductor (b) there is a net negative surface energy.

While the terms we have discussed so far present a fair approximation to $\mathcal{F}_S$, this form of the Helmholtz energy does not allow one to discuss which particular two-dimensional lattice of normal regions gives the mixed state of lowest Helmholtz energy; this is due to the linearity of the equation one obtains by trying to minimize $\mathcal{F}_S$ with respect to Δ. One must accordingly include the term of next highest power in Δ, which in this case will be proportional to Δ^4. The expression for the difference in Helmholtz energies between the superconducting and normal states will then be

$$\mathcal{F}_S - \mathcal{F}_N = \int \left\{ a|\Delta(\mathbf{r})|^2 + \frac{1}{2}b|\Delta(\mathbf{r})|^4 + \frac{\mathbf{H}^2}{8\pi} \right.$$
$$\left. + c\left|\left(\nabla - \frac{2ie\mathbf{A}}{\hbar c}\right)\Delta(\mathbf{r})\right|^2 \right\} d\mathbf{r} \quad (7.10.3)$$

with a, b, and c temperature-dependent constants and where the possibility of a complex Δ has been allowed for in the spirit of the discussion of the phase of the superconducting wavefunction given at the end of Section 7.9. Minimization of $\mathcal{F}_S$ with respect to both $\mathbf{A}(\mathbf{r})$ and $\Delta(\mathbf{r})$ yields the *Ginzburg–Landau equations*, which in principle enable one to calculate $\mathbf{A}$ and Δ as functions of position and of the constants a, b, and c. These constants can be evaluated in terms of the parameters of the homogeneous material when $\mathcal{F}_S$ is derived from the BCS Hamiltonian using some rather difficult procedures first applied by Gorkov. It is then possible to remove all but one of these parameters from appearing explicitly in the Helmholtz energy by working in terms of the dimensionless quantities in which $\mathbf{H}$ is measured in units proportional to H_c, distances are measured in units of the London penetration depth λ, and in which $\psi(\mathbf{r})$ is the ratio of $\Delta(\mathbf{r})$ to its value in the homogeneous material. The Ginzburg–Landau equations then take on the form

$$\left\{ 1 - |\psi|^2 - \left[\frac{1}{i\kappa}\nabla - \mathbf{A}\right]^2 \right\} \psi = 0 \quad (7.10.4)$$

$$|\psi|^2 \mathbf{A} + \nabla \times (\nabla \times \mathbf{A}) = \frac{1}{2i\kappa}(\psi^*\nabla\psi - \psi\nabla\psi^*) \quad (7.10.5)$$

where κ is λ/ξ_0. The fact that κ is the only parameter of the material to enter these equations confirms the idea that it is this quantity alone that determines whether a superconductor will be of type I or type II.

The detailed solution of Eqs. (7.10.4) and (7.10.5) leads to a variety of qualitatively correct predictions of the behavior of type II superconductors. The lower critical applied field H_{c1} at which it first becomes energetically favorable for a thread of normal material to exist in the superconductor can be shown to be less than H_c, the bulk critical field calculated from Δ, provided $\kappa > 1/\sqrt{2}$. Similarly the upper critical field H_{c2} below which a regular two-dimensional triangular lattice of threads of normal material

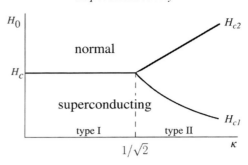

Figure 7.10.5. This diagram shows which of the three "phases" of a superconductor has the lowest free energy for a given Ginzburg–Landau parameter κ and applied field H_0.

forms the state of lowest energy can be calculated to be approximately $\sqrt{2}\kappa H_c$. This is illustrated in the "phase diagram" of Fig. 7.10.5, in which the state of the superconductor is shown as a function of the Ginzburg–Landau parameter κ and the applied magnetic field H_0. The existence of a superconducting surface layer in type II materials up to an applied field H_{c3} equal to about $1.7 H_{c2}$ may also be shown to follow from Eqs. (7.10.4) and (7.10.5).

7.11 High-temperature superconductivity

The technological promise of superconductivity is so rich that there has been a continual search for materials with higher critical temperatures. In high-voltage transmission lines, ohmic resistance losses consume about one percent of the power carried for every 100 km traveled, and even the best electric motors made of nonsuperconducting materials waste as heat several percent of the energy they use. If the need for refrigeration could be eliminated or reduced, the economic benefits flowing from the adoption of superconducting materials would be substantial.

In the decades before 1986, the progress made in this search was slow, for reasons that we can appreciate from the BCS expression (7.5.10) for the critical temperature. We might think that we could increase T_c by making the lattice more rigid, and thereby increasing ω_D. Unfortunately the interaction term V would simultaneously be reduced, as we see from Eq. (7.2.2), and little advantage would be gained. The remaining component of expression (7.5.10) is the density of states $D(\mu)$, and this can be increased by using transition metals and choosing the most favorable crystal structure. In this way a transition temperature of 23 K was achieved in Nb_3Ge. However, if we travel too far along the path of increasing the density of states we find a new obstacle. The electron–phonon interaction reduces the phonon frequencies through the process shown in Fig. 6.2.2, and a large electronic density of states enhances this effect. As we saw in the case of the Peierls transition described in Section 6.3, softening of the phonon modes eventually leads to a lattice instability.

Among other possible routes to high-temperature superconductivity we might look for pairing that involves a stronger force than arises from the electron–phonon interaction. One possibility could be a pairing between an electron and a hole in a material in which electrons and holes are present in equal numbers and with similar masses. However, systems of bound electron–hole pairs have a tendency to form spatially inhomogeneous structures in which the electric charge density or spin density varies periodically in space. Superconductivity is not favored in these spatially modulated structures.

It was thus a delightful surprise when, in 1986, Bednorz and Müller discovered that superconductivity occurred at 35 K in a ceramic compound of lanthanum, barium, copper, and oxygen. This delight was magnified the following year with the revelation by Chu and Wu that replacement of the lanthanum by yttrium raised T_c to 92 K. This allowed the use of liquid nitrogen, which boils at 77 K, for cooling. Many other ceramic compounds were subsequently found to be high-temperature superconductors. A feature common to most, but not all, of these was that they contained parallel layers of CuO_2, each layer separated from its neighbors by ionizable metallic atoms. Another characteristic was the increase in T_c that occurred when each single CuO_2 layer was replaced by two contiguous layers, and then by three, although a further increase to four layers produced a decrease in T_c.

The original motivation of Bednorz and Müller for looking at conducting oxides was the thought that the electron–phonon interaction could be strengthened if the copper ions were in nearly unstable positions in the lattice. The *Jahn–Teller* theorem states that, except in some special circumstances, a degeneracy in electronic states can be lifted by a distortion in which an atom moves to a less symmetric position. Since lifting the degeneracy moves the energies of the states apart, the lowest-lying state will be reduced in energy, and the system is unstable. The existence of nearly degenerate states on the copper ions could thus enhance the electron–lattice interaction.

The structure of the CuO_2 planes is as shown in Fig. 7.11.1, with each copper atom having four oxygen neighbors. The oxygen atoms are hungry for electrons, and succeed in removing not only the single electron in the outer 4s state of copper, but also one of the electrons from the filled 3d-shell. The copper is thus doubly ionized to Cu^{2+}. One might then expect to have a

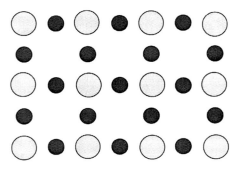

Figure 7.11.1. In many high-temperature superconductors there are planes of copper atoms (large spheres) and oxygen atoms (small spheres) arranged in this way.

metal, since the uppermost band should be only half filled. However, the mutual Coulomb repulsion of the 3d electrons now intervenes and splits this band. It does so by arranging the spins of the holes on the Cu sites in an antiferromagnetic ordering, in which the spin on every Cu site is aligned antiparallel to that of each of its four Cu neighbors. The size of the unit cell of the CuO_2 lattice is effectively doubled, and so the size of the first Brillouin zone is correspondingly halved. An energy gap now appears between the new first Brillouin zone and the new second Brillouin zone. The number of holes is now exactly equal to the capacity of the first Brillouin zone, making the compound an insulator. A filled valence band, consisting of states predominantly located on the oxygens, is separated by an appreciable gap from the empty conduction band in which electrons, had they been present, would be mostly concentrated on the copper sites. This is the situation for the undoped material, which is commonly referred to as a *Mott insulator*.

The antiferromagnetism, like all types of ordering, can be destroyed if the temperature is raised sufficiently. It can also be destroyed by increasing the number of holes in the CuO_2 layer, as this eliminates the one-to-one correspondence between holes and lattice sites. The doping process that adds extra holes can be achieved by either replacing some of the atoms with those of lower valency or modifying the amount of oxygen in the compound. A doping level (measured as the number of extra holes per CuO_2 unit) of a few percent is generally sufficient to destroy antiferromagnetism. As the number of holes is increased beyond this point one enters what is known as the *pseudogap* region. Here the electronic specific heat does not fit a picture of electrons as a gas or liquid of quasiparticles, but shows some traits characteristic of superconductors. A further increase in doping takes us into the superconducting state, which is often optimized at a doping level of about

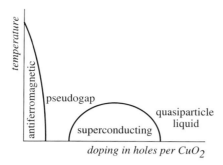

Figure 7.11.2. The generic phase diagram of a typical high-temperature superconductor.

20 percent. The generic phase diagram for high-temperature superconductors thus appears as in Fig. 7.11.2.

The existence of *planes* of copper and oxygen atoms suggests that the lower effective dimensionality of the system may be the factor that overcomes the difficulties that had been predicted in reaching high transition temperatures. In fact, there had been suggestions as early as 1964 that one-dimensional organic molecules might be a possible route to room-temperature superconductivity. One aspect of low effective dimensionality is that the screening of the Coulomb interaction is reduced. This has the two opposite effects of increasing the strength of the electron–lattice interaction, which should favor superconductivity, and increasing the mutual repulsion of the electrons, which should inhibit the formation of Cooper pairs and thus disfavor superconductivity.

The response of the system to this combination of a stronger indirect force of attraction (which may be due to effects other than the electron–phonon interaction) and a stronger direct repulsion could be a contributory factor in encouraging the electrons (or holes) to form Cooper pairs having d-wave symmetry rather than the isotropic s-wave symmetry of the elemental superconductors. The spins would again be antiparallel, but now the wavefunction would vanish as the particle separation tends to zero, reducing the energy cost of the strong short-range repulsion. Convincing experimental support for d-wave pairing has been found in a number of ingenious experiments. One of the most impressive pieces of evidence is seen in the beautiful picture on the cover of this book, which shows a scanning-tunneling-microscope image of the surface of a high-temperature superconductor. One of the copper atoms in a CuO_2 plane just below the surface has been replaced by a zinc atom. Scattering of the d-wave quasiparticles from the zinc atom results in a distribution that reflects the structure of the superconducting state. In d-wave superconductors the gap parameter Δ is no longer a constant, but is anisotropic, falling to zero in the four azimuthal directions where $\cos 2\phi$ vanishes. These are the directions in which the impurity wavefunction, and the tunneling currents associated with it, extend farthest.

While the standard BCS analysis can be readily modified to handle d-wave pairing, there are many other pieces of evidence to suggest that the BCS formalism cannot be applied to high-temperature superconductors without more drastic modification. Examples include the temperature dependence of the specific heat, the very small coherence length, and the large value of the ratio $2\Delta(0)/kT_c$. In Section 7.5 the BCS prediction for this ratio was shown to be 3.50, but the experimental result for $HgBa_2Ca_2Cu_3O_8$ is 12.8 in the direction that maximizes Δ. It is clear that superconductivity is a phenomenon that can appear in a number of forms, and whose complete explanation requires a correspondingly wide range of theoretical approaches.

Problems

7.1 Would you characterize the BCS theory as a mean field theory in the sense discussed at the beginning of Section 3.11? If so, at what stage is the mean field approximation introduced?

7.2 An alternative approach to the finite-temperature theory of Section 7.5 involves minimizing the Helmholtz energy, $\langle \mathcal{H}_{BCS}\rangle - TS$, with respect to both $x_\mathbf{k}$ and $f_\mathbf{k}$, with the entropy given by

$$S = -2k \sum_\mathbf{k} [f_\mathbf{k} \ln f_\mathbf{k} + (1-f_\mathbf{k})\ln(1-f_\mathbf{k})].$$

Show that this approach leads to the same expression for $\Delta(T)$ as given in Eq. (7.5.8).

7.3 Anderson's pseudospin formulation of the BCS theory starts by transforming from the pair creation operators $b_\mathbf{k}^\dagger = c_{\mathbf{k}\uparrow}^\dagger c_{-\mathbf{k}\downarrow}^\dagger$ to operators defined by $2s_x(\mathbf{k}) = b_\mathbf{k}^\dagger + b_\mathbf{k}$; $2s_y(\mathbf{k}) = i(b_\mathbf{k}^\dagger - b_\mathbf{k})$; $2s_z(\mathbf{k}) = 1 - n_{\mathbf{k}\uparrow} - n_{-\mathbf{k}\downarrow}$. Verify that in units where $\hbar = 1$ these operators have the commutation properties of spins as defined by Eq. (3.10.11).

7.4 Verify that when the transformation of Problem 7.3 is substituted in the BCS Hamiltonian (7.2.4) one finds a result of the form

$$\mathcal{H}_{BCS} = -\sum_\mathbf{k} \mathbf{H}(\mathbf{k}) \cdot \mathbf{s}(\mathbf{k})$$

where **H** is an "effective pseudomagnetic field" given by

$$\mathbf{H}(\mathbf{k}) = \left(\sum_{\mathbf{k}'} V_{\mathbf{k}\mathbf{k}'} s_x(\mathbf{k}'), \sum_{\mathbf{k}'} V_{\mathbf{k}\mathbf{k}'} s_y(\mathbf{k}'), 2\hat{\mathcal{E}}_\mathbf{k} \right).$$

[The energy of this system can now be minimized by arguing in analogy with the theory of domain walls in ferromagnets.]

7.5 Prove that no material can have a macroscopic magnetic susceptibility more negative than $-1/4\pi$. [Hint: consider a long cylinder held parallel to an applied field, and plot B_z as a function of position.]

7.6 Show that the fourth-order off-diagonal terms omitted from Eq. (7.3.8) have a negligible expectation value in the BCS ground state.

7.7 How does the electronic specific heat of a superconductor vary with temperature T as T tends to zero?

7.8 Provide the missing steps in the calculation that leads from Eq. (7.7.4) to Eq. (7.7.5).

7.9 The operator that describes the spin magnetic moment of the electron gas is

$$M = \mu_B \sum_\mathbf{k} (n_{\mathbf{k}\uparrow} - n_{\mathbf{k}\downarrow}).$$

In the ground state of the BCS superconductor, the magnetization vanishes. What is the minimum energy needed to create a state for which the expectation value of M is $2\mu_B$?

7.10 An exception to our general statement that an electron annihilation operator $c_\mathbf{k}$ only acts in partnership with a creation operator $c_\mathbf{k}^\dagger$ arises if we introduce a positron to the system. Consider a positron at rest in a BCS superconductor at zero temperature annihilating with an electron to produce two photons of total momentum $\hbar\mathbf{k}_1$. What is the difference between the total energy they would have in the superconductor and that in a normal metal?

7.11 Now raise the temperature in problem 7.10 to T_1, so that Δ is reduced to $\Delta(T_1)$. There are now two possible answers to the problem. What are they, and what is the ratio of the probabilities of occurrence of these two answers?

7.12 Calculate the electronic specific heat $C(T)$ for a BCS superconductor in the limit as $T \to T_c$, and hence find the ratio of $C(T_c - \epsilon)$ to $C(T_c + \epsilon)$ as $\epsilon \to 0$.

7.13 How does the penetration depth λ vary with temperature in a BCS superconductor as $T \to T_c$? Express your answer in terms of T, T_c, and $\lambda(T = 0)$.

Chapter 8

Semiclassical theory of conductivity in metals

8.1 The Boltzmann equation

We now have at our command many of the ingredients of the theory of the conduction of heat and electricity. In Section 3.9 we considered the heat current operator for phonons in a lattice, and in Section 4.6 we calculated the velocity of Bloch electrons and their dynamics in applied fields. The missing ingredients of the theory of the transport of heat or electricity, however, are the statistical concepts necessary to understand such irreversible processes. In this chapter we shall adopt the simplest attitude to these statistical problems, and begin with a discussion of the *probable occupation number* of a given phonon mode or Bloch state.

Let us start by considering a system of independent phonons or electrons. We know that we can define operators n_q and n_k whose eigenvalues are integers. If, for instance, there are three phonons of wavenumber $\mathbf{q}'$ present then the expectation value $\langle n_{\mathbf{q}'} \rangle$ of the operator $n_{\mathbf{q}'}$ will be equal to three. If the crystal is not in an eigenstate of the Hamiltonian, however, then $\langle n_{\mathbf{q}'} \rangle$ may take on some nonintegral value. If we wish to discuss the thermal conductivity of the lattice we should have to interpret the idea of a temperature gradient, and this must certainly involve some departure from the eigenstates of the lattice. We are thus obliged to consider linear combinations of different eigenstates as describing, for example, a lattice with a temperature gradient. Because there will be many different combinations of eigenstates that all give the appearance of a crystal with a temperature gradient we shall only have a very incomplete knowledge of the state of any particular crystal. We can, however, discuss the average results f_q and f_k that we expect to find when we make measurements of n_q and n_k, respectively, and can proceed to use semi-intuitive methods to derive equations that will govern their variation in time.

If we are to include the possibility of temperature gradients we shall have to allow $f_\mathbf{q}$ and $f_\mathbf{k}$ to be functions of position as well as of wavenumber. This appears self-contradictory, in that phonon and Bloch states are not localized, and so one cannot attempt to specify even the approximate location of a particle if one knows its wavenumber exactly. We must thus sacrifice some precision in determining the wavenumber if we allow $f_\mathbf{q}$ and $f_\mathbf{k}$ to vary with position. This need not be a serious limitation provided we restrict ourselves to slowly varying functions in r-space. If for instance we consider a temperature that varies as

$$T(\mathbf{r}) = T_0 + T_1 \cos(\mathbf{q}_T \cdot \mathbf{r})$$

then we must make $\mathbf{q}_T$ small enough that the concept of a local temperature is valid. We must certainly have a high probability that all the excitations have many collisions in traveling a distance q_T^{-1} so that they are properly "thermalized."

We now consider the *equation of continuity* for the function f, which can be the probable occupation number of either electrons or phonons. This relates the rate of change of f in the absence of collisions to the number of particles leaving an element of volume of six-dimensional k-r space. By a simple generalization of the usual three-dimensional version we find

$$\frac{\partial f}{\partial t} = -\frac{\partial}{\partial \mathbf{r}} \cdot \left(f \frac{d\mathbf{r}}{dt}\right) - \frac{\partial}{\partial \mathbf{k}} \cdot \left(f \frac{d\mathbf{k}}{dt}\right).$$

Upon differentiation this becomes

$$\frac{\partial f}{\partial t} = -\mathbf{v} \cdot \frac{\partial f}{\partial \mathbf{r}} - \frac{d\mathbf{k}}{dt} \cdot \frac{\partial f}{\partial \mathbf{k}} - f\left[\frac{\partial}{\partial \mathbf{r}} \cdot \mathbf{v} + \frac{\partial}{\partial \mathbf{k}} \cdot \frac{d\mathbf{k}}{dt}\right]. \tag{8.1.1}$$

The term in brackets vanishes both for phonons and for Bloch electrons in applied fields. For phonons this is obvious, since $\mathbf{v} = \partial \omega / \partial \mathbf{q}$, and does not depend on position while $\mathbf{k} = \mathbf{q}$, the wavenumber, and is constant. For electrons

$$\frac{d\mathbf{k}}{dt} = \frac{e}{\hbar c}(\mathbf{E}c + \mathbf{v} \times \mathbf{H}),$$

and the fact that $\hbar \mathbf{v} = \partial \mathcal{E}/\partial \mathbf{k}$ ensures this result. What remains of Eq. (8.1.1) constitutes the *Liouville equation*. We now specialize to consider only the *steady state*, in which $\partial f/\partial t$ vanishes, but we add to the right-hand side a

8.1 The Boltzmann equation

term to allow for changes in f due to collisions. We then have a form of the *Boltzmann equation*, which states

$$\left(\frac{\partial f}{\partial t}\right)_{\text{collisions}} - \mathbf{v} \cdot \frac{\partial f}{\partial \mathbf{r}} - \frac{d\mathbf{k}}{dt} \cdot \frac{\partial f}{\partial \mathbf{k}} = 0.$$

The spatial variation in f may be attributed solely to a variation in temperature, giving

$$\frac{\partial f}{\partial \mathbf{r}} = \frac{\partial f}{\partial T} \nabla T.$$

We then have the following equations for the phonon and electron systems:

$$\left(\frac{\partial f_\mathbf{q}}{\partial t}\right)_{\text{collisions}} = \frac{\partial \omega}{\partial \mathbf{q}} \cdot \nabla T \left(\frac{\partial f_\mathbf{q}}{\partial T}\right) \qquad (8.1.2)$$

$$\left(\frac{\partial f_\mathbf{k}}{\partial t}\right)_{\text{collisions}} = \frac{1}{\hbar} \frac{\partial \mathcal{E}}{\partial \mathbf{k}} \cdot \nabla T \left(\frac{\partial f_\mathbf{k}}{\partial T}\right) + \frac{e}{\hbar c} (E c + \mathbf{v} \times \mathbf{H}) \cdot \left(\frac{\partial f_\mathbf{k}}{\partial \mathbf{k}}\right). \qquad (8.1.3)$$

In general these two equations are coupled, since in the presence of the electron–phonon interaction the scattering probability for the electrons depends on the phonon occupation function, and conversely.

Before proceeding to an investigation of the solution of these equations in particular circumstances, we remind ourselves of the fact that applied electric fields and temperature gradients are usually rather small. If one is measuring thermal conductivity, for example, one usually wishes to determine this quantity as a function of temperature, and no accurate result could be obtained if the two ends of the sample were at widely different temperatures. The resulting currents of heat and electricity are consequently linear in $\mathbf{E}$ and ∇T. This leads us to a *linearization* of Eqs. (8.1.2) and (8.1.3). We expand f in powers of $\mathbf{E}$ and ∇T, and keep only the first two terms so that

$$f \simeq f^0 + f^1.$$

Here f^0 will represent the distribution in the equilibrium situation, and will be given by the Bose–Einstein function for phonons and the Fermi–Dirac function for electrons. By definition $(\partial f^0/\partial t)_{\text{collisions}}$ vanishes. We also note that for electrons f^0 depends on $\mathbf{k}$ only through the function $\mathcal{E}_\mathbf{k}$, and so

$$\frac{\partial f_\mathbf{k}^0}{\partial \mathbf{k}} = \frac{\partial f^0}{\partial \mathcal{E}} \frac{\partial \mathcal{E}}{\partial \mathbf{k}}.$$

It thus follows that the term $\mathbf{v}\times\mathbf{H}\cdot(\partial f^0/\partial\mathbf{k})$ also vanishes. On neglect of terms of second order of smallness we are then left with

$$\left(\frac{\partial f_{\mathbf{q}}^1}{\partial t}\right)_{\text{collisions}} = \mathbf{v}_{\mathbf{q}}\cdot\nabla T\left(\frac{\partial f_{\mathbf{q}}^0}{\partial T}\right) \tag{8.1.4}$$

$$\left(\frac{\partial f_{\mathbf{k}}^1}{\partial t}\right)_{\text{collisions}} = \frac{e}{\hbar c}\mathbf{v}_{\mathbf{k}}\times\mathbf{H}\cdot\frac{\partial f_{\mathbf{k}}^1}{\partial \mathbf{k}} + \mathbf{v}_{\mathbf{k}}\cdot\left(\nabla T\frac{\partial f^0}{\partial T} + e\mathbf{E}\frac{\partial f^0}{\partial\mathcal{E}}\right). \tag{8.1.5}$$

This is as far as we can conveniently go in simplifying the Boltzmann equations without specializing to a consideration of specific models and situations. We shall now proceed to investigate a few of these and attempt to see how some of the simple geometrical ideas such as the mean free path can be rescued from the complexity of the Boltzmann equation.

8.2 Calculating the conductivity of metals

We now specialize to consider the electrical conductivity of a metal in which the electrons are scattered elastically by a random array of n impurities. We argue that if the positions of the scattering centers are not correlated in any way, then we can neglect coherent scattering by the array of impurities as a whole, and assume that the scattering probability between Bloch states is just the scattering probability due to a single impurity multiplied by n. This argument will be valid if the density of the impurities is low, as we can then consider our calculation as finding the first term of an expansion of the resistivity ρ in powers of the impurity density n. While the function $\rho(n)$ turns out not to be analytic at $n=0$, the derivative $d\rho/dn$ does exist at this point, and so for small enough n we can write the scattering probability following Eq. (4.7.3) as

$$Q(\mathbf{k},\mathbf{k}') = n\left(\frac{2\pi}{\hbar}\right)|T_{\mathbf{k}\mathbf{k}'}|^2\delta(\mathcal{E}_{\mathbf{k}} - \mathcal{E}_{\mathbf{k}'}).$$

The rate of change of $f_{\mathbf{k}}$ will be the average net number of electrons entering state $\mathbf{k}$ from all other states $\mathbf{k}'$. Making allowances for the Exclusion Principle, which will prevent an electron from entering the state $\mathbf{k}$ if it is already occupied, we find

$$\left(\frac{\partial f_{\mathbf{k}}}{\partial t}\right)_{\text{collisions}} = \sum_{\mathbf{k}'}[f_{\mathbf{k}'}Q(\mathbf{k},\mathbf{k}')(1-f_{\mathbf{k}}) - f_{\mathbf{k}}Q(\mathbf{k}',\mathbf{k})(1-f_{\mathbf{k}'})].$$

8.2 Calculating the conductivity of metals

It was shown in Section 4.7 that $Q(\mathbf{k}, \mathbf{k}')$ was equal to $Q(\mathbf{k}', \mathbf{k})$, and so

$$\left(\frac{\partial f_\mathbf{k}}{\partial t}\right)_{\text{collisions}} = \sum_{\mathbf{k}'} Q(\mathbf{k}, \mathbf{k}')(f_{\mathbf{k}'} - f_\mathbf{k}).$$

Because energy is conserved $Q(\mathbf{k}, \mathbf{k}')$ vanishes unless $\mathcal{E}(\mathbf{k}) = \mathcal{E}(\mathbf{k}')$, and as $f_\mathbf{k}^0$ depends only on $\mathcal{E}(\mathbf{k})$ we immediately verify that

$$\sum_{\mathbf{k}'} Q(\mathbf{k}, \mathbf{k}')(f_{\mathbf{k}'}^0 - f_\mathbf{k}^0) = 0.$$

We may thus write

$$\left(\frac{\partial f_\mathbf{k}}{\partial t}\right)_{\text{collisions}} = \left(\frac{\partial f_\mathbf{k}^1}{\partial t}\right)_{\text{collisions}}$$

$$= \sum_{\mathbf{k}'} Q(\mathbf{k}, \mathbf{k}')(f_{\mathbf{k}'}^1 - f_\mathbf{k}^1).$$

In the absence of a temperature gradient the linearized Boltzmann equation (8.1.5) then becomes

$$\sum_{\mathbf{k}'} Q(\mathbf{k}, \mathbf{k}')(f_{\mathbf{k}'}^1 - f_\mathbf{k}^1) - \frac{e}{\hbar c} \mathbf{v}_\mathbf{k} \times \mathbf{H} \cdot \frac{\partial f_\mathbf{k}^1}{\partial \mathbf{k}} = e\mathbf{E} \cdot \mathbf{v}_\mathbf{k} \frac{\partial f_\mathbf{k}^0}{\partial \mathcal{E}}. \quad (8.2.1)$$

The changes, $f_\mathbf{k}^1$, that occur in f due to the action of the electric field are, of course, confined to the region of k-space in the vicinity of the Fermi surface. This is evident from the presence of the factor $\partial f_\mathbf{k}^0/\partial \mathcal{E}$ on the right-hand side of (8.2.1). For Fermi–Dirac statistics

$$\frac{\partial f_\mathbf{k}^0}{\partial \mathcal{E}} = \frac{\partial}{\partial \mathcal{E}}\left(\frac{1}{\exp[(\mathcal{E}_\mathbf{k} - \mu)/kT] + 1}\right)$$

$$= -\frac{\exp[(\mathcal{E}_\mathbf{k} - \mu)/kT]}{kT(\exp[(\mathcal{E}_\mathbf{k} - \mu)/kT] + 1)^2},$$

which may be rewritten in the form

$$\frac{\partial f_\mathbf{k}^0}{\partial \mathcal{E}} = -\frac{f_\mathbf{k}^0(1 - f_\mathbf{k}^0)}{kT}.$$

In the limit of low temperatures

$$\frac{\partial f_{\mathbf{k}}^0}{\partial \mathcal{E}} \to -\delta(\mathcal{E}_{\mathbf{k}} - \mathcal{E}_F), \quad (8.2.2)$$

the Fermi energy $\mathcal{E}_F$ being equal to the chemical potential μ at zero temperature. It is then clear that $f_{\mathbf{k}}^1$ is too rapidly varying a function for convenience in computation. We also note that the statement that $f_{\mathbf{k}}^1$ is linear in $\mathbf{E}$ means that the total change in $f_{\mathbf{k}}$ due to the field (E_x, E_y, E_z) is the sum of the changes $f_{\mathbf{k},x}^1, f_{\mathbf{k},y}^1$, and $f_{\mathbf{k},z}^1$ that would be produced by the three components of $\mathbf{E}$ acting separately. It is then convenient to define quantities $\Lambda_{\mathbf{k},x}, \Lambda_{\mathbf{k},y}$, and $\Lambda_{\mathbf{k},z}$ that satisfy

$$f_{\mathbf{k},x}^1 = -eE_x \Lambda_{\mathbf{k},x} \frac{\partial f_{\mathbf{k}}^0}{\partial \mathcal{E}}$$

and similarly for $f_{\mathbf{k},y}^1$ and $f_{\mathbf{k},z}^1$. More briefly we state that a vector $\Lambda_{\mathbf{k}}$ exists such that

$$f_{\mathbf{k}}^1 = -e\mathbf{E} \cdot \Lambda_{\mathbf{k}} \frac{\partial f_{\mathbf{k}}^0}{\partial \mathcal{E}} \quad (8.2.3)$$

and which is independent of $\mathbf{E}$. Equation (8.2.1) then becomes

$$-e\mathbf{E} \cdot \sum_{\mathbf{k}'} Q(\mathbf{k}, \mathbf{k}') \left(\Lambda_{\mathbf{k}'} \frac{\partial f_{\mathbf{k}'}^0}{\partial \mathcal{E}} - \Lambda_{\mathbf{k}} \frac{\partial f_{\mathbf{k}}^0}{\partial \mathcal{E}} \right)$$

$$+ \frac{e}{\hbar c} \left(\mathbf{v}_{\mathbf{k}} \times \mathbf{H} \cdot \frac{\partial}{\partial \mathbf{k}} \right) e\mathbf{E} \cdot \Lambda_{\mathbf{k}} \frac{\partial f_{\mathbf{k}}^0}{\partial \mathcal{E}} = e\mathbf{E} \cdot \mathbf{v}_{\mathbf{k}} \frac{\partial f_{\mathbf{k}}^0}{\partial \mathcal{E}}.$$

Now since $f_{\mathbf{k}'}^0 = f_{\mathbf{k}}^0$ for elastic scattering we may take the derivatives of $f_{\mathbf{k}}^0$ outside the summation. We also note that

$$\left(\mathbf{v}_{\mathbf{k}} \times \mathbf{H} \cdot \frac{\partial}{\partial \mathbf{k}} \right) \frac{\partial f_{\mathbf{k}}^0}{\partial \mathcal{E}} = 0,$$

which allows us to cancel terms in $e\mathbf{E}$ and $\partial f_{\mathbf{k}}^0/\partial \mathcal{E}$ when we substitute back into (8.2.1). We are left with the equation

$$\sum_{\mathbf{k}'} Q(\mathbf{k}, \mathbf{k}')(\Lambda_{\mathbf{k}} - \Lambda_{\mathbf{k}'}) + \frac{e}{\hbar c} \left(\mathbf{v}_{\mathbf{k}} \times \mathbf{H} \cdot \frac{\partial}{\partial \mathbf{k}} \right) \Lambda_{\mathbf{k}} = \mathbf{v}_{\mathbf{k}}. \quad (8.2.4)$$

8.2 Calculating the conductivity of metals

This bears no reference to the Fermi energy or to the temperature, and thus may be expected to lead to a smoothly varying function as its solution for the quantity $\Lambda_\mathbf{k}$.

Our object in solving the Boltzmann equation in this instance is the evaluation of the electrical conductivity tensor $\boldsymbol{\sigma}$ in terms of the band structure and the scattering probabilities. We have already noted that the current density due to a single Bloch electron is equal to $e\mathbf{v}_\mathbf{k}/\Omega$, and we may then write the total current density $\mathbf{j}$ as

$$\mathbf{j} = \Omega^{-1} \sum_\mathbf{k} e\mathbf{v}_\mathbf{k} f_\mathbf{k}$$

$$= \Omega^{-1} \sum_\mathbf{k} e\mathbf{v}_\mathbf{k} f_\mathbf{k}^1,$$

since in equilibrium no net current will flow. With use of (8.2.3) this becomes

$$\mathbf{j} = -\frac{e^2}{\Omega} \sum_\mathbf{k} \mathbf{v}_\mathbf{k}(\boldsymbol{\Lambda}_\mathbf{k} \cdot \mathbf{E}) \frac{\partial f_\mathbf{k}^0}{\partial \mathcal{E}}.$$

As $\boldsymbol{\sigma}$ is defined by

$$\mathbf{j} = \boldsymbol{\sigma} \cdot \mathbf{E}$$

we see that

$$\boldsymbol{\sigma} = -\frac{e^2}{\Omega} \sum_\mathbf{k} \mathbf{v}_\mathbf{k} \boldsymbol{\Lambda}_\mathbf{k} \frac{\partial f_\mathbf{k}^0}{\partial \mathcal{E}}. \tag{8.2.5}$$

This expression includes a summation over the two spin directions.

At this point we can make a connection with some pictorial concepts of transport theory by considering the conductivity of the simplest possible system – the free-electron gas in the absence of a magnetic field. In this case the conductivity tensor is by symmetry diagonal and

$$\sigma_{xx} = \sigma_{yy} = \sigma_{zz} = \tfrac{1}{3} \operatorname{Tr} \boldsymbol{\sigma}$$

$$= -\frac{e^2}{3\Omega} \sum_\mathbf{k} \mathbf{v}_\mathbf{k} \cdot \boldsymbol{\Lambda}_\mathbf{k} \frac{\partial f_\mathbf{k}^0}{\partial \mathcal{E}}.$$

Also by symmetry $\mathbf{v}$ and $\boldsymbol{\Lambda}$ must always be parallel to $\mathbf{k}$ and of uniform magnitude over the Fermi surface. Because thermal energies are small compared with the typical Fermi energy we adopt (8.2.2) and replace $\partial f_\mathbf{k}^0/\partial \mathcal{E}$

by $-\delta(\mathcal{E}_\mathbf{k} - \mathcal{E}_F)$ to find

$$\sigma_{xx} = \frac{e^2 v \Lambda}{3\Omega} \sum_\mathbf{k} \delta(\mathcal{E}_\mathbf{k} - \mathcal{E}_F)$$

$$= \frac{e^2 v \Lambda}{3\Omega} D(\mathcal{E}_F).$$

We recall that the energy density of states $D(\mathcal{E})$ is proportional to $\sqrt{\mathcal{E}}$ in the free electron gas, and that N, the total number of electrons, is given by

$$N = \int_0^{\mathcal{E}_F} D(\mathcal{E}) \, d\mathcal{E}.$$

Thus

$$D(\mathcal{E}_F) = \frac{3N}{2\mathcal{E}_F} = \frac{3N}{mv^2}$$

and

$$\sigma_{xx} = \frac{Ne^2 \Lambda}{\Omega m v}.$$

This is similar to a formula of elementary kinetic theory which expresses the conductivity in terms of a *mean free path* λ or a *relaxation time* τ, defined such that τ^{-1} is the probability per unit time of an electron having a collision in which it loses any momentum gained from the electric field. In such a theory one argues that at any instant an average electron has been traveling for a time τ since its last collision, and hence has an average extra velocity of $eE\tau/m$. The system thus has a conductivity

$$\sigma = \frac{Ne^2 \tau}{\Omega m},$$

which is equivalent to our previous expression when Λ is identified with τv. Looking back to Eq. (8.2.5) we see that in contrast to these simple ideas the solution of the Boltzmann equation in the more general circumstances of a metal with a nonspherical Fermi surface cannot, in general, be expressed in terms of a relaxation time. It is, however, an extremely useful approximation to make when the detailed nature of the scattering is not thought to be central to the problem under investigation. This may, for example, be the

8.2 Calculating the conductivity of metals

case in multilayered systems, such as metallic or semiconducting superlattices. These consist of a sequence of layers of two or more different materials, with each layer perhaps a few nanometers thick. At the interface between two different layers, there is in general a mismatch in the electronic band structure. Disorder due to interdiffusion of atoms across the interface may also lead to significant scattering. A lack of precise information about the form of the interface scattering can then make it unnecessary to worry about the lesser errors introduced by the relaxation-time approximation, which corresponds to writing

$$\left(\frac{\partial f_\mathbf{k}}{\partial t}\right)_{\text{collisions}} = -\frac{f_\mathbf{k}^1}{\tau}$$

in the Boltzmann equation. When this approximation is not valid one must return to a calculation of $\Lambda_\mathbf{k}$, which is known as the *vector mean free path* of the Bloch electrons, it being a natural generalization of the λ of kinetic theory.

When the system is spatially inhomogeneous on a macroscopic scale, there will be spatial derivatives in the Boltzmann equation. As a consequence, the distribution function will depend on $\mathbf{r}$ as well as on $\mathbf{k}$, in mild violation of the Uncertainty Principle as discussed at the beginning of this chapter. The simplest such system is a thin film, which also happens to be an important subject in such practical applications as the technology of integrated circuits. The conducting paths connecting different circuit elements are films of Cu or Al with thicknesses ranging from a few to a few tens of nanometers. In order to model the transport properties of such thin films correctly, we must include scattering at the boundaries.

With the film thickness d along the z-axis and with an electric field along the x-axis, the linearized Boltzmann equation is

$$v_z \frac{\partial f_\mathbf{k}^1(z)}{\partial z} + eEv_x \frac{\partial f_\mathbf{k}^0}{\partial \mathcal{E}} = -\frac{f_\mathbf{k}^1(z)}{\tau}.$$

For simplicity, we are here using the relaxation-time approximation and are also assuming that the film is homogeneous in the xy plane. We again remove the inconveniently rapidly varying part of $f_\mathbf{k}^1$ by defining a new smoother function $h_\mathbf{k}(z)$, just as we did in Eq. (8.2.3), and writing

$$f_\mathbf{k}^1(z) = h_\mathbf{k}(z) \frac{\partial f_\mathbf{k}^0}{\partial \mathcal{E}}.$$

The Boltzmann equation in terms of $h_\mathbf{k}(z)$ is then

$$\frac{\partial h_\mathbf{k}(z)}{\partial z} + \frac{h_\mathbf{k}(z)}{v_z \tau} = -\frac{v_x e E}{v_z}.$$

In order to solve this equation we shall need to stipulate the boundary conditions for $h_\mathbf{k}(z)$. A simple set of such conditions, which seems to work rather well in most practical applications, was given by Fuchs and Sondheimer. Their idea was that an electron incident on a boundary has a probability S of being specularly reflected. In such a reflection, the momentum of the electron parallel to the surface is conserved, and the momentum perpendicular to the boundary changes sign but preserves its magnitude. The energy of the electron is conserved. In addition, there will be nonspecular scattering due to imperfections at the surface, and so we assume that the electron has a finite probability $D = 1 - S$ of being diffusively reflected in a process in which only energy is conserved as the electron moves away from the surface after the reflection. In general, we may not know precisely how the electron is scattered diffusively. Usually one assumes that the distribution after a diffuse reflection is the equilibrium distribution for electrons moving away from the boundary. This is convenient for subsequent calculations of the total current, since the electrons that have scattered diffusively do not contribute to the total current. With these Fuchs–Sondheimer boundary conditions, the solutions will be different for electrons traveling in the positive and negative z-directions. Let us denote by $h_\mathbf{k}^+(z)$ and $h_\mathbf{k}^-(z)$ the distribution functions of electrons of wavevector $\mathbf{k}$ and position z with $v_z > 0$ and $v_z < 0$, respectively. We can then write

$$\frac{\partial h_\mathbf{k}^\pm(z)}{\partial z} \pm \frac{h_\mathbf{k}^\pm(z)}{|v_z|\tau} = \mp \frac{v_x e E}{|v_z|}.$$

The general solution for $h_\mathbf{k}^\pm(z)$ is

$$h_\mathbf{k}^\pm(z) = -e v_x E \tau [1 - F_\mathbf{k}^\pm e^{\mp z/(\tau |v_z|)}].$$

The coefficients $F_\mathbf{k}^\pm$ are to be determined using the boundary conditions. These are

$$h_\mathbf{k}^+(z=0) = S h_\mathbf{k}^-(z=0)$$
$$h_\mathbf{k}^-(z=d) = S h_\mathbf{k}^+(z=d).$$

We can then calculate the current density and conductance of the system. Here we state only the qualitative result without going through the details. If the specularity coefficients at the boundaries are not unity, there is current lost near the boundaries due to diffuse reflections. If the thickness d is less than or of the order of the elastic mean free path, the effect is an apparent increase in the resistivity of the material. As the system thickness d becomes greater than the elastic mean free path, the increase in resistivity becomes less and less important, and the resistivity approaches that of bulk material. As an example, in Cu the mean free path at room temperature is of the order of 20 nm. Therefore, the increase in resistivity can be quite appreciable for films of thickness 10 nm or less. We note that if $S = 1$, so that the electrons are reflected perfectly at the boundaries, a film of any thickness will have the same effective resistivity as bulk material since in this case there is no reduction in current density due to diffuse reflection at the boundaries. In real applications, the specularity coefficient S is generally quite low (close to zero) at the interface between a metal and either vacuum, metal, or insulator because of roughness and diffusion across the interface.

8.3 Effects in magnetic fields

The presence of the magnetic field term in the Boltzmann equation is a great complication, and makes the evaluation of the conductivity tensor a difficult task. The k-vectors of the Bloch electrons now follow orbits around the Fermi surface as described in Section 4.6, until they are scattered to some new k-state to start their journey again. At the same time they are accelerated by the electric field, but then have their extra velocity reversed as the magnetic field changes their direction of motion. We shall here outline in the briefest possible manner the formal solution for the conductivity tensor, and then indicate a few qualitative conclusions that may be drawn from it.

Let us first abbreviate the Boltzmann equation (8.2.4) by noting that the left-hand side is a sum of two terms, each of which represents an operator acting on Λ. We could thus rewrite (8.2.4) as

$$S\Lambda_{\mathbf{k}} - i\omega W \Lambda_{\mathbf{k}} = \mathbf{v}_{\mathbf{k}} \tag{8.3.1}$$

with S and W operators and ω the cyclotron frequency, eH/mc. In the absence of a magnetic field the vector mean free path is then the solution of

$$S\Lambda_{\mathbf{k}}^0 = \mathbf{v}_{\mathbf{k}}.$$

If we define a new operator T equal to $S^{-1}W$ then we can rewrite (8.3.1) as

$$(1 - i\omega T)\Lambda_{\mathbf{k}} = \Lambda_{\mathbf{k}}^0. \tag{8.3.2}$$

The operator T has a complete set of eigenfunctions and real eigenvalues τ_l, and so we can solve by expanding $\Lambda_{\mathbf{k}}^0$ in terms of these. This allows us to invert (8.3.2), substitute the solution for $\Lambda_{\mathbf{k}}$ into the expression (8.2.5) for the conductivity, and find an expression of the form

$$\sigma_{\nu\mu} = \sum_l \frac{\gamma_{\nu\mu}^{(l)}}{1 - i\omega\tau_l}, \tag{8.3.3}$$

where the $\gamma_{\nu\mu}^{(l)}$ are numbers that depend on the band structure, the direction of $\mathbf{H}$, and the form of the scattering, but are independent of the magnitude of $\mathbf{H}$. If we choose axes so that when $\mathbf{H} = 0$ the conductivity tensor is diagonal it happens that the diagonal $\gamma_{\nu\nu}^{(l)}$ are real while the off-diagonal $\gamma_{\nu\mu}^{(l)}$ are pure imaginary. One can then make simplifications of the form

$$\sigma_{\nu\nu} = \sum_l \frac{\gamma_{\nu\nu}^{(l)}}{1 + \omega^2\tau_l^2}; \quad \sigma_{\nu\mu} \atop (\nu\neq\mu) = \sum_l \frac{i\omega\tau_l \gamma_{\nu\mu}^{(l)}}{1 + \omega^2\tau_l^2}. \tag{8.3.4}$$

An added complication that has to be considered is that in an experiment the current is constrained by the boundaries of the sample to lie in a certain direction. The total electric field acting on the sample must then be in such a direction that $\boldsymbol{\sigma}\cdot\mathbf{E}$ conforms with the sample geometry. There are thus electric fields set up in the sample whose direction is not within the control of the experimenter, and it is these that must be measured when a given current is flowing. One thus measures the resistivity tensor, $\boldsymbol{\rho}$, which is the inverse of the conductivity tensor.

The diagonal elements of $\boldsymbol{\sigma}$ are functions of ω^2, and are thus unchanged when the magnetic field is reversed. The same is true for the diagonal elements of $\boldsymbol{\rho}$. One defines the *magnetoresistance* $\Delta\rho_{\nu\nu}$ as the increase in $\rho_{\nu\nu}$ as a function of the magnitude and orientation of $\mathbf{H}$. This quantity is proportional to ω^2 at low fields. When $\mathbf{H}$ is in the ν-direction then $\rho_{\nu\nu}(\omega) - \rho_{\nu\nu}(0)$ is said to be the longitudinal magnetoresistance, while the transverse magnetoresistance is measured with $H_\nu = 0$. The fact that the magnitude of $\mathbf{H}$ occurs only in the combination $\omega\tau_l$ gives rise to the approximation known as *Kohler's rule*. One argues that if $Q(\mathbf{k}, \mathbf{k}')$ were increased in the same proportion as $\mathbf{H}$ the products $\omega\tau_l$ would be unchanged, and the magnetoresistance would remain the same proportion of the zero-field resistance.

8.3 Effects in magnetic fields

Thus one should be able to find some function F such that for a wide range of impurity concentration

$$\Delta\rho_{vv} = \rho_{vv}(0) F\left(\frac{H}{\rho_{vv}(0)}\right).$$

In practice deviations occur from this rule, as it is not possible to alter the scattering without altering various other factors such as the electron velocities.

The off-diagonal elements of $\boldsymbol{\sigma}$ are odd functions of ω, and thus are changed in sign when the magnetic field is reversed. The off-diagonal elements of $\boldsymbol{\rho}$, on the other hand, are neither odd nor even in ω. In the particular case where $\mathbf{H}$ is perpendicular to $\mathbf{j}$ the presence of these terms constitutes the *Hall effect*. Let us now change to a coordinate system in which $\mathbf{H}$ is in the z-direction and $\mathbf{j}$ in the y-direction, and expand ρ_{xy} in powers of H_z, so that

$$\rho_{xy}(\omega) = \rho_{xy}(0) + RH_z + SH_z^2 + \cdots.$$

In this expression R is known as the *Hall coefficient*, while the term in H_z^2 is responsible for the so-called *transverse-even voltage*, which does not change sign when the magnetic field is reversed. One sometimes calls S the transverse-even coefficient for low fields.

Some interesting effects occur at high fields, where from (8.3.4) it appears at first glance that all elements of $\boldsymbol{\sigma}$ become small. This will be the case unless one of the τ_l should be equal to zero, in which case the diagonal elements of $\boldsymbol{\sigma}$ will tend to a constant value $\tau_{vv}^{(0)}$. To see when this situation will arise we look back to the definitions of T and W in (8.3.2), (8.3.1), and (8.2.4). The presence of a term equivalent to $\mathbf{H} \times \partial/\partial\mathbf{k}$ in W (and hence in T) gives the obvious answer that T acting on a *constant* always vanishes. However, if we are to expand $\boldsymbol{\Lambda}^{(0)}$ in eigenfunctions of T we do not expect to find any constant vector as a component of $\boldsymbol{\Lambda}^{(0)}$, since by symmetry there is no preferred direction on an orbit of the type δ in Fig. 4.6.5. However, when we look at the open orbits γ, we see that it would be possible for $\boldsymbol{\Lambda}^{(0)}$ to be a different constant on each part of the orbit without violating any symmetry requirements. Physically we could say that the magnetic field is incapable of reversing the velocity of the electrons on these orbits, and so they contribute an anomalously large amount to the conductivity in high magnetic fields. The transverse magnetoconductivity at high fields can thus give information about the topology of the Fermi surface. We also note that symmetry does not exclude constant components of $\Lambda_\mathbf{k}^0$ in the direction of $\mathbf{H}$, as shown in Fig. 8.3.1, from contributing to the current. This means that when $\mathbf{H}$ is in the z-direction, σ_{zz} will again tend to a constant at large magnetic fields.

We note, incidentally, that the free-electron metal with an isotropic scattering probability is a pathological case and is of little use as a model in which to understand magnetoresistance. This arises from the fact that $\mathbf{\Lambda}^{(0)}$ is composed of only two eigenfunctions of T in this case, and they both share the same eigenvalue τ, which we can identify with the relaxation time defined as in Section 8.2. Then, because the conductivity is of the form

$$\boldsymbol{\sigma} = \sigma_0 \begin{pmatrix} \dfrac{1}{1+\omega^2\tau^2} & \dfrac{-\omega\tau}{1+\omega^2\tau^2} & 0 \\ \dfrac{\omega\tau}{1+\omega^2\tau^2} & \dfrac{1}{1+\omega^2\tau^2} & 0 \\ 0 & 0 & 1 \end{pmatrix},$$

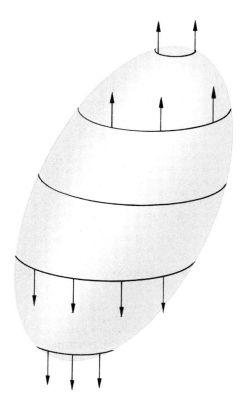

Figure 8.3.1 The component of the vector mean free path in the direction of the applied magnetic field contributes a term to the conductivity tensor that always causes the longitudinal magnetoresistance to saturate.

8.4 Inelastic scattering and the temperature dependence of resistivity

the resistivity tensor is

$$\rho = \sigma_0^{-1} \begin{pmatrix} 1 & \omega\tau & 0 \\ -\omega\tau & 1 & 0 \\ 0 & 0 & 1 \end{pmatrix}$$

and *shows no magnetoresistance*. One must thus at least generalize to the case of two parabolic bands to find a nonvanishing magnetoresistance.

8.4 Inelastic scattering and the temperature dependence of resistivity

The theory of the electrical conductivity of metals presented in Section 8.2 was based on the assumption that only elastic scattering occurred between Bloch states. At temperatures different from zero this assumption will not be valid, since then the electron–phonon interaction will cause electrons to be scattered between Bloch states with the emission or absorption of phonons. The matrix elements that appear in the expression for the scattering probability will be those of the electron–phonon interaction, and are functions of the occupation of the phonon states as well as of the Bloch states. We shall write the scattering probability between states to first order as

$$P(1,2) = \frac{2\pi}{\hbar} |\langle 1|\mathcal{H}_{e-p}|2\rangle|^2 \delta(\mathcal{E}_1 - \mathcal{E}_2) \tag{8.4.1}$$

where now $|1\rangle$ and $|2\rangle$ are descriptions of all the $n_{\mathbf{k}}$ and all the $n_{\mathbf{q}}$ of particular many-body states, and $\mathcal{E}_1$ and $\mathcal{E}_2$ the corresponding energies. In lowest order

$$\mathcal{H}_{e-p} = \sum_{\mathbf{k}\mathbf{k}'} M_{\mathbf{k}'\mathbf{k}} c_{\mathbf{k}'}^\dagger c_{\mathbf{k}} (a_{\mathbf{q}} + a_{-\mathbf{q}}^\dagger)$$

with $\mathbf{q} = \mathbf{k}' - \mathbf{k}$, suitably reduced to lie within the first Brillouin zone. This interaction changes the state of the metal by scattering the electron from $\mathbf{k}$ to $\mathbf{k}'$ and either absorbing a phonon of wavenumber $\mathbf{q}$ or emitting one of wavenumber $-\mathbf{q}$. The corresponding Born approximation for the scattering probability is found on substitution in (8.4.1) to be

$$P(\mathbf{k},\mathbf{k}') = \frac{2\pi}{\hbar} |M_{\mathbf{k}\mathbf{k}'}|^2 \langle c_{\mathbf{k}}^\dagger c_{\mathbf{k}'} c_{\mathbf{k}'}^\dagger c_{\mathbf{k}} a_{\mathbf{q}}^\dagger a_{\mathbf{q}} \rangle \delta(\mathcal{E}_{\mathbf{k}'} - \mathcal{E}_{\mathbf{k}} - \hbar\omega_{\mathbf{q}})$$

$$= \frac{2\pi}{\hbar} |M_{\mathbf{k}\mathbf{k}'}|^2 \langle n_{\mathbf{k}}(1-n_{\mathbf{k}'})n_{\mathbf{q}}\rangle \delta(\mathcal{E}_{\mathbf{k}'} - \mathcal{E}_{\mathbf{k}} - \hbar\omega_{\mathbf{q}})$$

when a phonon is absorbed and

$$P(\mathbf{k}, \mathbf{k}') = \frac{2\pi}{\hbar} |M_{\mathbf{k}\mathbf{k}'}|^2 \langle n_{\mathbf{k}}(1 - n_{\mathbf{k}'})(1 + n_{-\mathbf{q}}) \rangle \delta(\mathcal{E}_{\mathbf{k}'} - \mathcal{E}_{\mathbf{k}} + \hbar\omega_{-\mathbf{q}})$$

when a phonon is emitted. The collision term in the Boltzmann equation for the electrons (8.1.5) will then be of the form

$$\left(\frac{\partial f_{\mathbf{k}}}{\partial t}\right)_{\text{collisions}} = -\sum_{\mathbf{k}'} \frac{2\pi}{\hbar} |M_{\mathbf{k}\mathbf{k}'}|^2 \{f_{\mathbf{k}}(1 - f_{\mathbf{k}'})[f_{\mathbf{q}}\delta(\mathcal{E}_{\mathbf{k}'} - \mathcal{E}_{\mathbf{k}} - \hbar\omega_{\mathbf{q}}) \\
+ (1 + f_{-\mathbf{q}})\delta(\mathcal{E}_{\mathbf{k}'} - \mathcal{E}_{\mathbf{k}} + \hbar\omega_{-\mathbf{q}})] - f_{\mathbf{k}'}(1 - f_{\mathbf{k}}) \\
\times [f_{-\mathbf{q}}\delta(\mathcal{E}_{\mathbf{k}'} - \mathcal{E}_{\mathbf{k}} + \hbar\omega_{-\mathbf{q}}) + (1 + f_{\mathbf{q}})\delta(\mathcal{E}_{\mathbf{k}'} - \mathcal{E}_{\mathbf{k}} - \hbar\omega_{\mathbf{q}})]\}.$$

This very complicated expression can be simplified by a number of steps, but still remains difficult to interpret even within the framework of the free-electron model. One customarily assumes the phonon distribution to be in equilibrium so that $f_{\mathbf{q}}$ may be replaced by the Bose–Einstein distribution $f_{\mathbf{q}}^0$. A numerical solution of the Boltzmann equation with $\mathbf{H} = 0$ then shows that the mean free path Λ, defined as before, has a "hump" in it at the Fermi surface (Fig. 8.4.1). This reflects the fact that it is no longer possible to eliminate the chemical potential μ from the Boltzmann equation.

We shall leave the details of such a calculation to the more specialized texts, and simply examine the qualitative nature of the scattering and the

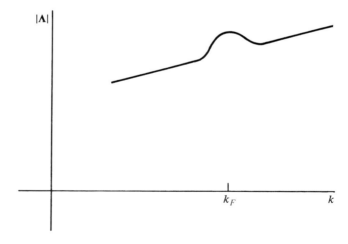

Figure 8.4.1 The inelastic nature of the scattering of electrons by phonons causes the mean free path to be a few per cent greater within the thermal thickness of the Fermi surface than elsewhere.

8.4 Inelastic scattering and the temperature dependence of resistivity

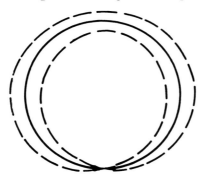

Figure 8.4.2. Scattering of an electron by absorption or emission of a phonon takes the electron onto one of the two surfaces defined by the energy-conservation relation $\mathcal{E}_{\mathbf{k}'} = \mathcal{E}_{\mathbf{k}} \pm \hbar\omega_{\mathbf{q}}$.

electrical resistance it causes. Firstly we note that an electron in state $\mathbf{k}$ is not scattered onto a surface of constant energy $\mathcal{E}_{\mathbf{k}'} = \mathcal{E}_{\mathbf{k}}$, but onto one of two surfaces slightly displaced in energy from it, so that $\mathcal{E}_{\mathbf{k}'} = \mathcal{E}_{\mathbf{k}} \pm \hbar\omega_{\mathbf{q}}$ (Fig. 8.4.2). If we were to make the approximation that the scattering is elastic, we should misrepresent the effect of the Exclusion Principle, in that some phonon emission processes are forbidden because of the low energy of the final electron state. However, only electrons within about kT of the Fermi surface can be scattered, and they would have to lose more than kT in energy to be scattered into a region where $f_{\mathbf{k}'}$ was close to unity. Since we do not expect to find many phonons present in equilibrium with energy more than kT, we can argue that the approximation only introduces a small error. Secondly we recall that the scattering matrix element will be proportional to the change in density in the sound wave, i.e., to $qy_{\mathbf{q}}$, where $y_{\mathbf{q}}$ is the amplitude. For a harmonic oscillator the average potential energy is half the total, and so

$$\tfrac{1}{2}(n_{\mathbf{q}} + \tfrac{1}{2})\hbar\omega_{\mathbf{q}} \sim \tfrac{1}{2}m\omega_{\mathbf{q}}^2 y_{\mathbf{q}}^2,$$

or

$$qy_{\mathbf{q}} \propto q\left(\frac{n_{\mathbf{q}}}{\omega_{\mathbf{q}}}\right)^{1/2}$$

$$\propto (qn_{\mathbf{q}})^{1/2}$$

for long waves, for which $\omega_{\mathbf{q}} \propto q$. Thus at low temperatures, when long waves are most important, we can make the approximation

$$Q(\mathbf{k}, \mathbf{k}') \propto qf_{\mathbf{q}}^0 \delta(\mathcal{E}_{\mathbf{k}} - \mathcal{E}_{\mathbf{k}'})$$

where

$$q = |\mathbf{k}' - \mathbf{k}|.$$

In calculating $\Lambda_\mathbf{k}$ from the Boltzmann equation we should then write

$$\sum_{\mathbf{k}'} q f_\mathbf{q}^0 \delta(\mathcal{E}_\mathbf{k} - \mathcal{E}_{\mathbf{k}'})(\Lambda_\mathbf{k} - \Lambda_{\mathbf{k}'}) \propto \mathbf{v}_\mathbf{k}.$$

The principal contribution to the difference between $\Lambda_\mathbf{k}$ and $\Lambda_{\mathbf{k}'}$ will be a change in direction, and one may then argue that the contribution of $\Lambda_\mathbf{k} - \Lambda_{\mathbf{k}'}$ to the sum will be about the same as $\Lambda_\mathbf{k}(1 - \cos\theta_{\mathbf{k}\mathbf{k}'})$, where θ is the angle between $\Lambda_\mathbf{k}$ and $\Lambda_{\mathbf{k}'}$. For low temperatures it will only be small θ that will be important, and then

$$1 - \cos\theta_{\mathbf{k}\mathbf{k}'} \simeq \tfrac{1}{2}\theta_{\mathbf{k}\mathbf{k}'}^2$$

$$\propto q^2.$$

Equation (8.2.1) is thus of the form

$$\Lambda_\mathbf{k} \sum_{\mathbf{k}'} q^3 f_\mathbf{q}^0 \delta(\mathcal{E}_\mathbf{k} - \mathcal{E}_{\mathbf{k}'}) \propto \mathbf{v}_\mathbf{k}.$$

The delta-function in energy restricts the sum to a surface in k-space. Changing the sum to an integral we find

$$\Lambda_\mathbf{k} \int q^3 f_\mathbf{q}^0 q \, dq \propto \mathbf{v}_\mathbf{k}.$$

Since $f_\mathbf{q}^0$ is a function of $\hbar\omega_\mathbf{q}/kT$, which for small q is proportional to q/T, we find the integral over q to be proportional to T^5; then Λ, and hence the conductivity, is proportional to T^{-5}. At high temperatures, on the other hand, the Bose–Einstein function $f_\mathbf{q}^0$ may be approximated by $kT/\hbar\omega_\mathbf{q}$. The detailed shape of the scattering probability is then unimportant, as the temperature only enters the Boltzmann equation through the function $f_\mathbf{q}^0$. If $f_\mathbf{q}^0$ is proportional to the temperature then Λ must vary as T^{-1}. The resistivity of a pure metal should thus be proportional to T^5 at low temperatures and to T at high temperatures. This prediction appears to be verified experimentally for the simple metals, but not for transition metals such as nickel, palladium, platinum, rhenium, and osmium. In these elements electron–electron

8.4 Inelastic scattering and the temperature dependence of resistivity

scattering appears to play a major role in limiting the current and in leading to a resistivity that varies as T^2 rather than T^5.

When both impurities and phonons are present the total probability of scattering from a Bloch state will be approximately the sum of the two scattering probabilities taken separately. This is so because the inelastic scattering by phonons will connect the state **k** with final states **k**′ which are different from those entered by elastic scattering. The two processes must be considered incoherent, and one adds the scattering probabilities rather than the scattering amplitudes. This leads to *Matthiessen's rule*, which expresses the idea that the electrical resistivity can be considered as a sum of two independent parts, one of which is a function of the purity of the metal and the other a function of temperature characteristic of the pure metal; i.e.,

$$\rho = \rho_i + \rho_0(T).$$

The addition of further impurities to a metal is then predicted to displace the curve of $\rho(T)$ and not to alter its shape (Fig 8.4.3).

In practice deviations of a few percent occur from this rule as a consequence of a variety of effects. Adding impurities, for example, may change the phonon spectrum, the electron–phonon interaction, or even the shape of the Fermi surface, while raising the temperature introduces the "hump" of Fig. 8.4.1 in Λ, which, in turn, changes the resistance.

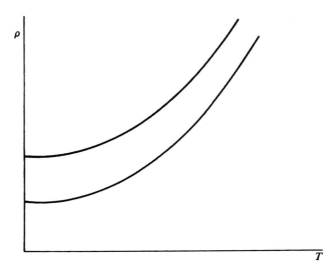

Figure 8.4.3 Matthiessen's rule predicts that addition of impurities to a metal has the effect of increasing the resistivity ρ by an amount that does not depend on the temperature.

8.5 Thermal conductivity in metals

One recognizes a metal not only by its large electrical conductivity but also by its large thermal conductivity. This indicates that the electrons must play an important role in the transport of heat – a fact which is at first surprising when one remembers the small heat capacity of the electron gas. The importance of the electrons lies in their long mean free paths and in the high velocities with which they travel, which more than compensate for their small heat capacity when compared with the phonon system.

Just as the electric current density was calculated from the expression

$$\mathbf{j} = \frac{e}{\Omega} \sum_{\mathbf{k}} \mathbf{v_k} f_\mathbf{k}, \qquad (8.5.1)$$

so one can write the energy current density $\mathbf{u}'$ as

$$\mathbf{u}' = \Omega^{-1} \sum_{\mathbf{k}} \mathcal{E}_\mathbf{k} \mathbf{v_k} f_\mathbf{k}.$$

This, however, is *not* the same thing as the heat current density $\mathbf{u}$ of the electrons, as can be seen by picturing the arrival at one end of a piece of metal of an electron of zero energy. As the only unoccupied k-states would be those with energy close to $\mathcal{E}_F$, all the thermally excited electrons would have to donate a small amount of their thermal energy to the new arrival, with the net result that the electron gas would be cooled; a zero-energy electron thus carries a large amount of coldness! We consequently have to measure $\mathcal{E}_\mathbf{k}$ relative to some carefully chosen reference energy. This is the same problem that we encountered in Section 3.3. There we saw that the appropriate zero of energy is the chemical potential μ; adding an electron of energy μ to the metal does not change the temperature of the system. This can be stated in thermodynamic terms by noting that

$$\frac{\partial \mathcal{F}}{\partial N} = \mu,$$

with $\mathcal{F}$ the Helmholtz energy. We accordingly write the heat current density due to the electrons as

$$\mathbf{u} = \Omega^{-1} \sum_{\mathbf{k}} (\mathcal{E}_\mathbf{k} - \mu) \mathbf{v_k} f_\mathbf{k}. \qquad (8.5.2)$$

8.5 Thermal conductivity in metals

The flow of heat resulting from the presence of any combination of fields and temperature gradients may now be calculated from the linearized Boltzmann equation (8.1.5). The thermal conductivity, for example, may be found by putting **H** and **E** equal to zero to obtain the equation

$$\left(\frac{\partial f_{\mathbf{k}}^1}{\partial t}\right)_{\text{collisions}} = \mathbf{v_k} \cdot \nabla T \left(\frac{\partial f_{\mathbf{k}}^0}{\partial T}\right). \tag{8.5.3}$$

Strictly speaking, the thermal conductivity tensor $\boldsymbol{\kappa}$ is defined by the equation

$$\mathbf{u} = -\boldsymbol{\kappa} \cdot \nabla T \tag{8.5.4}$$

under the condition that **j** rather than **E** be equal to zero. This only adds a very small correction, however (Problem 8.5), and so we shall neglect it here. We then proceed in analogy with the discussion of electrical conductivity given in Section 8.2. This time we define a vector mean free path by putting

$$f_{\mathbf{k}}^1 = (\mathcal{E}_{\mathbf{k}} - \mu) \frac{\nabla T}{T} \cdot \boldsymbol{\Lambda}_{\mathbf{k}} \frac{\partial f_{\mathbf{k}}^0}{\partial \mathcal{E}}. \tag{8.5.5}$$

Then since

$$\frac{\partial f_{\mathbf{k}}^0}{\partial T} = -\left(\frac{\mathcal{E} - \mu}{T}\right) \frac{\partial f^0}{\partial \mathcal{E}},$$

we find that when we consider only elastic scattering all the arguments used in Section 8.2 apply and we end up with the identical equation for $\boldsymbol{\Lambda}_{\mathbf{k}}$. Since from Eqs. (8.5.2) and (8.5.5) the heat current density is

$$\mathbf{u} = \Omega^{-1} \sum_{\mathbf{k}} (\mathcal{E}_{\mathbf{k}} - \mu)^2 \mathbf{v_k} \frac{\nabla T}{T} \cdot \boldsymbol{\Lambda}_{\mathbf{k}} \frac{\partial f_{\mathbf{k}}^0}{\partial \mathcal{E}}, \tag{8.5.6}$$

then from Eq. (8.5.4)

$$\boldsymbol{\kappa} = -\frac{1}{\Omega T} \sum_{\mathbf{k}} \mathbf{v_k} \boldsymbol{\Lambda}_{\mathbf{k}} (\mathcal{E}_{\mathbf{k}} - \mu)^2 \frac{\partial f_{\mathbf{k}}^0}{\partial \mathcal{E}}. \tag{8.5.7}$$

The fact that for elastic scattering it is identically the same $\boldsymbol{\Lambda}_{\mathbf{k}}$ that occurs in Eq. (8.5.7) as occurred in the Eq. (8.2.5) that defined the electrical conductivity leads us to the remarkable *Wiedemann–Franz law*. We recall from

Section 2.1 that a summation over allowed values of **k** can be replaced by an integration over k-space, so that for any well behaved function $A(\mathbf{k})$

$$\sum_{\mathbf{k}} A(\mathbf{k}) = \frac{2\Omega}{(2\pi)^3} \iint A(\mathbf{k}) \, d\mathcal{E} \, \frac{dS_{\mathbf{k}}}{\hbar v_{\mathbf{k}}}, \tag{8.5.8}$$

the first factor of 2 arising from the sum over spin directions, and $dS_{\mathbf{k}}$ representing an element of area of a surface in k-space of constant energy $\mathcal{E}$. If the functions $\mathbf{v}_{\mathbf{k}}$ and $\mathbf{\Lambda}_{\mathbf{k}}$ do not have any unusual kinks at the Fermi surface, then when Eq. (8.5.7) is changed into a double integral in the manner of Eq. (8.5.8) the result may be factorized to obtain

$$\boldsymbol{\kappa} = -\frac{1}{4\pi^3 T} \int \frac{\mathbf{v}_{\mathbf{k}} \mathbf{\Lambda}_{\mathbf{k}}}{\hbar v_{\mathbf{k}}} \, dS_{\mathbf{k}} \int (\mathcal{E} - \mu)^2 \frac{\partial f^0}{\partial \mathcal{E}} \, d\mathcal{E}. \tag{8.5.9}$$

This is possible because the presence of the term $df^0/\partial \mathcal{E}$ has the consequence that the integrand is only appreciable within the thickness kT of the Fermi surface. Since for a typical metal at room temperature $kT/\mathcal{E}F$ has a magnitude of 10^{-2} or less, then $\mathbf{v}_{\mathbf{k}}$ and $\mathbf{\Lambda}_{\mathbf{k}}$ can to a good approximation be considered independent of energy when the integration over $\mathcal{E}$ is performed. Similar arguments applied to Eq. (8.2.5) yield

$$\boldsymbol{\sigma} = -\frac{e^2}{4\pi^3} \int \frac{\mathbf{v}_{\mathbf{k}} \mathbf{\Lambda}_{\mathbf{k}}}{\hbar v_{\mathbf{k}}} \, dS_{\mathbf{k}} \int \frac{\partial f^0}{\partial \mathcal{E}} \, d\mathcal{E}. \tag{8.5.10}$$

The function $-\partial f^0/\partial \mathcal{E}$ has the shape shown schematically in Fig. 8.5.1(a), while $-(\mathcal{E} - \mu)^2 \partial f^0/\partial \mathcal{E}$ is the double-humped curve of Fig. 8.5.1(b). Both functions may be integrated by extending the limits to $\pm\infty$, and one finds the results 1 and $(\pi kT)^2/3$, respectively. Comparison of Eqs. (8.5.9) and (8.5.10) then yields the Wiedemann–Franz law, which states that

$$\boldsymbol{\kappa} = LT\boldsymbol{\sigma} \tag{8.5.11}$$

where L is known as the *Lorenz number*, and in our simple calculation is equal to

$$L_0 = \frac{\pi^2 k^2}{3e^2}, \tag{8.5.12}$$

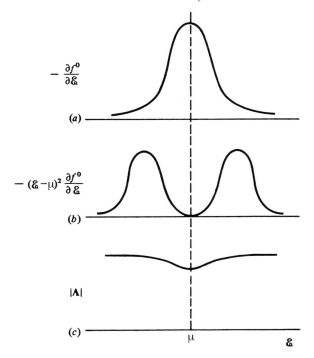

Figure 8.5.1 (a) The expression for the electrical conductivity contains the factor $\partial f^0/\partial \mathcal{E}$, and thus samples the electron distribution in the immediate vicinity of the Fermi surface. (b) The thermal conductivity of a metal, on the other hand, involves the double-humped function $(\mathcal{E}_k - \mu)^2 \partial f^0/\partial \mathcal{E}$, and measures Λ immediately below and above the Fermi energy. (c) If the mean free path varies appreciably over the energy range from $\mu - kT$ to $\mu + kT$ then the Lorenz number deviates from its usual value.

which has the value 2.45×10^{-8} V^{2}K^{-2}. We deduce from this result that κ is directly proportional to T at low enough temperatures, since the electrical conductivity then tends to a constant in the absence of superconductivity.

The Wiedemann–Franz law is obeyed to within a few percent by most good metals at most temperatures. Deviations occur whenever the mean free path becomes a function of energy, as may be seen by a glance back at Fig. 8.5.1. The electrical conductivity contains the factor $-\partial f^0/\partial \mathcal{E}$, and thus reflects the mean free path at the Fermi energy μ. The thermal conductivity, on the other hand, contains the factor shown in Fig. 8.5.1(b), and thus measures Λ_k at energies slightly below and slightly above μ. If Λ_k had a dip in it at the energy $\mathcal{E} = \mu$, as indicated in Fig. 8.5.1(c), then the calculation of σ would sample a lower value of $|\Lambda_k|$ than would the calculation of κ, and the Lorenz number L would show a positive deviation from the value L_0 that we previously derived. We should expect deviations in particular in alloys that exhibit the

Kondo effect discussed in Chapter 11, as well as in any situation where inelastic scattering can occur. Because an electron loses or gains energy $\hbar\omega_q$ in scattering by phonons, one generally observes deviations of L from L_0 as the temperature is increased from zero, and the electron–phonon interaction gradually takes over from the impurity potentials as the dominant scattering mechanism. At temperatures much greater than the Debye temperature it will be the case that $kT \gg \hbar\omega_q$ for all phonons, and it again becomes a good approximation to consider the scattering as elastic. Then the Lorenz number returns to its ideal value, L_0.

8.6 Thermoelectric effects

In the preceding section we calculated the heat current density $\mathbf{u}$ that results in a metal when a temperature gradient exists. It is a simple matter to extend this calculation to find the heat current density caused by the application of an electric field alone. Substitution of Eq. (8.2.3) in Eq. (8.5.2) yields the result

$$\mathbf{u} = -\frac{e}{\Omega} \sum_{\mathbf{k}} (\mathcal{E}_{\mathbf{k}} - \mu) \mathbf{v}_{\mathbf{k}} \frac{\partial f^0}{\partial \mathcal{E}} \Lambda_{\mathbf{k}} \cdot \mathbf{E}. \qquad (8.6.1)$$

The existence of this heat current is known as the *Peltier effect*. The Peltier coefficient Π is defined as the ratio of the heat current to the electric current induced in a sample by a weak electric field in the absence of temperature gradients.

Let us write Eq. (8.6.1) in the form

$$\mathbf{u} = \boldsymbol{\phi} \cdot \mathbf{E} \qquad (8.6.2)$$

and evaluate the components of the tensor $\boldsymbol{\phi}$ by changing the sum over k-states into a double integral in the manner of Eq. (8.5.8). We find

$$\boldsymbol{\phi} = -\frac{e}{4\pi^3} \int (\mathcal{E} - \mu) \frac{\partial f^0}{\partial \mathcal{E}} d\mathcal{E} \int \frac{\mathbf{v}_{\mathbf{k}} \Lambda_{\mathbf{k}}}{\hbar v_{\mathbf{k}}} dS_{\mathbf{k}}. \qquad (8.6.3)$$

We then immediately see that $\boldsymbol{\phi}$ is a very small quantity, for if we make the assumption that the integral over $dS_{\mathbf{k}}$ is independent of energy then the whole expression vanishes, as $(\mathcal{E} - \mu)(\partial f^0/\partial \mathcal{E})$ is an odd function of $\mathcal{E} - \mu$. This is illustrated schematically in Fig. 8.6.1, which shows the electron distribution

8.6 Thermoelectric effects

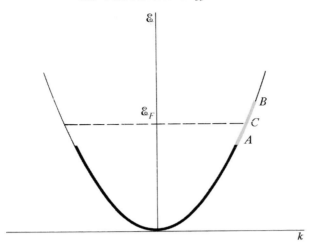

Figure 8.6.1 Although all the electrons in the region between A and B carry an electric current of approximately the same magnitude in the same direction, the heat current carried by those between A and C is almost exactly cancelled by those between C and B.

shifted in k-space by the action of the electric field. By symmetry the net electric and thermal currents are carried by those electrons between the points marked A and B. Because all these electrons are moving with approximately the Fermi velocity they all contribute a similar amount to the electric current. This is not the situation with the heat current, however, for those electrons between A and C have energies less than μ, and carry a current of coldness that almost cancels the positive heat current of the electrons between C and B. We thus have to take into account the energy dependence of $\mathbf{v_k}$ and $\Lambda_\mathbf{k}$, and so we make a Taylor expansion to first order of the integral over $dS_\mathbf{k}$ in Eq. (8.6.3). We write

$$\int \frac{\mathbf{v_k} \Lambda_\mathbf{k}}{\hbar v_\mathbf{k}} dS_\mathbf{k} \simeq \left[\int \frac{\mathbf{v_k} \Lambda_\mathbf{k}}{\hbar v_\mathbf{k}} dS_\mathbf{k}\right]_{\mathcal{E}=\mu} + (\mathcal{E} - \mu)\left[\frac{d}{d\mathcal{E}}\int \frac{\mathbf{v_k} \Lambda_\mathbf{k}}{\hbar v_\mathbf{k}} dS_\mathbf{k}\right]_{\mathcal{E}=\mu}$$

and substitute this in Eq. (8.6.3). The energy integral that survives is identical to that in Eq. (8.5.9), and so we find

$$\phi = \frac{e}{4\pi^3} \frac{(\pi kT)^2}{3} \left[\frac{d}{d\mathcal{E}}\int \frac{\mathbf{v_k} \Lambda_\mathbf{k}}{\hbar v_\mathbf{k}} dS_\mathbf{k}\right]_{\mathcal{E}=\mu}. \qquad (8.6.4)$$

The fact that Eq. (8.5.10) can be written in the form

$$\int \frac{\mathbf{v_k}\mathbf{\Lambda_k}}{\hbar v_k} dS_k = \frac{4\pi^3}{e^2} \boldsymbol{\sigma}$$

is sometimes made use of to put Eq. (8.6.4) in the form

$$\boldsymbol{\phi} = \frac{(\pi kT)^2}{3e} \frac{\partial \boldsymbol{\sigma}}{\partial \mu}. \tag{8.6.5}$$

The derivative $\partial\boldsymbol{\sigma}/\partial\mu$ is taken to mean the rate of change of conductivity with Fermi energy when it is assumed that the scattering and band structure remain constant. It is very important to realize that $\partial\boldsymbol{\sigma}/\partial\mu$ cannot be found simply by adding more electrons to the metal and measuring the change in resistance; as we saw in Section 6.4, certain kinks in the band structure and in the scattering are linked to the position of the Fermi energy, and would be altered by the addition of extra electrons.

Since $\mathbf{u} = \boldsymbol{\phi} \cdot \mathbf{E}$ and $\mathbf{E} = \boldsymbol{\sigma}^{-1} \cdot \mathbf{j}$, we can use Eq. (8.6.5) to write the Peltier coefficient (defined by $\mathbf{u} = \boldsymbol{\Pi} \cdot \mathbf{j}$) in the form

$$\boldsymbol{\Pi} = \frac{L_0 e T^2}{\mu} \frac{\partial \ln \boldsymbol{\sigma}}{\partial \ln \mu}, \tag{8.6.6}$$

with L_0 the ideal Lorenz number given in Eq. (8.5.12). For a material with a scalar conductivity (as, for example, a cubic crystal) the dimensionless quantity

$$\xi = \frac{\partial \ln \sigma}{\partial \ln \mu} \tag{8.6.7}$$

is a useful measure of the Peltier effect. It is generally of the order of magnitude of unity (Problem 8.7) but is extremely sensitive to the type of scattering. Alloys exhibiting the Kondo effect, for instance, have anomalously large Peltier coefficients as a consequence of the strong energy dependence of the scattering due to magnetic impurities.

In the same way that we calculated the heat current caused by an applied electric field we can investigate the electric current that results from the presence of a temperature gradient. On substituting Eq. (8.5.5) into Eq. (8.5.1) we find

$$\mathbf{j} = \frac{e}{\Omega T} \sum_{\mathbf{k}} (\mathcal{E}_\mathbf{k} - \mu) \mathbf{v_k} \frac{\partial f^0}{\partial \mathcal{E}} \mathbf{\Lambda_k} \cdot \nabla T. \tag{8.6.8}$$

8.6 Thermoelectric effects

In an isolated piece of metal charge will move to one end of the sample until an electric field **E** is built up that is just sufficient to induce an equal and opposite current that cancels that given by Eq. (8.6.8). The presence of this field constitutes the *Seebeck effect*; the ratio of **E** to ∇T is known as the *absolute thermoelectric power* or *thermopower* of the metal. If we write Eq. (8.6.8) in the form

$$\mathbf{j} = -\mathbf{\psi} \cdot \nabla T \qquad (8.6.9)$$

then the thermopower **S** is equal to $\boldsymbol{\sigma}^{-1} \cdot \boldsymbol{\psi}$.

We fortunately do not need to spend much time analyzing the Seebeck effect, for the thermopower **S** is related to the Peltier coefficient Π in a very simple way. Comparison of $\boldsymbol{\psi}$ [as defined by Eqs. (8.6.8) and (8.6.9)] with $\boldsymbol{\phi}$ [as defined by Eqs. (8.6.1) and (8.6.2)] yields the result

$$\boldsymbol{\phi} = T\boldsymbol{\psi}$$

or more generally

$$\Pi = T\tilde{\mathbf{S}} \qquad (8.6.10)$$

with $\tilde{\mathbf{S}}$ the transpose of **S**. A relationship of this kind was first derived by Lord Kelvin by arguments that are still appealing, but alas, no longer respectable. It is now thought of as an example of one of the *Onsager relations* that form the basis of the macroscopic theory of irreversible processes.

From Eqs. (8.6.6) and (8.6.10) one can write

$$\mathbf{S} = \frac{L_0 eT}{\mu} \left(\frac{\partial \ln \boldsymbol{\sigma}}{\partial \ln \mu} \right).$$

For a cubic metal this becomes

$$S = \frac{L_0 eT}{\mu} \xi. \qquad (8.6.11)$$

As μ is generally a few electron volts and ξ is of the order of magnitude of unity one finds that thermopowers in metals at room temperature have magnitudes of a few microvolts per kelvin. In semimetals such as bismuth, where μ is small, the thermopower is correspondingly larger. The proportionality of S to the absolute temperature suggested by Eq. (8.6.11) is dependent on ξ being independent of T. Because impurity scattering and phonon scattering may lead to completely different expressions for ξ it is not uncommon for even the sign of S to change as the temperature of the sample is changed. Thus dilute AuMn alloys have thermopowers that at the lowest temperatures are positive (i.e., a *negative* value of ξ, the electronic charge e being considered negative in Eq. (8.6.11)), but which become negative as the temperature is raised above a few kelvin.

Before leaving the topic of thermoelectric effects we should briefly consider an effect that occurs when phonon Umklapp processes are rare. One mechanism we have considered by which an electron can be scattered is the electron–phonon interaction, a phonon being created that carries off some of the momentum of the electron. We have implicitly been assuming that the momentum carried by this phonon is rapidly destroyed, either by a phonon Umklapp process or by scattering by lattice imperfections and impurities. In mathematical terms we have been uncoupling the Boltzmann equation for the phonons (Eq. (8.1.2)) from that for the electrons (Eq. (8.1.3)) by assuming the phonon relaxation time τ_{ph} to be very short. If this assumption is not valid then we must include in our computation of the heat current density the contribution of the perturbed phonon distribution. This added contribution to the Peltier coefficient (and hence also to the thermopower) is said to be due to *phonon drag*, the phonons being thought of as swept along by their interaction with the electrons. Such effects are negligible at very low temperatures (when there are few phonons available for "dragging") and at high temperatures (when phonon Umklapp processes "anchor" the phonon distribution) and thus cause a "hump" in the thermopower of the form shown in Fig. 8.6.2 at temperatures in the neighborhood of $\Theta/4$.

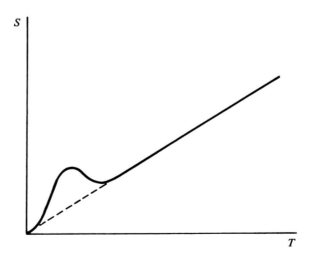

Figure 8.6.2 In very pure specimens the phenomenon of phonon drag may contribute appreciably to the thermopower at temperatures well below the Debye temperature.

Problems

8.1 When a certain type of impurity is added to a free-electron gas of Fermi energy $\mathcal{E}_F$ it is found that $|T_{\mathbf{kk'}}|^2$ is approximately constant, so that

$$Q(\mathbf{k}, \mathbf{k'}) \simeq \text{constant} \times \delta(\mathcal{E}_\mathbf{k} - \mathcal{E}_{\mathbf{k'}}).$$

How does the electrical conductivity, σ, of this system vary with $\mathcal{E}_F$? [Hint: In Eq. (8.2.4) the summation over $\mathbf{k'}$ may be replaced by an integral over k'-space. That is

$$\sum_\mathbf{k} \rightarrow \frac{\Omega}{8\pi^3} \int d\mathbf{k}.$$

Also

$$\int d\mathbf{k} \rightarrow \int \frac{dS}{|\partial \mathcal{E}/\partial \mathbf{k}|} \int d\mathcal{E},$$

where dS is an element of a surface of constant energy.]

8.2 With another type of impurity one finds that the scattering probability of Problem 8.1 is modified to the form

$$Q(\mathbf{k}, \mathbf{k'}) = \theta(\mathbf{k} - \mathbf{k'})\delta(\mathcal{E}_\mathbf{k} - \mathcal{E}_{\mathbf{k'}}),$$

where

$$\theta(\mathbf{k} - \mathbf{k'}) = \begin{cases} \text{constant} & \text{if } |\mathbf{k} - \mathbf{k'}| \leq t \\ 0 & \text{if } |\mathbf{k} - \mathbf{k'}| > t \end{cases}$$

and $t \ll k_F$, the Fermi radius. How does σ vary with $\mathcal{E}_F$ now?

8.3 The probability of an electron being scattered from $\mathbf{k}$ to $\mathbf{k'}$ with the emission of a phonon $\mathbf{q}$ of energy $\hbar\omega_\mathbf{q}$ is proportional to $f_\mathbf{k}(1 - f_{\mathbf{k'}})(1 + f_\mathbf{q})$, while the probability of the reverse process occurring, in which an electron scatters from $\mathbf{k'}$ to $\mathbf{k}$ with the absorption of a phonon is proportional to $f_{\mathbf{k'}}(1 - f_\mathbf{k})f_\mathbf{q}$. Are these expressions equal in equilibrium?

8.4 A monovalent simple cubic metal has the lattice potential

$$2V(\cos gx + \cos gy + \cos gz),$$

and a magnetoresistance which is found to saturate for all directions of the applied magnetic field **B**. The effect of a small strain on this metal, however, is to cause the magnetoresistance no longer to saturate for certain directions of **B**. Estimate V, explaining your reasoning carefully.

8.5 Express the exact thermal conductivity κ' (defined as the solution of Eq. (8.5.4) when $\mathbf{j} = 0$) in terms of $\boldsymbol{\sigma}$, **S**, and the approximate value κ given by Eq. (8.5.4) when $\mathbf{E} = 0$.

8.6 Is it meaningful to discuss "the limiting value of S/T for pure silver as $T \to 0$," where S is the thermopower? If not, why not?

8.7 Calculate the thermopowers of the metals defined in Problems 8.1 and 8.2.

8.8 An alternative approach to the kinetic theory mentioned in Section 8.2 argues that each electron receives an extra velocity of $e\mathbf{E}\tau/m$ by the time it has a collision, and thus has an average drift velocity over a long time of $e\mathbf{E}\tau/2m$. How can this apparent contradiction be resolved?

8.9 The room-temperature resistivity of Cu is about $2.0\,\mu\Omega$ cm. Use a simple free-electron gas model to find a relaxation time τ that gives you this resistivity at the right density of conduction electrons. Next, calculate the resistance of a thin Cu film by applying an electric field in the plane of the film. Use the Fuchs–Sondheimer boundary conditions to obtain a solution for the Boltzmann equation from which you can calculate the total current. What is the apparent resistivity of the film at thickness $d = 0.1\lambda$, $d = 0.5\lambda$, $d = \lambda$, and $d = 5\lambda$ (with λ the mean free path) for specularity coefficients $S = 0$, $S = 0.2$, and $S = 0.8$?

Chapter 9
Mesoscopic physics

9.1 Conductance quantization in quantum point contacts

In Chapter 8 we discussed the Boltzmann equation and the approach to describing transport properties, such as electrical conductivity, that it provides. In general, this approach works very well for most common metals and semiconductors, but there are cases where it fundamentally fails. This happens, for example, when the wave nature of the electron manifests itself and has to be included in the description of the scattering. In this case, interference may occur, which can affect the electrical conduction. We recall that the Boltzmann equation describes the electron states only through a dispersion relation of the Bloch states of an underlying perfect crystal lattice, a probability function, and a scattering function that gives the probability per unit time of scattering from one state to another. All these quantities are real, and do not contain any phase information about the electron states. Consequently, no wave-like phenomena can be described. The question then arises as to when the phase information is important. This really boils down to a question of length scales. We have earlier talked about the mean free path of an electron, which is roughly the distance it travels between scattering events. A simple example is given by scattering off static impurities that have no internal degrees of freedom. In this case the electron scattering is elastic, since an electron must have the same energy before and after a scattering event. Furthermore, in the presence of impurity scattering the phase of an electron wavefunction after a scattering event is uniquely determined by the phase before the scattering event. The wavefunction will in general suffer a phase shift as a consequence of the scattering, but this phase shift is not random and can be calculated for any wavefunction given the impurity potential. In view of this, we will now be more careful and specifically talk about the *elastic mean free path* ℓ_e as (roughly) the distance an electron travels between elastic scattering events. Note that the elastic mean free path is only weakly temperature dependent (through the temperature dependence of the Fermi distribution of the electrons).

Inelastic scattering, on the other hand, randomizes the phase of the electron wavefunction. A good example is provided by electron–phonon scattering. Consider such an event within the framework of perturbation theory in the electron–phonon interaction. In the scattering process, an initial electron Bloch state and phonon couple through some interaction. For a while, there will then be some complicated intermediate state made up of a multitude of electron Bloch states and phonon modes. Eventually the system settles into a direct product of another electron Bloch state and phonons consistent with energy and crystal momentum conservation. The electron in the intermediate state can have any energy for a time consistent with the Uncertainty Principle, but the time that the electron spends in intermediate states is not well specified. When the final electron Bloch state emerges, its phase is then unrelated to the initial phase of the electron wavefunction. Thus inelastic scattering inherently makes the phase before and after the scattering event incoherent. In the presence of scattering that breaks the phase coherence, it is useful to introduce a phase breaking length ℓ_ϕ. We can think of this as the distance an electron will travel while maintaining phase coherence.

Since inelastic scattering is typically much more strongly temperature dependent than elastic scattering (again, think of electron–phonon scattering), one can change the phase breaking length by varying the temperature. If we make the phase breaking length comparable to, or even smaller than, the system size, we enter an area where new phenomena, due to manifestations of the wave nature of the electron states, can occur. This is the area of mesoscopic physics. "Meso" means something like "in the middle" or "intermediate," and mesoscopic systems are larger than microscopic systems, which are of the order of maybe a Bohr radius and where we only have a few particles. Macroscopic systems contain perhaps 10^{23} particles and are very much larger than ℓ_e and ℓ_ϕ.

For a specific example, which will serve as a useful and illustrative model for making the transition from macro to meso, we consider a conductor of length L, width W, and thickness d. We start by taking L, W, and d all much greater than both ℓ_e and ℓ_ϕ. This system is depicted in Fig. 9.1.1. In a standard experiment, to which we will return several times, we inject a current into the system by connecting it to a potential difference across two terminals, e.g., 1 and 6. We then measure the resistance by noting the voltage drop along some section of the system at a fixed net current through the system. Let us assume that we inject a current into terminal 1, and draw

9.1 Conductance quantization in quantum point contacts

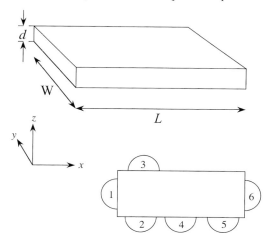

Figure 9.1.1. A system of length L, width W, and thickness d, with attached terminals numbered as shown.

the current out at terminal 6. We can then measure the voltage across two other terminals, such as 4 and 5. It is a property of a macroscopic system that the resistance we measure across probes 4 and 5 is the same regardless of whether we inject the current into probes 1, 2 or 3. In fact, the measured resistance is the same between any two probes separated by the same distance along the flow of the current. This is precisely due to the lack of phase coherence of the electron states. When we measure the voltage between, say, probes 4 and 5, we are measuring the lowest cost in energy to remove an electron from probe 5 and inject it at probe 4. This energy is due to the difference in electrochemical potential between the probes. When we remove an electron from probe 5, its phase is completely random because of the small ℓ_ϕ, and we have no way of figuring out whence (i.e., from which probe) the electron came. Similarly, an electron injected at probe 4 rapidly loses its phase and so is indistinguishable from other electrons at the electrochemical potential at that probe.

Now we take our model system into the mesoscopic regime. We do this by shrinking d and W until both are less than the phase breaking length ℓ_ϕ. As these lengths shrink, the energy states for motion along these directions will become discrete, with the separations between allowed energy eigenvalues growing as W^{-2} and d^{-2}, respectively. For example, if we consider periodic boundary conditions, the electron energies can be written as

$$\mathcal{E}(k_x, n, l) = \frac{\hbar^2 k_x^2}{2m_e} + \frac{2\pi^2 \hbar^2 n^2}{W^2 m_e} + \frac{2\pi^2 \hbar^2 l^2}{d^2 m_e},$$

where n and l are integers. We can then separate the energy values according to which sub-band n and l they belong. For the moment we assume that d is small enough that only the lowest sub-band $l = 0$ is occupied, and that only a small number of sub-bands $n \neq 0$ are occupied. In real mesoscopic systems, d can be of the order of one nanometer, with W ranging from perhaps a few nanometers to a few hundred nanometers. We also for now restrict our system to having only two external terminals, a source (S) and a drain (D), at which current is injected and drawn out, respectively, as in Fig. 9.1.2.

These kinds of system are rather easy to fabricate at the interface (known as a *heterojunction*) between two different semiconductors such as GaAs and GaAlAs grown by molecular beam epitaxy. Electrons (which are supplied by donors implanted some distance away from the heterojunction) are confined to move in the plane of the heterojunction. The source and drain can be made by doping heavily with donors in some small regions. The conducting channel connecting the source and the drain can be controlled by evaporating small metallic gates in the region between them, as shown schematically in Fig. 9.1.3. By applying a negative potential, or gate voltage V_G, to the gates,

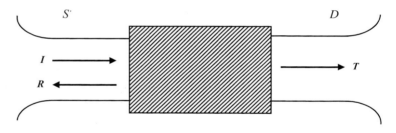

Figure 9.1.2. Schematic of a simple two-terminal device. An electron injected in one terminal has a probability R of being reflected and a probability T of being transmitted through the device to the other terminal.

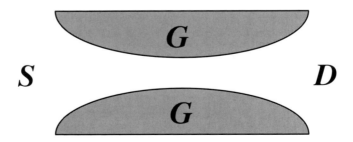

Figure 9.1.3. Schematic top view of a two-terminal device with gate electrodes.

the channel between source and drain through which the conduction electrons must pass can be made as narrow as we please. Heterojunctions can be grown cleanly enough to make the elastic mean free path of the order of or larger than the dimensions of the device. By cooling down to liquid-helium temperatures, the same can be achieved for the phase breaking length.

By measuring the voltage V_{SD} between source and drain as a function of the gate voltage V_G at fixed current I_{SD}, one can plot the conductance g in units of e^2/h as a function of V_G. In the extreme limit where $L \ll \ell_e$, the so-called ballistic limit in which electrons traverse the entire device without scattering elastically (or inelastically), one finds a remarkable result. The conductance shows a clear staircase-like behavior with steps at $g = Me^2/h$, with M an even integer. As the device is made longer and longer, the steps become noisier and noisier until they can no longer be discerned, even though each noisy curve is reproducible if the gate voltage is ramped up and down. This kind of behavior is sketched in Fig. 9.1.4.

Our first aim here is to understand the electrical conductance of this system. To this end, we first have to describe the electron states, and then how we drive a current through the system. Landauer pioneered mesoscopic physics with his insight that conduction should fundamentally be regarded as a scattering process in which we describe the electron states locally in our mesoscopic system and consider separately the means by which current is driven by an applied electrochemical potential difference. So what does this mean? Let us start with the electron states. We formally separate the mesoscopic part of the system, which we term the device and is the block of nominal length L, width W, and height d, from the terminals that are used to connect the device to the sources of electrochemical potential. Inside the device, the electron states maintain their phase coherence, since $L \ll \ell_\phi$. The

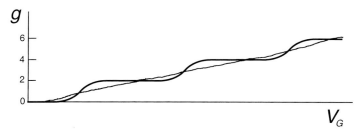

Figure 9.1.4. Conductance g (in units of e^2/h) vs. gate voltage V_G for a two-terminal ballistic device. At low temperatures (bold curve) the electrons pass through the device without any scattering, resulting in quantized conductance. At higher temperatures, the electrons scatter inelastically from phonons in the device, smearing out the quantization of the conductance (light curve).

two terminals of the system, S and D, are connected to very large reservoirs kept at thermodynamic equilibrium at electrochemical potentials $\mu + \Delta\mu$ and μ, respectively. These potentials are well defined deep in the reservoirs (far away from the actual device). The reservoirs inject electrons with energies up to the respective electrochemical potentials into the device through the terminals, and electrons flowing through the device exit it into the terminals and thereby enter the reservoirs. Dissipative processes within these reservoirs quickly thermalize electrons and randomize their phases, so that electrons entering the device from the reservoirs have random phases. Consequently, interference terms between *different* electron states average to zero and can be ignored. The natural way to describe the electrons that enter and leave the device is by using a basis of scattering states. In the scattering theory that we learn in basic quantum mechanics, we have well defined (asymptotic) incoming and outgoing electron states, which are connected by a scattering matrix, or S-matrix. The matrix elements $s_{\alpha\beta}$ of the S-matrix give the probability amplitude that an incoming electron in state β is scattered into the outgoing state α. Here, we have incoming states into the device from the reservoirs through the terminals, and these states can be scattered into outgoing states from the device into the terminals and the reservoirs by some potential in the device. The assertion that different states do not interfere with one another simplifies the discussion quite substantially, since we then do not have to work with complex transmission amplitudes (the elements of the S-matrix itself), but only the real probabilities of scattering from an incoming state to an outgoing one. The incoming and outgoing states are confined by some potential $V(y)$ (we can neglect the dependence on z since we are only considering the lowest sub-band of motion along z). For now, we are only considering a confining potential without any additional more complicated scattering inside the device itself. In the presence of the confining potential, the incoming and outgoing parts of the electron states can then locally be written as

$$\psi_{n,k}(x, y) = e^{\pm ikx} h_n(y), \quad (9.1.1)$$

where n is the sub-band index, and $h_n(y)$ is the transverse part of the wavefunction of the sub-band, or *channel*, n. This is a description of the electron states in the terminals – we do not attempt to describe the states in the device itself. The corresponding energy eigenvalue is

$$\mathcal{E}_{n,k} = \frac{\hbar^2 k^2}{2m} + \mathcal{E}_n.$$

9.1 Conductance quantization in quantum point contacts

Note that this is perfectly general, and given a good model for $V(y)$, we can determine $\mathcal{E}_n$. These states are solutions of the Schrödinger equation in the terminals, but are not appropriate scattering states. Those can be constructed by making linear combinations of the states given in Eq. (9.1.1). A state (s, m) incoming from the source into the device in channel m with unit amplitude is

$$e^{ik_{sm}x}h_{s,m}(y)$$

where we now also attach an index s to the sub-band wavefunction (for ease of notation we do not include the wavevector index). This will allow for generalization later, when the incoming sub-bands from different terminals are not necessarily the same. The probability that this state is scattered into an outgoing state (d, n) in the drain is then $T_{ds,nm}$, and the probability that the state is reflected into an outgoing state $e^{-ik_{sm}x}h_{s,m}(y)$ in the source is $R_{ss,mm}$. The scattering is elastic, so all states connected by $T_{ds,nm}$ and $R_{ss,mm}$ have the same energy eigenvalues. Note that in the absence of an external magnetic field there must exist an outgoing state in the same terminal with the same energy but opposite wavenumber.

We adopt the standard convention of positive incoming velocities, and at the same time change our convention about the sign of the charge on the electron. From now on, we shall make explicit the negative nature of this charge by writing it as $-e$, with e a positive quantity. The incoming current $i_{s,m}$ carried by the state (s, m) into the device is then

$$i_{s,m} = -ev_{s,m}(k) = -\frac{e}{\hbar}\frac{\partial \mathcal{E}_{s,m}(k)}{\partial k}\bigg|_{k=k_{sm}}.$$

This incoming state is scattered into states that are outgoing in the terminals and carry the current out of the device. Current conservation must be strictly obeyed, so the outgoing currents in all terminals add up to the incoming current. The transmitted outgoing current in state (d, n) is then $ev_{sm}T_{ds,nm}$, and the outgoing reflected current in state (s, m') is $ev_{sm}R_{ss,m'm}$.

Suppose now that we have applied an electrochemical potential difference $\Delta\mu$ between the source and the drain, so that source and drain are at electrochemical potentials $\mu + \Delta\mu$ and μ, respectively. With $f_{s,m}(\mathcal{E}_k)$ denoting the occupation numbers of the incoming states, the incoming current in the source is

$$I_s = -e\sum_{km} f_{s,m}(\mathcal{E}_k)v_{s,m}.$$

Similarly, the incoming current from the drain is

$$I_d = -e \sum_{kn} f_{d,n}(\mathcal{E}_k) v_{d,n}.$$

We can find the total current between source and drain by examining the current flowing near the source. Here, the total current consists of the difference between the *net* incoming current from the source (incoming from source minus current reflected into the source) and the part of the current from the drain that is transmitted to the source. The total current is then obtained by summing over all channels in source and drain:

$$I = -e \sum_{km} f_{s,m}(\mathcal{E}_k) v_{s,m} \left(1 - \sum_{m'} R_{ss,m'm}\right) + e \sum_{kn} f_{d,n}(\mathcal{E}_k) v_{d,n} \sum_{m'} T_{sd,m'n}.$$

Current conservation dictates that incoming current minus reflected current equals transmitted current,

$$1 - \sum_{m'} R_{ss,m'm} = \sum_{m'} T_{ds,m'm},$$

so we can write the total current as

$$I = -e \sum_{km} f_{s,m}(\mathcal{E}_k) v_{s,m} \sum_{m'} T_{ds,m'm} + e \sum_{kn} f_{d,n}(\mathcal{E}_k) v_{d,n} \sum_{m'} T_{sd,m'n}. \quad (9.1.2)$$

We first change the sums over wavevectors to integrals. When we do that, we have to insert the density of states in k-space, which in one dimension is just a constant, $1/2\pi$. We then change the integration variable to energy and insert a factor of $dk/d\mathcal{E}$ due to this change. Since the velocity is proportional to $d\mathcal{E}/dk$, we see that the density-of-states factor precisely cancels with the velocity in the integrals. This happens only because we can in mesoscopic physics consider separate, effectively one-dimensional, conducting channels. Because of this cancellation, all that is left in Eq. (9.1.2) are occupation numbers and transmission and reflection probabilities. All the specifics of the device, such as length and width, have disappeared. That makes the Landauer approach particularly powerful, simple, and beautiful.

9.1 Conductance quantization in quantum point contacts

The total current in the device is then

$$I = -\frac{e}{h}\int d\mathcal{E} \sum_m f_{s,m}(\mathcal{E}) \sum_{m'} T_{ds,m'm} + \frac{e}{h}\int d\mathcal{E} \sum_n f_{d,n}(\mathcal{E}) \sum_{m'} T_{sd,m'n}. \quad (9.1.3)$$

We now make the assumption that the driving force (in this case the electrochemical potential difference) is sufficiently small that in calculating the current we need only consider the leading term, which is proportional to $\Delta\mu$. We can then ignore the energy dependence of the transmission probabilities and evaluate them at the electrochemical potential μ. It is then convenient to define total transmission probabilities T_{sm} and T_{dn} for each channel by summing out the scattered channels. Thus

$$T_{sm} \equiv \sum_{m'} T_{ds,m'm}$$
$$T_{dn} \equiv \sum_{m'} T_{sd,m'n}, \quad (9.1.4)$$

where it is understood that the sums are evaluated at the electrochemical potential μ. Using Eq. (9.1.4) we then obtain for the total current

$$I = -\frac{e}{h}\sum_m \int d\mathcal{E}[f_{s,m}(\mathcal{E})T_{sm} - f_{d,m}(\mathcal{E})T_{dm}].$$

In the absence of an external magnetic field, the system is invariant under time reversal. This imposes constraints on the S-matrix, with the consequence that $T_{sm} = T_{dm} \equiv T_m$. This simplifies the expression for the total current, which now becomes

$$I = -\frac{e}{h}\sum_m T_m \int d\mathcal{E}[f_{s,m}(\mathcal{E}) - f_{d,m}(\mathcal{E})]$$
$$= -\frac{e}{h}\sum_m T_m \int_\mu^{\mu+\Delta\mu} d\mathcal{E} = -\frac{e}{h}\sum_m T_m \Delta\mu.$$

This is a remarkable and simple result: the total current is just the driving force times the transmission probability at the electrochemical potential times a universal constant. As we have discussed earlier, the voltage measured between source and drain is just the electrochemical potential difference, divided by the electron charge. The measured resistance is then

$$R_{sd} = -\Delta\mu/(eI) = \frac{h}{2e^2 \sum_m T_m},$$

where the additional factor of 2 comes from the two degenerate spin directions. For the case of a ballistic channel, $T_m \equiv 1$, so we obtain

$$R_{sd} = \frac{h}{e^2 M}, \qquad (9.1.5)$$

where M is the total number of channels (including spin degeneracy) connected by source and drain to the device. Equation (9.1.5) predicts a quantized conductance, just as is observed in measurements on quantum point contacts. The quantum of conductance is e^2/h, and the quantum number is M, the number of current-carrying channels in the system. For the conductance to be quantized, the channel has to be smaller than ℓ_e so that electrons traverse the device ballistically and $T_m = 1$. If the channel becomes longer than ℓ_e, or some scatterer is introduced into the channel, the transmission probabilities will in general be less than unity, and the conductance is no longer quantized. The steps start to degrade and become noisy, but the I–V curves are retraced if the gate voltage is swept up and down. If, on the other hand, the channel is made larger than ℓ_ϕ, there will be random inelastic scattering in the channel. The conductance will similarly not be quantized, but in this case there will be thermal noise on the I–V curve which will not be repeatable if the gate voltage is cycled.

One may wonder how it is that a ballistic device has a finite conductance. First of all, even though there is no scattering in the device itself, there must be inelastic scattering in the reservoirs in order for the electrons to thermalize. Each channel has a finite conductance due to this *contact resistance*, and a mesoscopic device with a finite number of channels cannot have infinite conductance. As the number of channels grows, so does the total conductance and we recover the perfect conductor (with infinite conductance) in the limit of an infinite number of channels.

9.2 Multi-terminal devices: the Landauer–Büttiker formalism

We hinted in the previous section at the fact that resistance in a mesoscopic device in general depends on which terminals are used as source and drain, and which terminals are used as voltage probes. In order to demonstrate this, we need to generalize the Landauer formula to multi-terminal devices. This generalization was developed by Markus Büttiker, and is a fundamental cornerstone of mesoscopic physics. A simple introductory example is given by a four-probe measurement, as depicted in Fig. 9.2.1. In such a measurement, the current flows between source and drain, and the electrochemical

9.2 Multi-terminal devices: the Landauer–Büttiker formalism

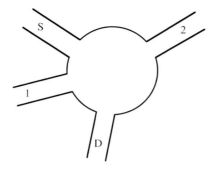

Figure 9.2.1. Schematic of a four-terminal device with current source (S) and drain (D), and two extra terminals 1 and 2.

potential difference between terminals 1 and 2 is measured. An ideal voltmeter has infinite internal resistance, and so the proper boundary condition to be imposed on terminals 1 and 2 is that the net current in (or out) of such terminals should be zero. On the other hand, there must be some well defined electrochemical potential associated with each of the probes 1 and 2 in order for a measurement of the electrochemical potential difference to make sense. Furthermore, electrons injected from one terminal into the system have in general finite probabilities of ending up at any of the other terminals. For example, electrons injected from the source may end up going into terminal 2. Since terminal 2 is also connected to a reservoir, just like the source and the drain, it too injects electrons into the system. The net current in or out of this terminal depends on the balance between incoming and outgoing currents. We can adjust this balance by changing the electrochemical potential of the reservoir to which this terminal is attached. In the end, we must then ensure that the electrochemical potentials at terminals 1 and 2 are self-consistently adjusted to yield zero net currents at these terminals.

Let us now formally state this for a general multi-terminal system with terminals $1, 2, 3, \ldots, N$. The terminals are in contact with reservoirs at well defined electrochemical potentials μ_i, with $i = 1, 2, \ldots, N$. Carriers are injected from the reservoir in all states with energies up to μ_i into terminal i. Electrons injected from the terminals have random phases and do not interfere with one another. An electron injected in state (i, m) (channel m in terminal i) has a probability $T_{ji,nm}$ of being scattered to outgoing state (j, n) (channel n in terminal j), and a probability $R_{ii,m'm}$ of being reflected into an outgoing state (i, m'). In looking solely for the linear response, we can ignore the energy dependence of the scattering and reflection probabilities and evaluate them at a common energy μ_0. We take this to be the lowest of the electrochemical potentials μ_i, $i = 1, 2, \ldots, N$. Then all states in the device with energies less than μ_0 are occupied and do not contribute to a net current in the device (why?). Terminal i then injects a current

$$I_{i,\text{injected}} = \sum_{m=1}^{M_i} -\frac{e}{h}(\mu_i - \mu_0),$$

where M_i is the total number of occupied channels at terminal i. The net current at terminal i is the difference between this injected current and the sum of the reflected current at the terminal and the current scattered into this terminal from other terminals. This net current is

$$I_i = -\frac{e}{h}\left[(M_i - R_{ii})\mu_i - \sum_{j(\neq i)} T_{ij}\mu_j\right]. \tag{9.2.1}$$

Here we have again used the definition $T_{ij} = \sum_{mm'} T_{ij,m'm}$ for the total transmission probability and have defined $R_{ii} = \sum_{mm'} R_{ii,m'm}$ for the reflection probability, even though this risks confusion with the resistance R_{ij}. Since the reference potential μ_0 is common to all reservoirs, it cancels out in Eq. (9.2.1). Current conservation dictates that the injected current in terminal i equal the reflected current and the total transmitted current to other terminals. In other words,

$$M_i = R_{ii} + \sum_{j(\neq i)} T_{ij}. \tag{9.2.2}$$

If we insert this into Eq. (9.2.1) we can write the net current I_i as

$$I_i = -\frac{e}{h}\sum_{j(\neq i)} T_{ij}(\mu_i - \mu_j). \tag{9.2.3}$$

Equations (9.2.2) and (9.2.3) express current conservation at a terminal and a relation between total current and driving forces. These are the mesoscopic versions of Kirchhoff's Laws.

Time reversal symmetry imposes constraints on the scattering matrix that connects incoming and outgoing states. In the absence of magnetic fields, we can reverse the directions of all incoming and outgoing states, and because of time reversal symmetry we must then have $T_{ij} = T_{ji}$. In the presence of a magnetic field, we can reverse the direction of all velocities if we also reverse the sign of the flux Φ penetrating the system, so in the more general case of an

9.2 Multi-terminal devices: the Landauer–Büttiker formalism

applied magnetic field, we have $T_{ij}(\Phi) = T_{ji}(-\Phi)$. This symmetry leads to a reciprocity theorem that relates resistances when the current and voltage probes are interchanged.

Armed with Eqs. (9.2.1) and (9.2.3) we can now go ahead and calculate the resistance $R_{ij,kl}$ due to a chemical potential difference $\mu_k - \mu_l$ between two terminals k and l when a current I flows from source (terminal i) to drain (terminal j). This amounts to solving the system of linear equations Eq. (9.2.1) for the electrochemical potentials μ_k and μ_l under the conditions that $I_m = 0$ for $m \neq i, j$, $I_i = -I_j \equiv I$, and $\mu_i = \mu_0 + \Delta\mu$, $\mu_j = \mu_0$. Once we have those, the resistance $R_{ij,kl}$ is obtained by just using Ohm's law. The voltage V between the terminals k and l is the electrochemical potential difference between the two terminals, divided by $-e$, and so $V = -\Delta\mu/e$, and $R_{ij,kl} = V/I$. The currents are proportional to the applied electrochemical potential difference, so it will cancel from the expression for the resistance. In the end, the resistance will be h/e^2 (which is a unit of resistance) times some combination of the transmission probabilities (which are dimensionless). We first work this out explicitly for the three-terminal case, and then sketch an approach for a general N-terminal case. So let us assume that we have a three-terminal device, with terminal 1 the source and terminal 3 the drain, and we are measuring a voltage between terminals 1 and 2. Thus, $I_1 = -I_3 = I$, and $I_2 = 0$. We will calculate the resistance $R_{13,12}$ due to the voltage between probes 2 and 1 with a current flowing from probe 1 to probe 3. Equations (9.2.1) then become

$$I = -\frac{2e}{h}[(M_1 - R_{11})\mu_1 - T_{12}\mu_2 - T_{13}\mu_3]$$

$$0 = -\frac{2e}{h}[(M_2 - R_{22})\mu_2 - T_{21}\mu_1 - T_{23}\mu_3] \quad (9.2.4)$$

$$-I = -\frac{2e}{h}[(M_3 - R_{33})\mu_3 - T_{31}\mu_1 - T_{32}\mu_2],$$

where the factor of 2 comes from summing over spin channels. Using Eq. (9.2.2), we can solve for μ_3 from the second line in Eq. (9.2.4) to obtain

$$\mu_3 = \frac{T_{21} + T_{23}}{T_{23}}\mu_2 - \frac{T_{21}}{T_{23}}\mu_1.$$

We insert this into the third line of Eq. (9.2.4), and after collecting some factors we obtain

$$-I = -\frac{2e}{hT_{23}}D(\mu_2 - \mu_1),$$

where we have defined $D = T_{31}T_{21} + T_{31}T_{23} + T_{32}T_{21}$. Then

$$R_{13,12} = -\frac{\mu_1 - \mu_2}{eI} = \frac{h}{2e^2}\frac{T_{23}}{D}.$$

Note that we could also have inserted the expression for μ_3 into the first line of Eq. (9.2.4) and obtained the same result. That is a consequence of having fewer independent chemical potentials than we have equations to solve.

The fact that the μ_i are overdetermined poses computational difficulties in the general case. Normally we would solve a set of linear equations like Eq. (9.2.3) by inverting a matrix, in this case the matrix

$$\hat{T}_{ij} \equiv \frac{e}{h}[(M_i - R_{ii})\delta_{ij} - T_{ij}]$$

(where it is understood that $T_{ii} = 0$). However, only the $N - 1$ electrochemical potential differences can be independent, and not the N electrochemical potentials at all reservoirs, and so the system of linear equations is singular, and the matrix cannot be inverted. The approach we take to get around this difficulty is first to note the obvious fact that setting all electrochemical potentials equal will yield a solution with all currents equal to zero. What is less obvious is that there may also be other sets of electrochemical potentials that yield zero currents. These are said to form the *nullspace* of the matrix $\hat{T}_{ij}$. Formally, the nullspace of an $N \times N$ singular matrix **A** consists of all N-dimensional vectors **x** such that $\mathbf{A} \cdot \mathbf{x} = 0$. Now these are most certainly not the solutions that we are interested in – in fact they are the problem rather than the solution! Technically, we need to separate out the "nullspace" from the domain of the linear mapping defined by the matrix $\hat{T}_{ij}$; that is, we need to weed out all the sets of electrochemical differences that yield all zero currents from those that yield the finite currents I_i that we have specified. A general and very powerful way of doing this (which can also be efficiently implemented on computers) is provided by a technique known as the *singular value decomposition* of the matrix $\hat{T}_{ij}$. This involves writing $\hat{T}_{ij}$ as a product of three matrices, one of which is diagonal and has only positive or zero elements (these are the *singular values*). The nontrivial solutions for the μ_i in terms of the currents can then, after some lengthy manipulations, be expressed in terms of another diagonal matrix that also has only positive or zero elements. Having thus formally obtained the electrochemical potentials as functions of the currents, one can calculate the resistance. While we have omitted some of the details in this description, we can nevertheless note that the general procedure has been as follows:

1. The current at the source is I and the current at the drain is $-I$.
2. The currents at all other terminals are set to zero.
3. We write down the linear equation (9.2.1) using these boundary conditions.
4. We solve this equation for the nontrivial electrochemical potentials by using singular value decomposition.
5. With electrochemical potentials obtained as functions of the applied current, we calculate the resistance between voltage probes.

The main point here is that the final expression for the resistance will involve transmission probabilities from all terminals i, j, k, and l in a combination that depends on which probes are source and drain, and which are voltage probes. This has the implication that the resistance will now depend on how the measurement is conducted and not just on a material parameter (resistivity) and geometry. As a consequence of the phase coherence in the mesoscopic system, an electron carries with it phase information from its traversal of the system, and the probability that an electron will reach one terminal depends on where it was injected.

The multi-terminal Landauer–Büttiker formula has become the standard approach for analyzing mesoscopic transport in areas ranging from quantum point contacts to the quantum Hall effect and spin-dependent tunneling transport. The basic physics underlying weak localization can also be understood from the point of view of the Landauer–Büttiker formalism, although in this case it does not easily lead to quantitative predictions. It is very intuitive, simple, and powerful. It also satisfies symmetries that lead to some specific predictions referred to as reciprocities, which have been verified experimentally. We will discuss some other consequences when we introduce a magnetic field in Chapter 10.

9.3 Noise in two-terminal systems

Any signal we ever measure has to be detected against a background of noise. Usually in practical applications, such as telecommunications, noise is a nuisance and we try to suppress it as much as possible. If the signal-to-noise ratio is low – it does not matter how large the actual signal amplitude is – the signal may be drowned in noise. But noise also contains information about the physical processes occurring in a system. Different processes have different kinds of noise, and by carefully analyzing the noise we can gather useful information. Here we look at some of the noise sources in a mesoscopic system and their characteristics. For simplicity, we here consider only two-terminal systems.

Noise is most conveniently analyzed in terms of its spectral density $S(\omega)$, which is the Fourier transform of the current–current correlation function,

$$S(\omega) = 2 \int_{-\infty}^{\infty} dt\, e^{i\omega t} \langle \Delta I(t + t_0, T) \Delta I(t_0, T) \rangle.$$

Here $\Delta I(t, T)$ is the time-dependent fluctuation in the current for a given applied voltage V at a given temperature T. In a two-terminal electric system there are two common noise sources. The first one is the thermal noise, or Johnson noise, of a device of conductance $G = 1/R$ at temperature T. It is due to thermal fluctuations, and can be derived using the fluctuation-dissipation theorem. For low frequencies (low, that is, compared with any characteristic frequency of the system, and such that $\hbar\omega \ll kT$), the Johnson noise has no frequency dependence, and is said to be white. The spectral density is given by

$$S = 4kTG.$$

The other common noise source is shot noise, which occurs when the current is made up of individual particles. When the transits of the particles through the device are uncorrelated in time, these processes are described as Poisson processes, and their characteristic noise is known as Poisson noise. The Poisson noise is also white for low frequencies, with a spectral density that is proportional to the current:

$$S_{\text{Poisson}} = 2eI.$$

Shot noise is a large contributor to noise in transistors, but in mesoscopic systems correlations can suppress it quite dramatically, giving it a spectral density much below that of Poisson noise. In the Landauer–Büttiker formalism, the current is due to transmission of electrons occupying scattering states. The Pauli principle forbids multiple occupancy of these states, which in two-terminal systems necessarily correlates the arrival of electrons to the source from a single scattering state.

Let us now look at this more quantitatively. At finite temperatures, the Landauer two-terminal formula is

$$I = \frac{e}{h} \int_0^{\infty} d\mathcal{E}\, [f_s(\mathcal{E}) - f_d(\mathcal{E})] T_n(\mathcal{E}),$$

9.3 Noise in two-terminal systems

where $f_s(\mathcal{E})$ and $f_d(\mathcal{E})$ are the Fermi distribution functions of states injected from the source and drain, respectively, and are given by

$$f_s(\mathcal{E}) = \frac{1}{1 + e^{(\mathcal{E}-\mu-eV)/kT}}$$

$$f_d(\mathcal{E}) = \frac{1}{1 + e^{(\mathcal{E}-\mu)/kT}}.$$

It is not a difficult exercise to evaluate the spectral density by inserting a current operator and using the relation between incoming and outgoing currents given by the Landauer formalism. With the linear response assumption (so that the transmission probabilities are taken to be independent of energy and evaluated at the common chemical potential) the result is

$$S = 2\frac{e^2}{h}\sum_n [2T_n^2 kT + T_n(1-T_n)eV \coth(eV/2kT)]. \quad (9.3.1)$$

This expression contains several interesting results. In the limit $eV/kT \to 0$ it reduces to the Johnson noise, and it is reassuring that we recover the central result of the fluctuation-dissipation theorem from the Landauer–Büttiker formalism. The second term is the shot noise. This one has some peculiar characteristics in a mesoscopic system. In the limit of zero temperature, the shot noise part of Eq. (9.3.1) becomes

$$S_{\text{shot}}(T \to 0) = 2eV \frac{e^2}{h} \sum_n T_n(1-T_n).$$

States for which $T_n = 1$ or $T_n = 0$ do not contribute to the shot noise. Shot noise represents fluctuations due to the uncorrelated arrivals of electrons. If $T_n = 1$, that transmission channel is "wide open," fully transmitting a steady stream of electrons without any fluctuations, which are suppressed by the Pauli Exclusion Principle. Similarly, if $T_n = 0$ there is no shot noise simply because there are no electrons at all arriving in that channel. The maximum noise that a single channel can contribute apparently occurs for $T_n = 0.5$, when the channel is half-way between closed and open, and there is maximum room for fluctuations.

This characteristic of the noise can be verified experimentally. Consider a two-terminal quantum point contact and its conductance as a function of gate voltage. Suppose the initial gate voltage is such that the conductance is quantized. There is then an integer number of channels for which $T_n = 1$

while for all others $T_n = 0$, and so the shot noise is zero. As the gate voltage is changed, the conductance moves towards a transition region where the conductance changes value. This happens when a new channel is opened or the highest-lying channel (in energy) is being pinched off. As this happens, the corresponding transmission probability goes from zero to unity (or vice versa), and the shot noise increases and goes through a maximum. As the conductance levels off on a new plateau, the shot noise vanishes. The effect of small, nonzero temperatures is just to round off the shot noise curve.

The shot noise in quantum point contacts has been measured, and the predictions described above have been verified, again demonstrating the simplicity and power of the Landauer–Büttiker formalism.

9.4 Weak localization

In general, the conductance of a metallic system increases monotonically as the temperature is reduced, but there are cases in which the conductance exhibits a maximum and then decreases as the temperature is reduced further. One such example is the Kondo effect, which will be the subject of Chapter 11. This phenomenon is the consequence of interactions between a local spin and the spins of conduction electrons. Systems that exhibit the Kondo effect are invariant under time reversal, and the conductance maximum is caused by the onset of strong interactions between conduction electrons and the local spin. There are, however, other systems that exhibit a maximum in the conductance but for which, in contrast to Kondo systems, the conductance maximum is closely related to issues of time reversal symmetry. One such example occurs when the conduction electrons in manganese are elastically scattered by impurities. As the temperature is decreased below a few kelvin the conductance decreases. Furthermore, at temperatures well below this conductance maximum the system exhibits *negative* magnetoresistance, which is to say that if an external magnetic field is applied the conductance *increases*. This indicates that the cause of the conductance maximum somehow depends on time reversal symmetry, and is destroyed if that symmetry is broken. Further evidence of this is given by the fact that if some small amount of gold is added, the magnetoresistance is initially positive as the external field is applied. Gold is a heavy element and has strong spin–orbit scattering, which destroys time reversal invariance by a subtle effect called the Aharonov–Casher effect.

This phenomenon of decreasing conductance, or increasing resistance, at low temperatures in the presence of time reversal symmetry is another manifestation of the long-range phase coherence of the electron wavefunctions. In

9.4 Weak localization

this case, the coherence leads to interference effects in which an electron interferes destructively with itself. In order for this to be possible at all, the phase breaking length must be long enough that the electrons diffuse through elastic scattering while maintaining their phase coherence for some reasonable distance. That is why the temperature has to be low in order for the effect, which is called weak localization, to be observable.

In principle, all the physics of weak localization is contained within the Landauer–Büttiker formalism. In this case, the transmission probabilities T_n must show some strong behavior for certain channels for which the interference effects must somehow reduce T_n. Note that while the Landauer–Büttiker formalism assumes that *different* electrons have no phase relations and so do not interfere with each other, it certainly leaves open the possibility that each electron state can interfere with itself on its path from one terminal to another. However, in the case of weak localization the Landauer–Büttiker formalism does not easily lend itself to practical calculations. In fact, in order to deal correctly with the problem, one has to use rather sophisticated many-body perturbation techniques. Instead, we will here give some more intuitive arguments for what lies behind weak localization.

We consider an electron as it traverses a mesoscopic system from source to drain. In this case, there is a rather high density of impurities, so the electron scatters frequently. As a consequence, the electron performs a random walk through the system. In many respects this is similar to the case for "normal" electron transport in which there is no phase coherence. The motion of a random walker can, at times long compared with a typical time between collisions, be described as a classical diffusion problem. According to the Einstein relation, the diffusion constant D_0 is proportional to the mobility, and hence to the conductance. In three dimensions, the probability that a particle has moved a net distance r in a time t is given by

$$P_3(r, t) = \frac{\exp(-r^2/4D_0 t)}{(4\pi D_0 t)^{3/2}}.$$

From this equation, we can find the probability amplitude that the particle returns to its original position. In general, there may be many different paths that take the electron back to the origin during some infinitesimal time interval dt at some time t. Let us for simplicity consider two such paths with probability amplitudes $\psi_1(r=0, t)$ and $\psi_2(r=0, t)$. To find the probability we have to take the squared modulus of the sum of the probability amplitudes, $|\psi_1(r=0, t)|^2 + |\psi_2(r=0, t)|^2 + \psi_1^*(r=0, t)\psi_2(r=0, t) + \psi_1(r=0, t)\psi_2^*(r=0, t)$. In "normal" macroscopic systems, the electron suffers inelastic collisions which randomize the phase along each path. Consequently, there is no phase relation between probability amplitudes $\psi_1(r = 0, t)$ and $\psi_2(r = 0, t)$, and the interference terms vanish as we add up contributions from all possible paths. In mesoscopic systems, the phase is preserved for all paths shorter than ℓ_ϕ, since then the electron returns to the origin within the phase breaking time $\tau_\phi = \ell_\phi/v_F$. Most of those paths, however, also have random relative phases and the interference terms vanish. However, there is now one class of paths for which the interference terms do not vanish. These are paths that are related by time reversal. In the absence of a magnetic field such paths have probability amplitudes that are precisely complex conjugates of each other, $\psi_1(r = 0, t) = \psi_2^*(r = 0, t)$, and these paths interfere constructively. This means that the particle has an *enhanced* probability of returning to the origin, compared with incoherent classical diffusion. As a consequence, the probability that the particle has moved some net distance in a time t is reduced. This reduces the diffusion constant, and, according to the Einstein relation, reduces the conductivity.

We can make the argument more quantitative by considering an electron wavepacket at the Fermi surface. In order for the wavepacket to be able to interfere it must have a spatial extent Δx of the order of its wavelength, $\Delta x \approx \lambda_F$. In a time δt, the wavepacket thus sweeps out a volume $\delta \Omega \approx \lambda_F^{d-1} v_F \delta t$, where d is the spatial dimensionality of the system. This wavepacket can interfere with itself provided it returns to the origin at some time t. The probability for this to happen is

$$P(r = 0, t) \approx \frac{\delta \Omega}{(4\pi D_0 t)^{d/2}} = \frac{\lambda_F^{d-1} v_F \delta t}{(4\pi D_0 t)^{d/2}}.$$

Each such event decreases the effective diffusion. We now need to sum over all such events. These can occur only at times less than the phase breaking time τ_ϕ, since for longer times the phases of the two time-reversed paths have been randomized and will no longer interfere. We must also insert some minimum time τ_0 below which there are on average no collisions, denying the electron any chance of returning. This lower limit is of the order of the elastic scattering time $\tau_e = \ell_e/v_F$. In three dimensions we put $d = 3$ to find

$$\int_{\tau_e}^{\tau_\phi} dt P_3(r = 0, t) \approx \frac{2\lambda_F^2 v_F [(\tau_e)^{-1/2} - (\tau_\phi)^{-1/2}]}{(4\pi D_0)^{3/2}}. \tag{9.4.1}$$

This probability that an electron can return to the origin with some memory of its original phase will be of roughly the same magnitude as the fractional

9.4 Weak localization

reduction in conductance $\Delta\sigma/\sigma$ caused by the interference of the time-reversed paths. If we substitute the Drude formula $D_0 = v_F \ell_e / d$ we find the fractional increase in the resistivity to be

$$\Delta\rho/\rho \approx [1 - (\tau_e/\tau_\phi)^{1/2}]/(k_F\ell_e)^2. \qquad (9.4.2)$$

By including the temperature dependence of the elastic relaxation time (which yields the temperature dependence of the elastic mean free path) and the phase breaking time we also get an estimate of the overall temperature dependence of the increase in resistivity.

In two dimensions, the result is more striking, since the denominator in the expression for $P_2(r = 0, t)$ decays only as t^{-1} rather than as $t^{-3/2}$. An integration analogous to that in Eq. (9.4.1) gives the fractional increase in resistivity to be

$$\Delta\rho/\rho \approx \frac{1}{k_F\ell_e} \ln\left(\frac{\tau_\phi}{\tau_e}\right), \qquad (9.4.3)$$

which can be quite significant if the ratio of phase breaking time to elastic relaxation time is large. In two dimensions, the probability of a random walker returning to the origin is much larger than in three dimensions, which leads to a dramatic enhancement in the resistivity.

Note that weak localization depends sensitively on the phase relation between time-reversed paths. If this relation is altered, the weak localization is in general suppressed. One way to achieve this is to apply an external magnetic field. This adds a so-called Aharonov–Bohm phase to the path of an electron. For closed paths, this phase is equal to the 2π times the number of flux quanta enclosed by the path, and the sign is given by the sense of circulation of the path. If the Aharonov–Bohm phase added to one path is $\Delta\phi$, then the time reversed path gets an added phase $-\Delta\phi$, and the interference term between these two paths now has an overall phase factor. Since different pairs of time-reversed paths will have different, and, on average random, phase factors, the net effect of the magnetic field is to wipe out rapidly the weak localization as the contributions to the increase in resistance from different pairs of paths are added together. Hence, the resistance of a system in the weak localization regime is observed to decrease ("negative magnetoresistance") as an external magnetic field is applied.

There are also other phenomena associated with phase coherence. One such is the existence of the so-called universal conductance fluctuations. If the conductance of a mesoscopic system is measured as a function of some external control parameter, for example a magnetic field or impurity configuration, there are seen to be fluctuations in the conductance. The specific fluctuations are sample-dependent and reversible as the external control parameter is swept up and down, but the magnitude is universal and is of the order of the conductance quantum e^2/h. The root cause of the fluctuations is interference between different paths between two points in the sample. As the control parameter is varied, the precise interference patterns change but do not disappear, in contrast to weak localization. The universal conductance fluctuations are, however, suppressed in magnitude by processes that destroy time-reversal invariance. In fact, one can use universal conductance fluctuations to detect the motion of single impurities in a sample, as this will lead to observable changes in the conductance fluctuation pattern.

9.5 Coulomb blockade

We close this chapter by briefly discussing a phenomenon known as *Coulomb blockade*. Although it does not *per se* depend on wavefunction coherence, it is intimately related to small (nano-scale) devices, and so one can make the case that it is a mesoscopic effect. It is a quantum phenomenon in that, while it does not depend directly on $\hbar$, it does rely for its existence on the quantized nature of electric charge. It reflects the fact that the capacitance of mesoscale devices can be so small that the addition of a single electron may cause an appreciable rise in voltage.

We consider a small metallic dot, like a tiny pancake, connected to two leads. We here explicitly take the connections to the leads to be *weak*. This means that there are energy barriers that separate the leads from the dot, and through which the electrons have to tunnel in order to get on and off the dot. This allows us, to a reasonable approximation, to consider the electron states on the dots as separate from the electron states in the leads. As a consequence, it is permissible to ask how many electrons are on the dot at any given time. If the leads had been strongly coupled to the dot, then electron eigenstates could simultaneously live both on the dot and in the leads, and we would not have the restriction that only an integer number of electrons could reside on the dot. Also, the resistances at the junctions with the dot have to be large enough that essentially only one electron at a time can tunnel on or off. The condition for this is that the junction resistances be large compared with the resistance quantum $R_Q = h/e^2$. The tunneling rates are then low enough that only one electron at a time tunnels. Note also that it is important that we consider the dot to be metallic, so that there is a fairly large number of electrons on the dot, and the available states form a continuum. Technically,

9.5 Coulomb blockade

we can then consider the operator for the number of electrons to be a classical variable, just as we did in the liquid-helium problem when we replaced the number operator $a_0^\dagger a_0$ for the condensate with the simple number N_0. Experimentally, the dots can also be made out of semiconductors, in which case one may have a very small number of electrons – of the order of ten or so – on the dots. In that case, the discrete spectrum of eigenstates on the dots has to be considered more carefully, and there may be interesting and complicated correlation effects between the electrons.

We now imagine that we connect the leads to some source of potential difference V, and we monitor the current that flows from one lead, through the dot, and into the other lead. What we find is that for most values of V the current is totally negligible, while for some discrete set of voltages the conductance through the dot is rather high, resulting in a conductance vs. bias voltage curve that looks rather like a series of evenly spaced delta-functions. It turns out that it is rather easy to come up with a qualitative picture that is even quantitatively rather accurate. We model the dot and the junctions according to Fig. 9.5.1, in which C_1 and C_2, and V_1 and V_2 are the capacitances and voltages across each junction, respectively. The total voltage applied by the voltage supply is $V = V_1 + V_2$, and the charge on each junction is $Q_1 = n_1 e = C_1 V_1$ and $Q_2 = n_2 e = C_2 V_2$, with n_1 and n_2 the number of electrons that have tunneled onto the island through junction 1, and the number of electrons that have tunneled off the island through junction 2, respectively. Because the tunneling rates across each junction may differ, Q_1 and Q_2 are not necessarily equal. The difference is the net charge Q on the island,

$$Q = Q_2 - Q_1 = -ne,$$

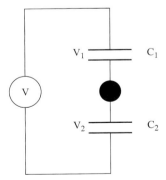

Figure 9.5.1. Equivalent electrostatic circuit of an island connected to a voltage source through two tunneling junctions.

with $n = n_1 - n_2$ an integer. The island itself has a capacitance $C_{tot} = C_1 + C_2$, obtained by grounding the external voltage sources and connecting a probe voltage source directly to the island.

The electrostatic energy of the system (junctions, island, and voltage source) consists of electrostatic energy $\mathcal{E}_s$ stored in the junctions, minus the work W done by the voltage source in moving charges across the junction. As an electron tunnels off the island through junction 2, there is a change in voltage across junction 1. Charge then flows to junction 1 from the voltage source, and the voltage source does work. Similarly, if an electron tunnels onto the island through junction 1, the voltage across junction 2 changes and charge flows from junction 2 to the voltage source. Coulomb blockade occurs if there is some minimum voltage V that has to be supplied in order to have an electron tunnel onto or off the island. If V is less than this threshold, no current can flow through the island.

In order to look at this quantitatively, we first note that we can write

$$V_1 = \frac{C_1 V_1 + C_2(V - V_2)}{C_{tot}} = \frac{C_2 V + ne}{C_{tot}} \tag{9.5.1}$$

$$V_2 = \frac{C_2 V_2 + C_1(V - V_1)}{C_{tot}} = \frac{C_1 V - ne}{C_{tot}}. \tag{9.5.2}$$

These two equations give us the voltage across one junction as an electron tunnels through the other junction. The electrostatic energy stored in the junctions is

$$\mathcal{E}_s = \frac{C_1 V_1^2 + C_2 V_2^2}{2} = \frac{C_1 C_2 (V_1 + V_2)^2 + (C_1 V_1 - C_2 V_2)^2}{2 C_{tot}} = \frac{C_1 C_2 V^2 + Q^2}{2 C_{tot}}.$$

If an electron of charge $-e$ tunnels out of junction 2, then the charge Q on the island increases by $+e$. According to Eq. (9.5.1), the voltage V_1 then changes by $-e/C_{tot}$. To compensate, a charge $-eC_1/C_{tot}$ flows from the voltage source. If we now consider n_2 electrons tunneling off the island through junction 2, the work done by the voltage source is then $-n_2 eVC_1/C_{tot}$. Next, we apply the same reasoning to electrons tunneling onto the island through junction 1. The result is that for n_1 electrons tunneling onto the island, the voltage source does an amount of work equal to $-n_1 eVC_2/C_{tot}$. For a system with a charge $Q = -ne = -n_2 e + n_1 e$ on the island, the total energy is then

$$\mathcal{E}(n_1, n_2) = \mathcal{E}_s - W_s = \frac{1}{2C_{tot}}[C_1 C_2 V^2 + Q^2] + \frac{eV}{C_{tot}}[C_1 n_2 + C_2 n_1].$$

We can now look at the cost in energy for having an electron tunnel onto or off an initially neutral island ($Q = 0$). If we change n_2 by $+1$ or -1, the energy changes by

$$\mathcal{E}_2^\pm = \frac{e}{C_{\text{tot}}}\left[\frac{e}{2} \pm VC_1\right].$$

Similarly, if we change n_1 by $+1$ or -1, the change in energy is

$$\mathcal{E}_1^\pm = \frac{e}{C_{\text{tot}}}\left[\frac{e}{2} \mp VC_2\right].$$

Because the first term in each of these expressions is inherently positive, the energy change is also positive when V is small, and the process of electron transfer will not occur. It will not be until we reach a threshold voltage of $V = e^2/2C_1$ and $V = e^2/2C_2$, respectively, that a reduction in energy will accompany the transfer. In other words, until the threshold voltage has been reached, the charge on the island cannot change, which means that no current flows through the system. This prevention of conduction by the requirement that the charging energy be negative is called Coulomb blockade. We note that for symmetric barriers, for which $C_1 = C_2$, the threshold voltage for an initially neutral island is $V = e/C_{\text{tot}}$.

Quantum dots can in principle be made into single-electron transistors, and logical circuits can be constructed with single-electron transistors as building blocks. However, for this to be practically useful, one has to ensure that the charging energy $e^2/2C \gg kT$, where T is the temperature. For a dot of size $20 \times 20\,\text{nm}^2$, which can be fabricated by electron beam lithography, the capacitance is of the order of 10^{-17} F, and then the phenomenon will be observable only at temperatures less than about 10 K. However, dots or atomic clusters of size 1 nm or less would have capacitances of the order of or less than 10^{-19} F, in which case the charging energy is of the order of electron volts. This opens up the possibility of very compact integrated circuits and computers.

Problems

9.1 It was stated at the end of the paragraph following Eq. (9.1.1) that "in the absence of an external magnetic field there must exist an outgoing state in the same terminal with the same energy but opposite wave-number." Why is this?

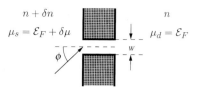

Figure P9.1. Schematic of the ballistic quantum point contact for Problem 9.2.

9.2 Calculate the conductance of a ballistic quantum point contact in a semiclassical two-dimensional electron gas. Assume that a barrier partitions the electron gas with Fermi energies μ_s and μ_d on each side of the barrier, respectively, with a corresponding difference δn in densities, as in Fig. P9.1. A constriction of width w lets electrons cross from one side to the other. First calculate the net flux through the constriction. This is due to the excess density δn at the source incident with speed v_F on the constriction, and averaged over angle ϕ of incidence. This will give you the current I through the constriction as a function of δn. The chemical potential difference is $\delta\mu = eV$, with V the source-to-drain voltage. In the expression for the conductance, $\delta n/\delta\mu$ can be taken to be the density of states of the two-dimensional electron gas.

9.3 A model of a smooth quantum point contact is the saddle-point potential,

$$V(x,y) = V_0 - \tfrac{1}{2}m\omega_x^2 x^2 + \tfrac{1}{2}m\omega_y^2 y^2,$$

where the curvature of the potential is expressed in terms of the frequencies ω_x and ω_y. This potential is separable, and one can solve for the transmission probabilities. With the reduced variable

$$\epsilon_n = 2\left[\frac{\mathcal{E} - (n+\tfrac{1}{2})\hbar\omega_y - V_0}{\hbar\omega_x}\right],$$

where n denotes the transverse channels, the transmission probabilities are

$$T_{nm} = \delta_{nm}\frac{1}{1+e^{-\pi\epsilon_n}}.$$

Plot T_{nn} as a function of $(\mathcal{E} - V_0)/(\hbar\omega_x)$ for different values of ω_y/ω_x ranging from <1 to >1 for the three lowest channels $n = 0, 1, 2$. Set up the expression for the total conductance, given these transmission probabilities. Under what conditions (at zero temperature) would you say that the conductance is quantized? [Hint: Calculate the maximum and minimum slope of the conductance vs. Fermi energy. How wide and flat are the plateau regions of the conductance? What would you require of ω_x and ω_y in order to say that the conductance is well quantized?] (This problem was posed by M. Büttiker.)

9.4 Derive the set of equations for a four-terminal system that correspond to those given for a three-terminal system in Eqs. (9.2.4).

9.5 A tunneling device with transmission probabilities $T_n \ll 1$ of resistance $100\,\Omega$ is to be connected in series with a $25\,\Omega$ resistor. The system, consisting of tunneling device and resistor, will operate at room temperature. Assume that the noise spectrum is white for any bandwidth under consideration. Under what conditions will the noise power of the system be dominated by Johnson and shot noise, respectively? [Hint: start with Eq. (9.3.1) and obtain an expression for the noise in the limit of $T_n \ll 1$ for the tunneling device. You must also add the Johnson noise from the resistor.]

9.6 Fill in the missing steps that lead from Eq. (9.4.1) to Eq. (9.4.2).

Chapter 10
The quantum Hall effect

10.1 Quantized resistance and dissipationless transport

The Hall effect has long been a standard tool used to characterize conductors and semiconductors. When a current is flowing in a system along one direction, which we here take to be the y-axis, and a magnetic field H is applied in a direction perpendicular to the current, e.g., along the z-axis, there will be an induced electrostatic field along the x-axis. The magnitude of the field E is such that it precisely cancels the Lorentz force on the charges that make up the current. For free electrons, an elementary calculation of the type indicated in Section 1.8 yields the Hall resistivity $\rho_H = -H/\rho_0 ec$, and apparently provides a measure of the charge density of the electrons. For Bloch electrons, as we saw in Section 8.3, the picture is more complicated, but ρ_H is still predicted to be a smoothly varying function of H and of the carrier density. In some circumstances, however, the semiclassical treatment of transport turns out to be inadequate, as some remarkable new effects appear.

In a two-dimensional system subjected to strong magnetic fields at low temperatures, the response is dramatically different in two respects. First, the Hall resistivity stops varying continuously, and becomes intermittently stuck at quantized values $\rho_H = -h/je^2$ for a *finite* range of control parameter, e.g., external magnetic field or electron density. In the integer quantum Hall effect, j is an integer, $j = 1, 2, \ldots$, and in the fractional quantum Hall effect, j is a rational number $j = q/p$, with p and q relative primes and p odd. (In addition, there exists a fractional quantum Hall state at $\nu = 5/2$ and possibly other related states. The physics of these is, however, very different from that of the "standard" odd-denominator fractional quantum Hall states and will not be discussed here.) Second, at the plateaus in ρ_H at which it attains these quantized values, the current flows without dissipation. In other words, the longitudinal part of the resistivity tensor is zero. The resistivity

10.1 Quantized resistance and dissipationless transport

and conductivity are both tensor quantities, and it happens that the longitudinal conductivity also vanishes at these plateaus. This may sound a little strange, but is a simple consequence of the dissipationless transport in two dimensions in crossed electric and magnetic fields.

As we shall see, these two observations, a quantized Hall resistance and dissipationless transport, can be understood if the system is incompressible (that is, it has an energy gap separating the ground state from the lowest excited state) and if there is disorder, which produces a range of localized states. Our first task will be to ask what causes the incompressibility and energy gap. In the integer quantum Hall effect, the energy gap (which is responsible for the incompressibility) is a single-particle kinetic energy gap due to the motion of single particles in an external field. It is not necessary to introduce electron–electron interactions in order to explain the integer quantum Hall effect. In the fractional quantum Hall effect, on the other hand, the energy gap and the ensuing incompressibility are entirely due to electron–electron interactions. This, and the absence of any small parameter in the problem that would permit a perturbation expansion, makes it a very difficult system to study theoretically.

The presence of disorder is necessary in order to explain the *plateaus* in the quantized Hall resistivity. Disorder gives us a range of energies within which states are localized, and as the Fermi energy sweeps through these states the Hall resistivity exhibits a plateau. In the integer quantum Hall effect, the disorder dominates over the electron–electron interactions. In the fractional effect the strengths are reversed. The fractional effect occurs only in samples that are very clean, and which consequently have a very high electron mobility. There is, of course, no sharp division between the integer and the fractional quantum Hall effect, and there is no magical amount of disorder at which the fractional quantum Hall effect is destroyed. Which plateaus, and thus which fractional or integer quantum Hall states, will be observed depends on how much disorder there is in the system and what the temperature is. If we start by imagining a very clean system in the limit of zero temperature, the Hall resistivity vs. control parameter will exhibit a series of plateaus corresponding to all fractional and integer Hall states, but the extent of each plateau becomes very small. As we start to add impurities to the system, the fractional quantum Hall states with the smallest energy gaps are destroyed, since the perturbations introduced by the disorder become larger than the smallest energy gaps. The corresponding plateaus disappear and neighboring plateaus grow in size. At sufficient disorder, all fractional quantum Hall plateaus have vanished and we are left with only the plateaus of the integer quantum Hall states. Similarly, increasing the temperature will destroy the quantum Hall effect, as it is, strictly speaking, a zero-temperature phenomenon. By this we mean that the quantization of the Hall conductance and vanishing of the longitudinal resistance are exact only in the limit of low temperatures. As the temperature is raised, the weakest fractional quantum Hall states will start to disappear. As the temperature is increased further, the integer quantum Hall states will eventually suffer the same fate.

10.2 Two-dimensional electron gas and the integer quantum Hall effect

We start by considering a two-dimensional gas of N noninteracting electrons in an external magnetic field and with no disorder. Let the area in the xy-plane be A and the magnetic field be $\mathbf{B} = B\hat{\mathbf{z}}$. The Hamiltonian of this system is simply

$$\mathcal{H}_0 = \frac{1}{2m^*} \sum_{j=1}^{N} \left[\mathbf{p}_j + \frac{e}{c} \mathbf{A}(\mathbf{r}_j) \right]^2,$$

where $\mathbf{A}(\mathbf{r}_j)$ is the vector potential at the position $\mathbf{r}_j$ of electron j, the charge on the electron is now taken to be $-e$, and m^* is the band mass of the electron, e.g., $m^* \approx 0.07 m_e$ in GaAs. The first thing we have to do is to fix a gauge for the vector potential, and there are two common choices for this depending on which symmetry we want to emphasize. The first choice is the so-called Landau gauge, $\mathbf{A} = Bx\hat{\mathbf{y}}$. This gauge is translationally invariant along the y-axis and so the single-particle eigenstates can be taken to be eigenstates of p_y. This choice of gauge is convenient for rectangular geometries with the current flowing along the y-axis. The other choice is the symmetric gauge $\mathbf{A} = \frac{1}{2} B(x\hat{\mathbf{y}} - y\hat{\mathbf{x}}) = \frac{1}{2} Br\hat{\boldsymbol{\phi}}$. As the last equality shows, this gauge is rotationally invariant about the z-axis, and the single-particle eigenstates can be taken to be eigenstates of the z-component of angular momentum. This choice of gauge is convenient for circular geometries (quantum dots) and is the gauge in which the Laughlin wavefunction for fractional quantum Hall states is most easily represented.

For now, we use only the Landau gauge, and with this choice the Hamiltonian $\mathcal{H}_0$ becomes

$$\mathcal{H}_0 = \frac{1}{2m^*} \sum_j \left[-\hbar^2 \frac{\partial^2}{\partial x_j^2} - \hbar^2 \frac{\partial^2}{\partial y_j^2} + 2\frac{\hbar}{i}\frac{e}{c} Bx_j \frac{\partial}{\partial y_j} + \frac{e^2}{c^2} B^2 x_j^2 \right]. \quad (10.2.1)$$

10.2 Two-dimensional electron gas and the integer quantum Hall effect

In the absence of potentials that break the translational invariance along the y-axis, we can write the single-particle states as

$$\psi_{kn}(x, y) = \phi_{kn}(x)e^{iky}. \qquad (10.2.2)$$

We apply periodic boundary conditions along a length L_y on the y-axis. The admissible values of k are then given by $k = 2\pi i_k/L_y$, with $i_k = 0, \pm 1, \pm 2, \ldots$ By applying the Hamiltonian (10.2.1) to the wavefunction (10.2.2) we obtain the single-particle Schrödinger equation

$$\left[-\frac{\hbar^2}{2m^*}\frac{d^2}{dx^2} + \frac{1}{2}m^*\omega_c^2(x - x_k)^2\right]\phi_{kn}(x) = \mathcal{E}_{kn}\phi_{kn}(x), \qquad (10.2.3)$$

where $\omega_c = eB/(m^*c)$ is the cyclotron frequency, and $x_k = -(\hbar c/eB)k$, which we write as $-\ell_B^2 k$, with ℓ_B the *magnetic length*, $\sqrt{\hbar c/eB}$. This is the characteristic length scale for the problem, and is about 10 nm for magnetic fields of 5 to 10 T. Equation (10.2.3) is, for each allowed value of k, the equation for a harmonic oscillator centered at the position $x = x_k$, and so the energy eigenvalues are

$$\mathcal{E}_{nk} = (n + \tfrac{1}{2})\hbar\omega_c \quad n = 0, 1, 2, \ldots \qquad (10.2.4)$$

Surprisingly, the energy eigenvalues do not depend on the momentum $\hbar k$ along the y-axis, but only on the index n, the so-called Landau level index, and all states with the same quantum number n form a Landau level. This means that there is a huge degeneracy in energy. The center points of the states are $x_{k_i} = -\ell_B^2 k_i$, and the centers of two neighboring states along the x-axis are separated by a distance $\Delta x = 2\pi \ell_B^2/L_y$. If the system has a width L_x we can fit $L_x/\Delta x$ states in one Landau level across this width. Each Landau level thus contains $L_x/\Delta x = L_x L_y/(2\pi \ell_B^2)$ states, which is the degeneracy of each Landau level. Another way to think of this is that each state occupies an area $2\pi \ell_B^2$, and the degeneracy is just the total area $A = L_x L_y$ divided by the area per state.

The degeneracy, or the area per state in units of $2\pi \ell_B^2$, leads us to define a very useful quantity, the filling factor ν, which is a conveniently scaled measure of the density of the system. The filling factor is defined as $\nu = 2\pi \ell_B^2 \rho$, with ρ now the number of electrons per unit area. Thus, when $\nu = 1$, all the states in the lowest Landau level $n = 0$ that lie within the area A are filled. Another way to look at the filling factor, which is especially useful when we deal with the fractional quantum Hall effect, is that it is a measure of the number of electrons per flux quantum. For this system the flux quantum is $\Phi_0 = hc/e$. It is double the flux quantum $\phi_0 = hc/2e$ introduced in Section 7.9 because we are now dealing with single electrons rather than electron pairs. The total flux piercing the system is $\Phi = BA = \Phi_0 BAe/(hc) = \Phi_0 A/(2\pi\ell_B^2)$, so the number of flux quanta N_{Φ_0} is $A/(2\pi\ell_B^2)$. Thus the number of electrons per flux quantum is $\rho A/N_{\Phi_0} = 2\pi\ell_B^2 \rho = \nu$.

With the single-particle energy spectrum given by Eq. (10.2.4), the density of states for the system of noninteracting particles consists of a series of δ-functions of weight $A/(2\pi\ell_B^2)$ at the energies $(n+\frac{1}{2})\hbar\omega_c$, as depicted in Fig. 10.2.1. If we plot the ground state energy $\mathcal{E}_0(\nu)$ of the N independent electrons as a function of filling factor ν we obtain a piecewise linear plot with slope $(n+\frac{1}{2})\hbar\omega_c$ and with discontinuities of magnitude $\hbar\omega_c$ in the slope at integer filling factors, as shown in Fig. 10.2.2. As we add more electrons to a system, we occupy states in the lowest Landau level n' that still has vacant states available. These states are all degenerate and each extra electron adds

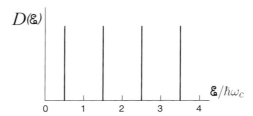

Figure 10.2.1. The density of states of a noninteracting two-dimensional electron gas in a magnetic field.

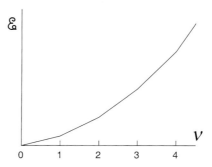

Figure 10.2.2. Ground-state energy $\mathcal{E}$ vs. filling factor ν for a noninteracting two-dimensional electron gas in a magnetic field. As the nth Landau level is being filled, the energy increases by $(n+\frac{1}{2})\hbar\omega_c$ per particle. When the nth Landau level is precisely filled, adding a new electron will require $(n+\frac{3}{2})\hbar\omega_c$, causing the slope of the curve to change discontinuously.

an energy $(n' + \frac{1}{2})\hbar\omega_c$. When the last available state in this Landau level has been filled, the next electron will need an energy $(n' + 1 + \frac{1}{2})\hbar\omega_c$, and so the slope increases discontinuously by $\hbar\omega_c$. Since the ground-state energy has angles at integral ν, this implies that the zero-temperature chemical potential,

$$\mu = \left(\frac{\partial \mathcal{E}_0}{\partial N}\right)_B,$$

has discontinuities at integer filling factors, as shown in Fig. 10.2.3. Finally, we use the fact that the isothermal compressibility κ is related to the chemical potential μ through

$$\kappa^{-1} = \rho^2 \frac{d\mu}{d\rho},$$

with the derivative taken at constant (here $T = 0$) temperature. At the integer filling factors, the slope of μ vs. ν approaches infinity, and so the compressibility vanishes there. The compressibility measures the energy cost of "squeezing" the system infinitesimally. The compression is created by exciting particles from just below the Fermi energy to just above the Fermi energy in order to make a long-wavelength density perturbation. For a compressible system, this costs only an infinitesimal energy. However, when the system is said to be incompressible, compressing the system infinitesimally requires a *finite* energy. This is what happens at integer filling factors: one set of Landau levels is completely filled, and particles can only be excited by crossing the energy gap $\hbar\omega_c$ to the next Landau level.

Let us now turn to the response of the system to a transverse electric field. In the absence of any external potential (including disorder), we can easily calculate the current carried by each single-particle state. The operator that

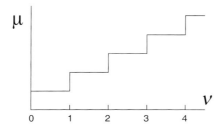

Figure 10.2.3. The chemical potential for the two-dimensional electron gas in a magnetic field has discontinuous jumps whenever a Landau level has been filled.

describes the current is

$$\mathbf{J} = -\frac{e}{m^*}\left(\mathbf{p} + \frac{e}{c}\mathbf{A}\right)$$

for an electron of charge $-e$. This operator can be thought of as being proportional to a derivative of the Hamiltonian with respect to the vector potential. This is a very useful observation, and we turn it into a formal device by introducing a fictitious vector potential $\mathbf{a} = -(q\Phi_0/L_y)\hat{\mathbf{y}} = -[qhc/(eL_y)]\hat{\mathbf{y}}$. Here q is a dimensionless parameter, Φ_0 is the flux quantum hc/e, and we have applied this fictitious vector potential along the y-axis in order to relate it most easily to J_y. We note that $\nabla \times \mathbf{a} = 0$, so $\mathbf{a}$ does not correspond to any physical magnetic field through the system. However, if we imagine making the system a loop in the yz-plane by tying together the ends along the y-direction, qhc/e could be due to a real magnetic field piercing the center of the loop with q flux quanta.

With this extra vector potential, the Hamiltonian is

$$\mathcal{H}(q) = \frac{1}{2m^*}\left(\mathbf{p} + \frac{e}{c}\mathbf{A} - q\frac{e}{c}\frac{\Phi_0}{L_y}\hat{\mathbf{y}}\right)^2.$$

Here we have explicitly indicated the parametric dependence on q. The current operator can then be written

$$J_y(\mathbf{r}, q) = -\frac{e}{m^*}\left[\mathbf{p} + \frac{e}{c}\mathbf{A} - \frac{qh}{L_y}\hat{\mathbf{y}}\right]\cdot\hat{\mathbf{y}} = \frac{eL_y}{h}\frac{\partial\mathcal{H}(q)}{\partial q}. \quad (10.2.5)$$

We now make an interesting observation. According to Eq. (10.2.5) we can evaluate the current in any state by forming the expectation value of the derivative of the Hamiltonian with respect to a fictitious vector potential in that state. If a state carries any current, this derivative must obviously be nonzero, and the eigenvalue spectrum must also depend on this fictitious vector potential. But this added vector potential is "pure gauge," which is to say that it does not correspond to any physical magnetic field and can be completely removed by a gauge transformation. Therefore, it should have no effect whatsoever on the spectrum of the system. The solution to this paradox lies in the fact that the vector potential $\mathbf{a}$ adds a phase to the electron wavefunction. This is the so-called Aharonov–Bohm phase $\phi_{AB} = -e/\hbar c \int \mathbf{a} \cdot d\mathbf{r}$. In the present case we integrate along the y-direction and obtain $\phi_{AB} = 2\pi q$. The phase of a single-particle electron wavefunction thus advances by $2\pi q$ as

10.2 Two-dimensional electron gas and the integer quantum Hall effect

it travels around the circumference L_y in the y-direction. This is precisely what happens if there is a flux $\Phi = -q\Phi_0$ piercing the center of the ring. In addition to this Aharonov–Bohm phase, the phase of the wavefunction also advances by kL_y, where k is the wavenumber along the y-direction. If q is an integer, $q = 0, \pm 1, \pm 2, \ldots$, nothing new is added and the standard wavenumbers $k_i = 2\pi i/L_y$ with $i = 0, \pm 1, \pm 2, \ldots$ satisfy the condition that the wavefunctions be single-valued. But if q is not an integer, the phase added to the wavefunction due to $\mathbf{a}$ as we go around the ring is not an integer times 2π. The wavenumbers k_i would then make the wavefunction multiple-valued. In order to avoid this, we have to adjust the wavenumbers so that $k_i L_y$ carries an extra phase that precisely cancels the phase due to the vector potential $\mathbf{a}$, and the allowed wavenumbers are now $k_i = 2\pi(i - q)/L_y$. In other words, the presence of the vector potential $\mathbf{a}$ changes the boundary conditions, unless $\mathbf{a}$ corresponds to an integer number of flux quanta piercing the system. Note that this effect hinges on the phase coherence of the wavefunction extending around the ring. If the localization length is much smaller than the circumference, the wavefunction will not run the risk of being multiple-valued. For example, the wavefunction can be localized at some position y_0 and decay exponentially with a decay length $\ell \ll L_y$ away from $y = y_0$. The same wavefunction is then single-valued as we go around the ring no matter what q is (except for some exponentially small corrections that we can ignore). The spectrum of $\mathcal{H}(q)$ then has no dependence on q and the wavefunctions cannot carry any current, which must obviously be the situation if the wavefunctions are localized. This is the case for disordered insulators.

We now apply this kind of argument specifically to a two-dimensional electron gas on a "ribbon" of width L_x along the x-axis and having a circumference L_y along the y-axis. We apply a field $\mathbf{E} = E_x \hat{\mathbf{x}}$ across the width of the ribbon and then calculate the Hall resistivity of the system. We use our trick from the previous paragraph of adding a fraction q of a flux quantum piercing the system to calculate the current density. The single-particle Schrödinger equation in the Landau gauge and in the absence of disorder is then

$$\mathcal{H}(q)\psi_\alpha(x, y) = \left\{ \frac{1}{2m^*} \left[\mathbf{p} + \frac{e}{c} B\left(x - \frac{q\Phi_0}{BL_y}\right)\hat{\mathbf{y}} \right]^2 + eE_x x \right\} \psi_\alpha(x, y)$$

$$= \mathcal{E}_\alpha \psi_\alpha(x, y), \tag{10.2.6}$$

where α represents an enumeration of the eigenstates. It shows what happens as we slowly add a fraction q of a flux quantum through the system – the electrons "march" to the right, moving their center points from $x_{k_i} = -k_i \ell_B^2 = -(2\pi i/L_y)\ell_B^2$ to $x_{k_i} + q\Phi_0/BL_y$. As we complete the addition of one unit of flux quantum through the system, the set of center points becomes mapped back onto itself. If there were no electric field present, the single-particle eigenstates for different values of q would all be equivalent, there would be no dependence of the spectrum of $\mathcal{H}$ on q, and thus no current. But the presence of the electric field changes this. There will now be a dependence of the spectrum of eigenvalues on q. We make this explicit by inserting single-particle states $\phi_{kn;q}(x)e^{iky}$ into Eq. (10.2.6) and completing the squares:

$$\mathcal{H}(q)\phi_{kn;q}(x)e^{iky} = \left\{ \frac{1}{2m^*} p_x^2 + \frac{1}{2} m^* \omega_c^2 \left[x + \left(x_k - \frac{q\Phi_0}{BL_y} + \frac{v_d}{\omega_c} \right) \right]^2 \right.$$
$$\left. - \frac{1}{2} m^* v_d^2 - \hbar k v_d + e \frac{E_x}{B} \frac{q\Phi_0}{L_y} \right\} \phi_{kn;q}(x)e^{iky},$$

where v_d is the classical drift velocity $v_d = cE/B$. The electric field introduces a dependence of the single-particle energies on wavenumber k, and hence on the center point x_k. This means that as we now add a fraction q of a flux quantum, the energy of the system changes. By virtue of the relation between single-particle energies and currents, the states carry currents in the presence of the electric field. Since the total energy of the system changes as we insert some flux, this apparently means that we must do work on the system in order to insert flux. Clearly, the work that we do in this process must be related to how the electron single-particle states march to the right and increase their energies. Imagine that we slowly insert precisely one flux quantum into the system. The single-particle states and their energies are all the same before and after inserting the flux quantum. But in the process, all *occupied* states moved one step over to the right, so that at the end of the process, we have transferred precisely one electron per occupied Landau level across the width L_x of the system. The cost in energy of this process is clearly $\Delta \mathcal{E} = neV = neE_x L_x$, where n is the number of occupied Landau levels.

We can work this out in more detail. Let the resistivity tensor of the system be ρ. If there is no dissipation, then the diagonal components of the resistivity tensor must vanish, and only the off-diagonal components are nonzero. Now we consider the electric field in the y-direction rather than the x-direction, and use Faraday's law to write

$$\frac{1}{c}\frac{d\Phi}{dt} = \frac{1}{c}\int d\mathbf{S} \cdot \frac{d\mathbf{B}}{dt} = \int_C d\boldsymbol{\ell} \cdot \mathbf{E}_y = \int_C d\ell \rho_{yx} j_x,$$

where C is a contour enclosing the flux quantum and j_x is the current density, equal to J_x/L_y, in the x-direction. If we now integrate this equation from $t = -\infty$ to $t = \infty$, we can relate the change in flux $\Delta\Phi$ to a transfer of charge along the x-axis:

$$\frac{1}{c}\Delta\Phi = \rho_{yx}\int_C d\ell \int dt\, j_x = \rho_{yx}\int dt\, J_x. \qquad (10.2.7)$$

Now choose $\Delta\Phi = \Phi_0$. Then the net charge transferred ($\int dt\, J_x$) is $-ne$, where n is the number of occupied Landau levels. Thus

$$\frac{1}{c}\Phi_0 = -\rho_{yx}ne,$$

so, using $\Phi_0 = hc/e$, we obtain

$$\rho_{yx} = -\frac{h}{ne^2}. \qquad (10.2.8)$$

While we have indeed derived a quantized Hall resistance for an ideal ribbon-like system using a rather sophisticated gauge-invariance argument, we could for the simple system above have taken a much simpler approach. We could have calculated the current carried by each single-particle state, summed up the result to get the total current, calculated the energy difference between the left-most and right-most occupied states, and we would have arrived at the same result. So why did we go through all this effort to calculate something we could have derived using very elementary techniques? The reason is that real systems are not ideal, but are composed of interacting electrons in the presence of disorder. The simple methods cannot be used in those cases. The gauge-invariance argument, on the other hand, is very powerful, and is independent of the details of the system. It allows us to turn now to a "real" system with disorder.

First, we summarize the main ingredients of the gauge-invariance argument that we use in this case: (a) only states that are extended through the system respond to the flux inserted; (b) there is a *mobility gap*, i.e., there is a finite energy gap separating the bands of current-carrying states from each other, so that the system remains dissipationless and the diagonal part of the resistivity tensor vanishes; and (c) if we add precisely one flux quantum, the eigenstates of the system before and after the flux quantum is inserted are equivalent. Therefore, if there is a change in energy as we slowly add one flux quantum, this change must be due to a different occupation of single-particle states within the same Landau level. It *cannot* be due to exciting electrons to higher Landau levels, since such processes must overcome the cyclotron energy and cannot occur adiabatically. It must be due to transferring n electrons from one side of the system to the other.

The remaining issues are how to relate n to the number of filled Landau levels, and the origin of the mobility gap. For the ideal noninteracting system in the absence of an external electric field, the density of states consists of a series of delta-functions at the energies $\mathcal{E} = (n + \frac{1}{2})\hbar\omega_c$, each of which has a weight $A/(2\pi\ell_B^2)$. As we add electrons to the system, the chemical potential will always be at one of these energies, except when an integer number of Landau levels are completely filled. Hence, there are almost always extended states just above and below the Fermi energy, without any energy gap separating them. As we add impurities to the system, extended states will start to mix due to scattering off the impurities. This both introduces dissipation (due to a finite probability of an incident electron being back-scattered) and also broadens each Landau level into a band. We will here assume that the bandwidth is smaller than the cyclotron energy so that each broadened Landau level is separated from the neighboring ones (see Fig. 10.2.4). It is generally then assumed that at the center of each broadened Landau level, there remains a small number of extended states that can carry current, while the states on each side of the center of the Landau level are localized. This is a crucial assumption. According to the localization theory of noninteracting particles, all electrons in two dimensions in the presence of *any* disorder should be localized, which seems contrary to the assumption we just made. What makes the difference is the presence of the magnetic field. There is strong theoretical and experimental evidence that the localization length of the localized states diverges as the center of the Landau level is approached. In a more pragmatic vein, we can also argue that it is an experimental fact that these systems do carry current, and so there must be *some* extended states present.

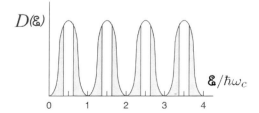

Figure 10.2.4. In the presence of disorder, the density of states shows that the Landau levels have broadened into bands. The shaded areas represent localized states that carry no current.

As most of the states in the original Landau levels become localized and shifted in energy away from the center of the Landau level, this gives us the mobility gap that we need – an energy range through which we can sweep the Fermi energy while an energy gap separates the occupied extended states from the unoccupied ones. While the Fermi energy lies in a band of localized states, the transport is dissipationless, since dissipation is due to scattering between current-carrying occupied and unoccupied states at the Fermi energy. Here, all occupied current-carrying states are well below the Fermi energy and cannot scatter to unoccupied states without a finite increase in energy.

Next, let us for simplicity assume that all localized states are in a region $-L_x/2 + \delta x < x < L_x/2 - \delta x$ and that the extended states occupy the regions $|x| > L_x/2 - \delta x$. As we now adiabatically insert a flux quantum through the system, the extended states in both regions march one step to the right. But that means that we must have effectively transferred one electron for each Landau level with its extended states below the Fermi energy across the region of localized states and across the width of the system – since all states in the localized region are initially occupied, there is no empty state to move into from the extended-state region at $-L_x/2 < x < -L_x/2 + \delta x$ unless the net effect is to transfer one electron across the band of localized states for each such Landau level. Therefore, for this example, the integer n in the gauge argument above is equal to the number of filled Landau levels even in the presence of (moderate) disorder. We should, however, point out that while this argument can still be strengthened a little bit, there is no general proof that n must be equal to the number of filled Landau levels, or, for that matter, nonzero. For example, in a strip of finite width, all the energy levels are discrete, and it is therefore impossible to move a charge adiabatically across the system without adding any energy during this process. Only in the limit of very wide strips do the energy levels form a continuum and make it possible to move charges adiabatically. Another approach to quantization will be presented in the next section, in which the current response to changes in the chemical potential is studied using the Landauer–Büttiker formalism. That approach has the advantage of being closer to real experiments in which the current response is measured in systems that are real, disordered, and of finite size.

10.3 Edge states

In the previous section, we implicitly attached a special significance to the edges in the physics of the quantized Hall conductance. For the ideal system with an electric field, we transferred one electron per Landau level from current-carrying edge states on one side of the system to current-carrying edge states on the other side. Furthermore, while there is an excitation gap in the *bulk* of the system, the edge states provided gapless excitations. It turns out that because of the strong magnetic field, there will always be gapless excitations of current-carrying states flowing along the perimeter of the system. In this section, we examine these edge states more closely. We will find that there is a very natural interpretation of the quantized Hall resistance using these edge states in a Landauer–Büttiker formalism.

We start by first giving a simple argument, due to Allan MacDonald, which demonstrates that in a bounded system, there must always be gapless excitations at the boundaries of the system. Consider a finite system with a density ρ^* at which the bulk is incompressible with a filling factor ν^*. The chemical potential μ then lies in the bulk excitation gap, i.e., we have to pay the price of the energy gap in order to add particles to the bulk of the system. We now imagine that we increase the chemical potential by an infinitesimal amount $\delta\mu$. In the bulk, the current density cannot change since $\delta\mu$ is infinitesimal and cannot overcome the mobility gap in the bulk. It follows that if there is a change in the current density as a response to $\delta\mu$, this change must be at the edges of the system. Charge conservation also requires that if there is a resulting change in the current along the edge, this change must be uniform along the edge. We can relate the change in current δI to the change in orbital magnetization density through

$$\delta I = \frac{c}{A}\delta M, \tag{10.3.1}$$

with A the total area of the system. This relation is nothing but the equation for the magnetic moment of a current loop. But we can write δM in terms of $\delta\mu$ using a Maxwell relation:

$$\delta M = \left.\frac{\partial M}{\partial \mu}\right|_B \delta\mu = \left.\frac{\partial N}{\partial B}\right|_\mu \delta\mu. \tag{10.3.2}$$

By combining Eqs. (10.3.1) and (10.3.2) we arrive at

$$\frac{\delta I}{\delta\mu} = c\left.\frac{\partial \rho}{\partial B}\right|_\mu. \tag{10.3.3}$$

When the filling factor is locked at a particular value ν^* then changing the magnetic field at fixed μ necessarily changes the density, since

$\partial \rho^*/\partial B = (\partial/\partial B)(\nu^*/2\pi\ell_B^2) = \nu^* e/2\pi\hbar c$. Then Eq. (10.3.3) shows that there is a corresponding current response to a change in the chemical potential. We conclude that: (a) there must be gapless excitations in the system (since there were states into which we could put more particles at an infinitesimal cost in energy); and (b) these excitations must be located at the edges of the system. Since all real systems are finite and inhomogeneous, the low-energy properties probed by experiments such as transport measurements must be determined by the gapless edge excitations.

Next, we discuss in more detail the origin of these gapless edge states. First of all, we may quite generally assume that there is some confining potential $V_{ext}(\mathbf{r})$ that keeps the electrons in the system. This potential is caused by electron–ion interactions and electron–electron interactions, but for simplicity we assume that we have noninteracting electrons confined by a potential $V_{ext}(x)$. In the center of the system the potential is flat, and we can here set $V_{ext}(x) = 0$, but as we approach the edges of the system, the potential bends upwards, providing a well that confines the electrons to the interior of the system. The nonzero gradient of the confining potential also causes the states near the edges to carry a finite current. From our earlier discussion about gauge invariance we related the derivative of the Hamiltonian with respect to a fictitious flux to the current operator:

$$\frac{\partial \mathcal{H}}{\partial q} \propto j_y.$$

We now take the expectation value of this relation in one of the eigenstates of $\mathcal{H}$. The result is

$$\frac{\partial \mathcal{E}_{nk}}{\partial k} = -\frac{\hbar L_y}{e} i_{nk},$$

where i_{nk} is the net current carried by the state $|nk\rangle$. This equation just relates the group velocity $(\partial \mathcal{E}/\partial k)$ to the current carried. In the interior, where the confining potential is flat, the eigenvalues are constant with respect to k and these states carry no net current. Near the edges, where the confining potential slopes upward, the eigenvalues change with k, giving rise to a finite velocity of the eigenstates, and hence a finite current carried by each state. Along one edge the current flows in the positive y-direction, and along the other edge, the current flows in the negative y-direction, since the gradient of $V_{ext}(x)$ has opposite signs at the two edges. The eigenvalues change because of the relation $x_k = -\ell_B^2 k$ between the center points x_k of the states and the wavenumber k. For example, in the limit of a very slowly rising potential

$$\frac{\ell_B}{\hbar\omega_c}\frac{dV_{\text{ext}}(x)}{dx} \ll 1,$$

and the eigenvalues are approximately $\mathcal{E}_{nk} \approx (n+\frac{1}{2})\hbar\omega_c + V_{\text{ext}}(x_k)$, so $d\mathcal{E}_{nk}/dk \approx \ell_B^2 dV_{\text{ext}}(x_k)/dx$. We briefly described these current-carrying edge states in terms of semiclassical skipping orbits in Section 1.8.

The theorem we discussed at the beginning of this section stated that the current-carrying states must be located at the edges. This lateral localization is due to the strong external magnetic field. For a confining potential $V_{\text{ext}}(x)$ that preserves translational invariance along the y-axis, the single-particle eigenstates can be labeled by the y-momentum k and can be constructed from basis states that are a product of e^{iky} and a function of $x - x_k$. All these basis states are localized in the x-direction about x_k on a scale given by ℓ_B. In the presence of the potential at the edge, linear combinations of these will form new energy eigenstates, which will also be localized in the x-direction. The strong magnetic field also prevents mixing of edge states at opposite edges, provided the separation between the edges is large compared with the magnetic length ℓ_B. Consider the effect of a local potential $V(x,y)$ on two well separated edge states $|1\rangle$ and $|2\rangle$. In perturbation theory, the mixing of the edge states depends on the matrix element $\langle 1|V|2\rangle$, which falls off roughly as $\exp(-d^2/\ell_B^2)$, where d is the separation between the edges.

In the language of the Landauer–Büttiker formalism, the lack of mixing between current-carrying edge states at opposite edges makes the transmission probabilities for edge states unity. To be more precise, it can be shown that the matrix element for back-scattering across a system in the presence of disorder is of order $\exp(-L_x/\xi)$, where ξ is a disorder-dependent length characterizing the extent of the edge state in the direction across the system. The suppression of back-scattering makes the Landauer–Büttiker formalism particularly well suited for systems in the integer quantum Hall regime, and also provides these systems with a very convenient framework for interpretation. Consider first a system with the Fermi energy midway between the centers of two Landau levels. In discussing bulk systems we argued that this places the Fermi energy in the mobility gap of the bulk states, so that in the bulk there are no current-carrying states at the Fermi energy. On the edge, however, there will be current-carrying states at the Fermi energy. We now connect the system to a source and a drain and apply an infinitesimal electrochemical potential energy difference between them. The current-carrying edge states injected at one terminal i cannot back-scatter but flow along their respective edges until they encounter the next terminal j along the edge with transmission probability $T_{ji} = 1$ and all other transmission probabilities zero. It is then a matter of simple algebra to conclude that: (a) the resistance between source and drain is $h/e^2 N$, with N the number of filled Landau levels; (b) the resistance between any two terminals along the same edge is zero; and (c) the resistance between any two terminals on opposite edges is $h/e^2 N$.

As we increase the Fermi energy, it will eventually approach the center of the next Landau level. There are now extended states all across the system that are mixed by impurity scattering. Back-scattering is therefore no longer prohibited, and the longitudinal resistance between terminals along the same edge attains a finite, nonzero value. At the same time, we are beginning to add edge states belonging to a new Landau level, and the Hall resistance and resistance between source and drain decreases. As soon as the Fermi energy has swept past this new Landau level, the bulk states are in a new mobility gap, there is no back-scattering, and the longitudinal resistance vanishes. At the same time, we have populated the edge states originating from this new Landau level, and the Hall resistance attains a new quantized value $h/e^2(N+1)$.

As with all transport phenomena, the simple linear theory of the integer quantum Hall effect fails at sufficiently large currents. The transport at a quantized plateau then ceases to be dissipationless, while the Hall resistance may or may not change appreciably from its quantized value. This can happen through a variety of mechanisms. When one increases the electro-chemical potential difference $\Delta\mu$ between source and drain it is observed that at some value of $\Delta\mu$ the longitudinal resistance starts to increase dramatically, and eventually becomes Ohmic, and thus linear in $\Delta\mu$. One loss mechanism involves coupling to phonons. As soon as the drift velocity exceeds the sound velocity, electrons can emit phonons, and dissipation occurs even in a system with no other source of disorder.

10.4 The fractional quantum Hall effect

As we stated earlier, the fractional quantum Hall effect is observed at very large magnetic fields in very clean systems. Here, the energy gap is caused by electron–electron interactions. In order to observe resistance plateaus of finite width, there must be some degree of disorder present in order to provide a mobility gap, but too much disorder has the contrary effect of quenching this necessary energy gap. The first theoretical evidence for an energy gap caused by electron–electron interactions came from numerical diagonalizations of small systems, which showed a downward dip in the ground state energy per particle near $\nu = \frac{1}{3}$. As the magnetic fields considered are of the order of 10 T, it is a good first approximation to assume that the cyclotron energy is much larger than any other energy scale. This means that we can restrict the basis states for electrons in the bulk of the two-dimensional sample to only the lowest Landau level. Hence, in the absence of external potentials, all single-particle states are completely degenerate. The problem becomes one of finding the ground state and elementary excitations of this system in which many electrons of equal unperturbed energy interact through a Coulomb potential that is screened only by the static dielectric constant of the material.

Almost all our understanding of the fractional quantum Hall effect comes from a bold variational trial wavefunction first proposed by Laughlin in 1983. He demonstrated that this wavefunction is incompressible at filling factors $\nu = 1/p = 1/(2m+1)$ with m an integer and that the quasiparticles at these fillings have fractional charge $\pm e/p = \pm e/(2m+1)$. Subsequent theoretical advances based on Laughlin's suggestion helped to establish why his wavefunction gives a good description of the ground state. This work showed that there is a low-energy branch of collective modes called magneto-rotons (named in analogy with rotons in liquid helium), and established that there is a hidden, so-called off-diagonal long-range order in the Laughlin ground state. This latter insight led to the development of effective field theories in which this order parameter and long-wavelength deviations from it are the central quantities. Subsequent pioneering work by Jain, also based on the Laughlin wavefunction and the off-diagonal long-range order it contains, showed that the fractional quantum Hall effect can be described as an integer quantum Hall effect of composite particles consisting of electrons bound to an even number of flux quanta. These entities are called composite fermions. Finally, it was pointed out that the spin degree of freedom in GaAs systems is very important, and leads to a new class of excitations with spin textures. This is at first counter-intuitive, since one is inclined to assume that in strong magnetic fields, the spin degree of freedom is frozen out. However, in GaAs, atomic and band-structure properties conspire to drive the effective Landé g-factor close to zero, rendering the spin contribution to the energy per electron much smaller than any other energy.

We will for now ignore the spin degree of freedom and consider N spin-polarized electrons in a magnetic field strong enough that we need only consider single-particle basis states in the lowest Landau level. It is convenient to work in the symmetric gauge, in which $\mathbf{A} = \frac{1}{2}(\mathbf{B} \times \mathbf{r})$, since then the system is rotationally invariant about the z-axis, and the z-component of total angular momentum, L_z, commutes with the Hamiltonian. Our task is then to find the best choice for the ground state of the degenerate system of electrons in eigenstates of L_z when the Coulomb interaction is turned on. It is conventional (even though it is not a little confusing!) to use the complex notation $z_j \equiv x_j - iy_j$ for the coordinates of the jth electron. In the lowest Landau level, the single-particle basis functions in the symmetric gauge are then written as

$$\phi_m(z_j) = \frac{1}{(2\pi\ell_B^2 2^m m!)^{1/2}} \left(\frac{z_j}{\ell_B}\right)^m e^{-|z_j|^2/4\ell_B^2}.$$

The probability densities of these states form circles about the origin with the peak density occurring at $r \simeq \ell_B\sqrt{2m}$. One can verify that $\phi_m(z_j)$ is an eigenstate of L_z with eigenvalue $\hbar m$.

The set of all N-particle Slater determinants composed of the lowest-Landau-level single-particle wavefunctions forms a basis in which we can expand the N-particle wavefunctions. For the special case of $\nu = 1$, we can write down the wavefunction by inspection. It is, except for a trivial normalization factor,

$$\Psi_1 = \prod_{i<j} (z_i - z_j) \prod_k \exp\left(-\frac{1}{4\ell_B^2}|z_k|^2\right). \tag{10.4.1}$$

This clearly satisfies the requirement imposed by the Pauli principle that the wavefunction vanish as $z_i \to z_j$ for any pair i, j. The fact that it vanishes linearly with the separation of the particles stems from its nature as a Slater determinant of N wavefunctions. As we saw in Chapter 2, this also has the effect of lowering the repulsive Coulomb energy. If we now turn to a fractional quantum Hall state at $\nu = 1/p = 1/(2m+1)$, we must impose the following conditions on a candidate for the ground-state wavefunction: (a) the wavefunction must be odd under interchange of the positions of two electrons; (b) it must contain a factor $\exp(-\sum_i |z_i|^2/4\ell_B^2)$; and (c) the wavefunction must be an analytic function that is an eigenstate of the total angular momentum. The second of these comes from the fact that all single-particle lowest-Landau-level wavefunctions contain the factor $\exp(-|z|^2/4\ell_B^2)$, and so must any Slater determinant constructed from them. The wavefunction must be analytic because it can be expanded in Slater determinants of lowest-Landau-level single-particle wavefunctions, each of which is a polynomial in the z_i. The simplest possible wavefunctions that satisfy these requirements are

$$\Psi_p = \prod_{i<j}(z_i - z_j)^p \prod_k \exp\left(-\frac{1}{4\ell_B^2}|z_k|^2\right),$$

with p an odd integer. These wavefunctions are of the same form as (10.4.1), except for the fact that the wavefunction now vanishes as an odd power p with the separation of two particles. This gives the particles even higher impetus to stay apart and avoid the Coulomb repulsion. They are also eigenfunctions of the z-component of total angular momentum, with eigenvalue $\frac{1}{2}\hbar N(N-1)p$.

This wavefunction is not easy to visualize, or even to make use of in calculating the energy of the system analytically or numerically. Instead, we shall avail ourselves of a neat trick to study the particle density described by these wavefunctions. The probability distribution of the electrons in Ψ_p is

$$|\Psi_p|^2 = \prod_{i<j}|z_i - z_j|^{2p} \prod_k \exp\left(-\frac{1}{2\ell_B^2}|z_k|^2\right)$$

$$= \exp\left(2p\sum_{i<j} \ln|z_i - z_j| - \frac{1}{2\ell_B^2}\sum|z_i|^2\right) = e^{-\mathcal{H}_p},$$

where we have defined

$$\mathcal{H}_p = -2p\sum_{i<j} \ln|z_i - z_j| + \frac{1}{2\ell_B^2}\sum|z_i|^2.$$

The fictitious Hamiltonian $\mathcal{H}_p$ actually describes a real, albeit classical, system. Consider a two-dimensional plasma, consisting of charges $-q$ at positions r_i in the plane, interacting with one another and with a neutralizing positive background charge density of density ρ. The total interaction between the particles is given by

$$V_p = -q^2 \sum_{i>j} \ln r_{ij}, \qquad (10.4.2)$$

with r_{ij} the distance between the charges. The logarithmic interaction comes from the fact that the charges are infinitely long rods. The interaction energy with the neutralizing background charge density can be shown to be

$$V_{\text{background}} = \frac{\pi}{2}\rho q^2 \sum_i r_i^2. \qquad (10.4.3)$$

The system described by Eqs. (10.4.2) and (10.4.3) is called a one-component classical two-dimensional plasma. In order to minimize its energy, the particles will spread out uniformly to reach the density ρ of the neutralizing background charge. In comparing $\mathcal{H}_p$ and Eqs. (10.4.2) and (10.4.3), it is clear that the probability density of Ψ_p corresponds to that of a classical one-component two-dimensional plasma with density $\rho_p = 1/(2\pi \ell_B^2 p)$ and charges $q^2 = 2p$. We can therefore infer that the wavefunction Ψ_p corresponds to a state with uniform electron charge density $\rho_p = 1/(2\pi \ell_B^2 p)$. This is the electron density at the fractional quantum Hall fillings $\nu = 1/p$. Detailed calculations of the electron–electron correlation functions have verified that Ψ_p describes a translationally invariant liquid, and not a solid. This is significant, because another contender for a fractional quantum Hall state is the so-called Wigner crystal, in which the electrons arrange themselves on a regular lattice in order to minimize the Coulomb repulsion. The Wigner crystal is, as the name implies, a crystal and not a liquid. However, numerical calculations have shown that the Laughlin wavefunction Ψ_p has the lower energy for filling factors ν greater than about $\frac{1}{7}$.

Numerical calculations have also verified that the Laughlin wavefunction is remarkably accurate. This was done by calculating numerically the ground-state energy by exact diagonalization and comparing that with the energy expectation value of the Laughlin wavefunction, and by also calculating the overlap of the numerically obtained ground state with the Laughlin state. The reason for this accuracy (in spite of the apparent simplicity of wavefunction) is very deep and is intimately connected with the extra powers with which the wavefunction vanishes as two electrons are brought together.

10.5 Quasiparticle excitations from the Laughlin state

The elementary excitations from the Laughlin state have the remarkable property that they have fractional charge. At first, this may seem really bizarre, and may lead one to suspect that quarks are somehow involved. This is of course not the case. The fractionally charged elementary excitations are not single particles in the sense that they can exist alone, but are displacements of the electron charge density such that the total local deficit or excess of charge adds up to a fraction of an electron charge. The local charge density is made up of a complicated correlated motion of the real electrons in the system, and in order to create a quasihole, we need to create a local charge deficit. Let us first consider how a filled Landau level ($\nu = 1$) responds to a flux tube carrying a total flux Φ inserted adiabatically through the center of the system. In the symmetric gauge, the single-particle wavefunctions in the lowest Landau level are

$$\phi_m(z) = \left(\frac{z}{\ell_B}\right)^m \exp\left(-\frac{|z|^2}{4\ell_B^2}\right),$$

where z is again equal to $x - iy$ and not the out-of-plane Cartesian coordinate, and we have omitted an uninteresting normalization constant. The extra flux Φ adds an Aharonov–Bohm phase of $\alpha = 2\pi\Phi/\Phi_0$ to each single-particle state as we encircle the origin. In order to preserve single-valuedness of the wavefunctions, we must then add a compensating phase, and the single-particle state now becomes

$$\phi_{m,\alpha}(z) = \left(\frac{z}{\ell_B}\right)^{m+\alpha} \exp\left(-\frac{|z|^2}{4\ell_B^2}\right).$$

If the flux Φ is precisely one unit of flux quantum, $\Phi = \Phi_0$, the mth single-particle wavefunction becomes the $(m+1)$th single-particle wavefunction. This is precisely analogous to our discussion earlier in the Landau gauge. If we started with a full Landau level with uniform charge density, then by inserting one unit of flux quantum we have expelled precisely one electron from the center of the system by pushing all electron charge uniformly out to the edges.

We use these ideas to create a charge deficit – a quasihole – in the Laughlin state. Since the change in the single-particle wavefunctions depends on fundamental principles such as gauge invariance and minimal coupling, the Laughlin wavefunction must respond in a very similar way by shifting $z_i^m \to z_i^{m+1}$ if the center of the system is pierced by a flux quantum. The result will be some deficit of charge. Laughlin used this observation to propose the following Ansatz wavefunction for a quasihole at the position z_0:

$$\Psi_p^+ = \prod_i (z_i - z_0) \prod_{i<j} (z_i - z_j)^p \prod_j \exp\left(-\frac{|z_j|^2}{4\ell_B^2}\right).$$

One can verify that this wavefunction has a component of angular momentum in the direction of the applied magnetic field equal to $\hbar[\frac{1}{2}N(N-1)p + N]$. There is now an extra zero at $z_i = z_0$ for each electron, so that all electrons are pushed away from this point, at which there is then a local deficit of charge relative to the ground state. We can calculate this charge deficit by using the plasma analogy again. Then

$$|\Psi_p^+|^2 = e^{-\mathcal{H}_p^+},$$

with

$$\mathcal{H}_p^+ = -2p\left(\frac{1}{p}\sum_i \ln|z_i - z_0| + \sum_{i<j} \ln|z_i - z_j|\right) + \sum_i \frac{|z_i|^2}{2\ell_B^2}.$$

Again with the identification $2p \to e^2$ we see that $\mathcal{H}_p^+$ corresponds to a classical two-dimensional one-component plasma with an extra repulsive phantom charge e^* that is fixed at $z = z_0$, and which is smaller in magnitude by a factor of $1/p$ than the other charges. Since this charge is repulsive, the plasma will respond by depleting charge $-e/p$ from around $z = z_0$. This charge is expelled to the edges of the system. Another way to understand that Ψ_p^+ must have electrons depleted from $z = z_0$ is the fact that we have inserted into the wavefunction a factor that vanishes as any electron approaches $z = z_0$. There must therefore be a charge deficiency near $z = z_0$. Since there is a *deficiency* of $-e/p$ near $z = z_0$, this means that there is a net extra charge of $+e/p$ relative to the ground state Ψ_p near $z = z_0$, so this must be a quasihole.

The quasiparticle, which has charge $-e/p$, can be constructed in an analogous manner. Instead of adding a flux tube at $z = z_0$ with extra flux, which pushes electrons away from the flux tube, we add a flux tube that depletes the flux at $z = z_0$ by one flux quantum. Technically, this is a little more complicated than creating the quasihole. We need to construct an operation that locally removes one flux quantum at $z = z_0$, thereby decreasing the z-component of total angular momentum by $\hbar N$ by moving all single-particle states in one step towards z_0, which we can without loss of generality take to be the origin. We can then phrase the task at hand in the following way: given the Laughlin function Ψ_p, how can we construct an operator that transforms all the polynomial factors z^m to z^{m-1}? A candidate for this operator is

$$S_m^\dagger \equiv \prod_i^N \left(2\ell_B^2 \frac{\partial}{\partial z_i} - z_0^*\right),$$

where it is understood that this operator acts only on the polynomial part of Ψ_p, leaving the exponential factor intact. Note that the single-particle state with zero angular momentum about z_0 is lifted to the next Landau level by this operator. Since we demand that our states have no components in Landau levels other than the lowest one, we interpret this as an annihilation. The same line of argument as for the quasihole can then be used to show that

$$\Psi_p^- = S_m^\dagger \Psi_p$$

corresponds to an excess charge $-e/p$ near $z = z_0$ in an otherwise uniform state. The extra charge is removed from the edges and moved to $z = z_0$ by "contracting" all single particle states towards $z = z_0$.

The elementary excitations in the Laughlin state are not only fractionally charged – they also can be thought of as having fractional statistics. A good way to start to clarify this concept is to consider regular particles – bosons or fermions – in three dimensions, where the spin-statistics theorem tells us that these are the only two classes of particles allowed. The bosons are described by a wavefunction that is symmetric as two particles exchange positions, and fermions are described by one that is antisymmetric under this exchange. It is useful to cast this in terms of an "exchange phase." We imagine a many-body system in three dimensions consisting of identical bosons or fermions, and keep all but two of the particles fixed at their positions. We then move one of the remaining two adiabatically in a counterclockwise rotation of π about the other, and perform a translation that puts one particle at the original position of the other, and vice versa. The net result is then to have interchanged the positions of two particles. During the operation, the phase of the wavefunction changes. This change in phase depends directly only on the positions of the two particles that we move. At the end of the operation, the change in phase is an even integer times π if the particles are bosons, and an odd integer times π if the particles are fermions. These are the only possibilities in three dimensions. In two dimensions, on the other hand, particles can be chosen to be fermions or bosons or anything in between (with respect to the statistics of the particles), so long as the change in statistics, *the change in exchange phase*, is compensated for by including some interaction between the particles.

Let us now try to apply this kind of reasoning to two quasiholes in a Ψ_p Laughlin state. First, we have to construct a viable candidate for a two-quasihole wavefunction. We might think that something like

$$\widetilde{\Psi_p^{+2}} = \prod_i (z_i - u)(z_j - w) \prod_{i<j} (z_i - z_j)^p \prod_i \exp\left(-\frac{|z_i|^2}{2\ell_B^2}\right), \qquad (10.5.1)$$

where u and w are the positions of the quasiholes of charge e/p, might do the trick. However, this wavefunction has a serious deficiency. Imagine moving one quasihole in a circle about the other, keeping all other charges (including the other quasihole) fixed. Since we are moving a charge e/p around a flux tube (the location of the other quasihole) carrying a unit flux quantum, the many-body wavefunction must pick up an Aharonov–Bohm phase of $2\pi/p$. However, the wavefunction given in Eq. (10.5.1) clearly picks up a phase that is only an integer times 2π, and so we must add something to the wavefunction to fix this difficulty. In order for the wavefunction to pick up the correct phase, it must be a function only of the difference of the coordinates of the quasiholes. This is also more generally required by the fact that the system must be translationally invariant. In addition, we must require that the wavefunction remain in the lowest Landau level. A trial wavefunction that satisfies these conditions is

$$\Psi_p^{2+} = (u-w)^{1/p} \prod_{i,j}(z_i - u)(z_j - w) \prod_{i<j}(z_i - z_j)^p$$

$$\times \prod_k \exp\left[-\frac{|z_k|^2}{4\ell_B^2}\right] \exp\left[-\frac{1}{4p\ell_B^2}(|u|^2 + |w|^2)\right]. \quad (10.5.2)$$

By using the by now familiar plasma analogy we can see that this wavefunction indeed corresponds to a uniform plasma interacting with two positive phantom charges e/p located at u and w, respectively, plus a term corresponding to the interaction between the two phantom charges themselves.

We are now in a position to examine the exchange phase and the statistics of the quasiholes. We note that the factor $(u-w)^{1/p}$ in Eq. (10.5.2) makes the wavefunction Ψ_p^{2+} multi-valued in the parameters u and w, and so we are prepared to believe that something odd can happen to the phase of the wavefunction. Indeed, if we now perform the exchange operation on the two quasiholes at u and w, we find that the phase of the wavefunction changes by π/p. Not only is the charge of the quasiholes a fraction set by the denominator in the filling factor, so is the exchange phase! The possibility of exchange phases and therefore particle statistics other than π and 2π is unique to two dimensions, and has led to use of the term *anyons* to describe particles of arbitrary exchange phase and statistics.

Just as the quasiholes obey fractional statistics, so too do the quasiparticles. The same arguments about translational invariance, Aharonov–Bohm phase, and analyticity can be applied to a two-quasiparticle wavefunction, but the algebra is a little more involved.

There is another way, due to Arovas and coworkers, to arrive at the exchange phase. Imagine that we start with a Laughlin state Ψ_p, and we then add a quasihole of charge $e^* = e/p$ at fixed external magnetic field. This means that we cannot easily write down the wavefunction for this state, since the quasiholes or quasiparticles we constructed previously involved adding or subtracting flux quanta, which necessarily changes the total external magnetic field. Next, we drag the quasihole around some closed contour $\mathcal{C}$ enclosing an area A. This gives an Aharonov–Bohm phase change

γ to the total wavefunction, with

$$\gamma = -\frac{e^*}{\hbar c}\oint_C \mathbf{A}\cdot d\boldsymbol{\ell} = -2\pi\,\frac{e^*}{e}\,\frac{\Phi}{\Phi_0},$$

where Φ is the total flux enclosed by the contour. Since there are p flux quanta per electron, we also have $\Phi = -2\pi p N_{\text{enc}}$, where $N_{\text{enc}} = -\rho A/e$ is the number of electrons enclosed by the contour. So the Aharonov–Bohm phase counts the charge enclosed by the contour. Next, we add another quasihole of charge e^* inside the contour, still keeping the total magnetic field fixed. As we now drag our first quasihole around the contour C, there is a net charge of $-eN_{\text{enc}} + e^*$, so the phase change of the wavefunction is now $\gamma' = -2\pi[N_{\text{enc}} - 1/p]$. That is, there is a phase contribution of $-2\pi/p$ due to the one quasiparticle encircling the other quasiparticle inside the contour. Since the exchange phase of two quasiholes is half of this, we conclude that the exchange phase is of magnitude π/p. The same argument can also be made for quasiparticles.

To show that the Hall resistivity is quantized in a Laughlin state, we can apply a gauge argument similar to the one we used in the integer quantum Hall effect. The difference is that we have to apply p flux quanta in order for the ground state to return to itself, if we consider a fractional quantum Hall state at $v = 1/p$ on a ribbon. This gives a Hall resistivity of $-ph/e^2$ (cf. Eq. (10.2.8)). Perhaps we can feel intuitively that we must have $\rho_H = -h/(ve^2)$ for the fractional quantum Hall effect, too. One factor of e in the denominator comes from the minimal coupling to the vector potential, and a factor of ev comes from the charge of the quasiparticles. The finite width of the plateaus comes again from disorder, which creates bands of localized states (in this case, quasiparticles and quasiholes, which are created as the filling factor is moved away slightly from $1/p$), in which we can pin the Fermi energy. The transport is dissipationless because of the excitation gap for extended states.

We have now seen that the Laughlin wavefunction gives a very good description of the ground state and its elementary excitations at filling factors $v = 1/p$, with p an odd integer. What about other quantum Hall fractions, such as $v = 2/5$? One can attempt to construct sequences of Laughlin states to describe these. For example, we can imagine that we start with the $v = 1/3$ state. We then further increase the external magnetic field. This will move the system to a lower filling, which we can attempt to describe as a Laughlin state with a relatively high density of quasiholes. Eventually, these holes may condense into a Laughlin state of quasiholes. However, the technical details of this sort of description rapidly become intractable. In addition, the energy gaps predicted by a simple application of this approach do not bear a very great resemblance to those deduced from experimental observations of the relative prominence of the various plateaus. It turns out that there is a much more convenient and simple description, based on *composite fermions*, which not only correctly describes the sequence of energy gaps of the ground states, but also the elementary and collective excitations above these. We discuss composite fermions in the last section of this chapter.

10.6 Collective excitations above the Laughlin state

Suppose we move the overall filling factor slightly off a Laughlin filling factor $v = 1/p$, for example by increasing or decreasing the strength of the external magnetic field. It is energetically favorable for the system to respond by creating quasiholes or quasiparticles, while keeping most of the electron density fixed at $v = 1/p$. These excitation energies, $\mathcal{E}^+$ and $\mathcal{E}^-$, consequently give us the change in slope of the ground state energy at a filling factor $v = 1/p$. Equivalently, the lowest-energy way to add or remove electrons from the system is to have the density excess or deficit break into quasiparticles or quasiholes, leaving most of the system unchanged at $v = 1/p$. The quasiholes and quasiparticles are charged excitations. We can also consider neutral excitations, which occur when the particle number and magnetic field are kept fixed. We can construct these from quasiholes and quasiparticles as long as we maintain charge neutrality by creating only quasihole–quasiparticle pairs. We can imagine that we create such a pair, and then move the localized quasihole and quasiparticle far apart. The situation is now reminiscent of the one we encountered in Section 2.7, when we considered the effect of creating an electron–hole pair in the three-dimensional electron gas. There we found that the operator $c^\dagger_{\mathbf{p+q}} c_{\mathbf{p}}$ created a satisfactory excitation when q was large, but that when q was small we needed to form the linear combination of the operators $c^\dagger_{\mathbf{p+q}} c_{\mathbf{p}}$ that created *density fluctuations*. The same circumstances arise in considering excitations above the Laughlin ground state. If we consider excitation energy as a function of wavevector $\mathbf{k}$, this quasihole–quasiparticle pair corresponds to the excitation energy at large wavevectors, $k\ell_B \gg 1$. The limit $k\ell_B \to 0$, on the other hand, gives the energy of a very long-wavelength density fluctuation, which must be made up of a linear combination of many quasihole–quasiparticle pairs. Since the system is incompressible, there must also be an energy gap at $k\ell_B \to 0$. We have already noted that at filling factors v smaller than about $1/7$, the ground state of the system is a Wigner crystal. This means that as we lower the filling factor down towards 1/7, the electron liquid of the Laughlin state must somehow freeze and transform into a crystal structure with a length scale of about ℓ_B. The analogy we make now is with the Peierls transition discussed in Section 6.3, where a phonon mode was softened until its frequency vanished, whereupon a permanent distortion occurred. We may expect a similar phenomenon here, and look for an excitation mode at $k \sim 1/\ell_B$ to become increasingly soft as the filling factor is reduced, and for its excitation energy to reach zero. When this happens, the translationally invariant liquid is no longer the ground state.

It turns out that there is in fact a minimum in the excitation energy of the Laughlin state as a function of wavevector at about $k \approx 1/\ell_B$. This minimum is very well described by a theory analogous to the theory that produced the so-called roton minimum in liquid ^{4}He shown in Fig. 3.4.2, and the excitations at this minimum are therefore appropriately called magneto-rotons. The core of the theory is a calculation of the energy expectation value of a variational *Ansatz*, or trial, wavefunction for an excited state corresponding to a density wave of wavevector $\mathbf{q}$. The obvious choice is

$$\Psi_{\mathbf{q}} = N^{-1/2} \rho_{\mathbf{q}} |\Psi_0\rangle,$$

where $\rho_{\mathbf{q}}$ is the Fourier transform of the density operator and $|\Psi_0\rangle$ the ground state whose energy is $\mathcal{E}_0$. The factor of $N^{-1/2}$, with N the particle number, has been inserted for convenience. The norm of this state, which is required to evaluate the expectation energy, is

$$s(\mathbf{q}) = N^{-1} \langle \Psi_0 | \rho_{\mathbf{q}}^\dagger \rho_{\mathbf{q}} | \Psi_0 \rangle,$$

which is also known as the static structure factor of the ground state $|\Psi_0\rangle$. This quantity can be measured directly by, for example, neutron scattering. The expectation value of the excitation energy of the state $\Psi_{\mathbf{q}}$ is then

$$\Delta(\mathbf{q}) = \frac{\langle \Psi_{\mathbf{q}} | \mathcal{H} - \mathcal{E}_0 | \Psi_{\mathbf{q}} \rangle}{s(\mathbf{q})} = \frac{N^{-1} \langle \Psi_0 | \rho_{\mathbf{q}}^\dagger [\mathcal{H}, \rho_{\mathbf{q}}] | \Psi_0 \rangle}{s(\mathbf{q})} = \frac{f(\mathbf{q})}{s(\mathbf{q})},$$

where $f(\mathbf{q})$ is called the oscillator strength. It is a measure of how much of the phase space available to excitations is filled by the mode under consideration. For the case of liquid helium, it is easy to derive the result

$$f(q) = \frac{\hbar^2 q^2}{2m},$$

10.6 Collective excitations above the Laughlin state

since the potential energy and the density operators commute with one another, while the kinetic energy and the density operators do not. This leads to the so-called Feynman–Bijl formula for the liquid He excitation energy:

$$\Delta(q) = \frac{\hbar^2 q^2}{2ms(q)}.$$

This equation says that the excitations are essentially free particle excitations renormalized by the structure factor $s(q)$ due to correlations between particles.

One can construct a very similar theory for collective excitations in the fractional quantum Hall effect. One complication, which makes the algebra too lengthy to reproduce here, is that we need to project all actions of operators on wavefunctions onto the lowest Landau level. At the end of the day, one arrives at an equation very similar to the Feynman–Bijl formula, except that the oscillator strength and the structure factor have to be replaced by quantities projected onto the lowest Landau level, $\bar{f}(\mathbf{q})$ and $\bar{s}(q)$. The corresponding equation for the collective mode excitation energy is then

$$\Delta(\mathbf{q}) = \frac{\bar{f}(\mathbf{q})}{\bar{s}(\mathbf{q})}.$$

One important difference between liquid helium and the fractional quantum Hall effect is that the latter has an excitation gap at $q \to 0$, while, as we saw in Fig. 3.4.2, the former does not. This implies that $\lim_{q \to 0} \Delta(q)$ is finite. One can show that $\bar{f}(q \to 0) \sim |q|^4$, which means that we must have $\bar{s}(q \to 0) \sim |q|^4$ in order to have a finite gap at $q \to 0$. Detailed calculation shows that this is indeed the case for *any* liquid ground state in the lowest Landau level, not just the Laughlin functions. For finite q the excitation energies $\Delta(q)$ for Laughlin states can be found by numerically evaluating the projected structure factor for the Laughlin wavefunction. The result is in very good agreement with direct numerical diagonalizations for the lowest-lying excited mode, and with experimental observations.

This approach to finding the magneto-roton collective modes is called a single-mode approximation, because it assumes that there is a single excitation mode for each q. The reason that this theory works for liquid helium lies in the fact that the symmetry of the wavefunction for bosons only allows for low-lying collective density modes. But this is not the case for a system of fermions, for which there can in principle be a continuum of single-particle excitations and, for the case of the quantum Hall effect, intra-Landau level excitations. This is where the existence of a gap comes in and saves the day.

The excitation gap in the fractional quantum Hall effect quenches out single-particle-like excitations and leaves only the low-lying collective modes.

10.7 Spins

So far, we have limited our discussion to a fully spin-polarized system. This seems reasonable, since in the strong magnetic fields used in experiments on quantum Hall systems one might expect the Zeeman spin splitting $g\mu_B B$, where g is the effective Landé factor and μ_B the Bohr magneton, to be large enough that the high-energy spin direction would be energetically inaccessible. However, two factors conspire to make the Zeeman splitting very low in GaAs, which is the material from which many quantum Hall devices are constructed. First of all, spin–orbit coupling in the GaAs conduction band effectively lowers the Landé factor to $g \approx 0.44$. Second, the low effective mass, $m^* \approx 0.067 m_e$, further reduces the ratio of spin-splitting energy to cyclotron energy to about 0.02, compared with its value of unity for free electrons. For magnetic fields of about 1–10 T, the Coulomb energy scale of the electron–electron interactions in GaAs is $e^2/(\epsilon \ell_B)$, with the static dielectric constant ϵ being about 12.4, and is of the same order as the cyclotron energy. As a first approximation, one should then set the Zeeman energy to zero, rather than infinity, since it is two orders of magnitude smaller than the other energy scales. As a consequence, the spin degree of freedom is governed by the electron–electron interactions, rather than by the Zeeman energy. This dramatically changes the nature of the low-energy bulk single-particle excitations near filling factors $\nu = 1/p$, with p odd, from single-particle spin-flips to charge-spin textures. In these objects, loosely called *skyrmions*, the spin density varies smoothly over a distance of several magnetic lengths, so that the system can locally take advantage of the exchange energy by having spins roughly parallel over distances of the order of a magnetic length. The magnetization has a finite winding number n, which is to say that if we encircle a skyrmion along some closed path, the magnetization direction will change by $2\pi n$, where n is an integer. The kind of spin texture excitations that make up skyrmions have been known to exist in other models of magnets, but what is remarkable about the skyrmions in the quantum Hall effect is that they carry charge, and the charge is equal to ne/p, with $1/p$ being the filling factor ν. This coupling between charge and spin is a direct consequence of the fact that these are two-dimensional systems in the presence of a strong external magnetic field. Let us suppose that the (bulk) filling factor is initially unity. If there is a region in space where the spin is slowly varying spatially, that changes the effective magnetic field $\mathbf{B}_{\text{eff}}(\mathbf{r}) = B\hat{z} + 4\pi \mathbf{M}(\mathbf{r})$, where $\mathbf{M}(\mathbf{r})$

10.7 Spins

is the magnetization density due to the varying spin density. But that would result in a concomitant change in local effective filling factor away from unity. The system desperately wants to maintain a filling factor of unity, so it responds by locally transferring some charge into the region of varying spin density to maintain an effective filling factor of unity. The net effect is a local accumulation (or deficit) in charge relative to the ground state.

We can make this argument more formal in the following way. Let us assume that the spin density varies slowly on the scale of ℓ_B for a quantum Hall system at bulk filling factor $\nu = 1/p$. The spin $\mathbf{S}_j$ of electron j sees an effective exchange field $\mathbf{b(r)}$ due to the spin density of the other electrons. This exchange field just expresses the fact that the electrons gain exchange energy by keeping their spins parallel. Formally, it is defined as the change in exchange-correlation energy as we change the direction of one electron's spin, while keeping the others fixed. In a mean-field approximation, we do not distinguish between the exchange fields of different electrons, but take the field at $\mathbf{r}$ to be a suitable average over the exchange fields in some neighborhood about $\mathbf{r}$. A model Hamiltonian expressing this coupling would be

$$\mathcal{H}_{\text{eff}} = -\sum_{j}^{N} \mathbf{b(r)} \cdot \mathbf{S_j}.$$

Imagine that we move a single electron adiabatically around a closed path $\mathcal{C}$ in real space, keeping all other electrons (and their spins) fixed. The electron will keep its spin aligned with the exchange field as we move it along $\mathcal{C}$ and trace out some path ω in spin-space. This path is the path drawn by a unit vector on the unit sphere as the vector moves through the same angles (θ, ϕ) as the spin along $\mathcal{C}$. This means that the electron wavefunction will acquire an extra phase, a so-called *Berry's phase*, which is analogous to the Aharonov–Bohm phase that a charge acquires as it moves through a region with a finite vector potential. The Berry's phase is $\Omega/2$, where Ω is the solid angle subtended by the path ω. This is illustrated in Fig. 10.7.1.

We must not forget that the electron also acquires an Aharonov–Bohm phase, since it is charged and moves through a region with a finite vector potential. There are thus two contributions to the added phase, the Berry's phase, and the Aharonov–Bohm phase. The electron cannot tell where the contributions to the phase come from, and we might as well replace the Berry's phase by adding some extra flux $\Delta\Phi$ in the region enclosed by $\mathcal{C}$, such that the Aharonov–Bohm phase due to this flux equals the Berry's

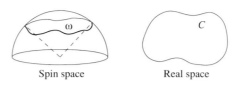

Figure 10.7.1. As an electron moves along a path C in real space with an inhomogeneous magnetic field, the direction of the electron's spin traces a path ω on the unit sphere in spin space. The solid angle subtended by the path ω is Ω.

phase. This means that

$$\Delta \Phi = \frac{\Omega}{4\pi} \Phi_0,$$

where Φ_0 is the flux quantum. But adding extra flux in the region enclosed by C will add extra charge, for we remember that the quasiholes and quasiparticles were generated by adding flux at the positions of the elementary excitations. A simple application of the flux argument from Section 10.5 shows that the extra charge induced is

$$\Delta Q = -ev \frac{\Delta \Phi}{\Phi_0} = -ev \frac{\Omega}{4\pi},$$

if the bulk Hall conductivity $\sigma_{xy} = -ve^2/h$.

We can construct simple trial functions for skyrmions of charge $\pm e$ for the case of $v = 1$. It is convenient to work in the symmetric gauge, and to consider the wavefunction of a skyrmion centered at the origin. The basic idea is to start with a filled, spin-polarized Landau level, which we write as $|\Psi_0\rangle = \Pi_m c_m^\dagger |0\rangle$, where $|0\rangle$ is the vacuum state, and $c_m^\dagger$ creates an up-spin electron of z-component of orbital angular momentum $\hbar m$ in the lowest Landau level. To construct a skyrmion hole we first remove the spin-up electron in the $m = 0$ state by use of the operator c_0, and then replace part of the amplitude of the other spin-up states with a component with spin down and with a value of m reduced to $m - 1$. We do this by operating with a product of terms of the form $(v_m + u_m b_m^\dagger c_{m+1})$ where $b_m^\dagger$ creates a down-spin electron of angular momentum $\hbar m$. The term $(v_0 + u_0 b_0^\dagger c_1)$, for example, replaces part of the spin-up electron amplitude in the $m = 1$ state by a contribution with spin down in the $m = 0$ state. It thus has the double effect of increasing the amount of spin-down component in the wavefunction and bringing closer to the origin some of the charge. This reduces somewhat the charge deficit near the origin due to the initial destruction of the spin-up electron with $m = 0$. The resulting wavefunction will then be

$$|\psi_-\rangle = \prod_{m=0}^{\infty} [u_m b_m^\dagger + v_m c_{m+1}^\dagger] |0\rangle. \qquad (10.7.1)$$

In a similar way we can create a skyrmion particle by first adding a down-spin electron in the $m = 0$ state by use of the operator $b_0^\dagger$, and then operating with terms like $(v_m - u_m b_{m+1}^\dagger c_m)$. This has the effect of replacing part of the amplitude of the spin-up states with a component with spin down and with a value of m increased to $m + 1$. It yields the skyrmion particle wavefunction

$$|\psi_+\rangle = \prod_{m=0}^{\infty} [-u_m b_{m+1}^\dagger + v_m c_m^\dagger] b_0^\dagger |0\rangle. \qquad (10.7.2)$$

In this case some of the charge is repelled from the origin by the increase in m, which reduces the increase of charge near the origin due to the initial creation of a spin-down electron with $m = 0$. The set of numbers u_m and v_m are to be determined by a minimization of the energy in a procedure similar to the one we used in the BCS theory of superconductivity in Section 7.3, subject to the normalizing constraint that $|u_m|^2 + |v_m|^2 = 1$. The signs of u_m and v_m are chosen to make $|\psi_+\rangle$ and $|\psi_-\rangle$ orthogonal. The spin texture of a skyrmion is illustrated in Fig. 10.7.2.

What we now want to do is to demonstrate that $|\psi_+\rangle$ and $|\psi_-\rangle$ are approximate energy eigenstates when we make the right choices of u_m and v_m with $u_0 \neq 0$ and u_m decaying as m increases so that the original spin-polarized Landau level is restored far away from the skyrmion. We must also verify that the energies of these states are lower than the single-electron quasi-electron or quasihole energies. It is a good exercise to verify that the spin

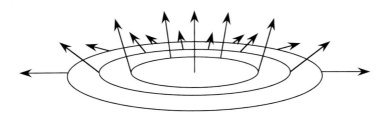

Figure 10.7.2. The spin orientation due to a skyrmion located at the origin.

densities of $|\psi_+\rangle$ and $|\psi_-\rangle$ point downwards at the origin and upwards as $m \to \infty$, and that the projection of the spin polarization on the xy-plane rotates by $\pm 2\pi$ along any path encircling the origin.

We start with the Coulomb Hamiltonian for the lowest Landau level, but include the Zeeman energy. We have

$$\mathcal{H} = \frac{1}{2} \sum_{\substack{m_1,m_2 \\ m_3,m_4}} V_{m_1 m_2 m_3 m_4} : [b^\dagger_{m_1} b_{m_2} + c^\dagger_{m_1} c_{m_2} - \delta_{m_1 m_2}][b^\dagger_{m_3} b_{m_4} + c^\dagger_{m_3} c_{m_4} - \delta_{m_3 m_4}] :$$

$$- g\mu_B B \sum_m [c^\dagger_m c_m - b^\dagger_m b_m].$$

Here $V_{m_1 m_2 m_3 m_4}$ are the matrix elements of the Coulomb interaction between angular-momentum single-particle states,

$$V_{m_1 m_2 m_3 m_4} = \int d^2 r_1 d^2 r_2 \phi^*_{m_1}(\mathbf{r}_1) \phi^*_{m_2}(\mathbf{r}_2) \frac{e^2}{\epsilon |\mathbf{r}_1 - \mathbf{r}_2|} \phi_{m_3}(\mathbf{r}_2) \phi_{m_4}(\mathbf{r}_1),$$

and $:[\ldots]:$ indicates *normal ordering*, which keeps all the creation operators to the left of the annihilation operators. We have not included a uniform positive background charge density, since that is not important here.

Next, we proceed with the Hartree–Fock reduction of the terms with four electron creation and annihilation operators. When we were studying the uniform electron gas in Section 2.4, our procedure was to group together pairs of electron creation and annihilation operators to make number operators, which we then replaced with their expectation values. In the present case, our wavefunction is no longer a simple Slater determinant, but is of the form given by Eqs. (10.7.1) and (10.7.2). Thus we must allow for the fact that not only terms like $\langle b^\dagger_m b_m \rangle$ and $\langle c^\dagger_m c_m \rangle$ but also those like $\langle b^\dagger_m c_{m\pm 1} \rangle$ will contribute. (If we had been considering skyrmions with winding numbers larger than unity, there could also in principle be other combinations, such as $\langle b^\dagger_m c_{m\pm n} \rangle$ with $n > 1$.)

The minimization of the expectation value of $\mathcal{H}$ proceeds in a manner very close to that used in the BCS theory to arrive at Eq. (7.3.12), which related the coefficient x_k to the gap parameter Δ through the relation

$$x_\mathbf{k} = \pm \frac{\mathcal{E}_\mathbf{k}}{2\sqrt{\mathcal{E}_\mathbf{k}^2 + \Delta_\mathbf{k}^2}},$$

but with $\Delta_{\mathbf{k}}$ self-consistently dependent on the interaction V and on x_k itself. For the skyrmion problem we have the very similar results

$$u_m = \frac{U_\pm^{sk}(m)}{\sqrt{\mathcal{E}_m^2 + |U_\pm^{sk}(m)|^2}},$$

$$v_m = \frac{\mathcal{E}_m}{\sqrt{\mathcal{E}_m^2 + |U_\pm^{sk}(m)|^2}},$$
(10.7.3)

but with rather more complicated definitions of the components. Here

$$\mathcal{E}_m = \tfrac{1}{2}\{\mathcal{E}_b(m) - \mathcal{E}_c(m) + [(\mathcal{E}_b(m) - \mathcal{E}_c(m))^2 + 4|U_\pm^{sk}(m)|^2]^{1/2}\},\quad (10.7.4)$$

and, with a standing for either b or c,

$$\mathcal{E}_a(m) = s_a g\mu_B B + U^H(m) + U_a^{ex}(m),\quad (10.7.5)$$

with $s_a = 1$ for $a = b$, and $s_a = -1$ for $a = c$. The interaction-energy terms U are

$$U^H(m) = \sum_{m_1} V_{mm_1 mm_1}[\langle b_{m_1}^\dagger b_{m_1}\rangle + \langle c_{m_1}^\dagger c_{m_1}\rangle - 1]$$

$$U_a^{ex}(m) = \sum_{m_1} V_{mm_1 m_1 m}\langle a_{m_1}^\dagger a_{m_1}\rangle \quad (10.7.6)$$

$$U_\pm^{sk}(m) = \sum_{m_1} V_{m,m_1,m_1\pm 1,m\pm 1}\langle b_{m_1}^\dagger c_{m_1\pm 1}\rangle,$$

with $U^H(m)$ the direct (Hartree) term, $U^{ex}(m)$ the exchange (Fock) term, and $U_\pm^{sk}(m)$ the skyrmion-specific terms that describe a spin-flip together with a change in angular momentum. The order parameters, i.e., the expectation values that contribute to the energy, are given by the relations

$$\langle b_m^\dagger b_m\rangle = |u_m|^2,$$

$$\langle c_m^\dagger c_m\rangle = |v_{m\mp 1}|^2, \quad (10.7.7)$$

$$\langle b_m^\dagger c_{m\pm 1}\rangle = u_m^* v_m.$$

The self-consistent equations (10.7.3) to (10.7.7) must be solved numerically. The result is that for any value of g, the hole-skyrmion and the particle-skyrmion have *lower* energies than the simple quasihole or quasiparticle, in which a regular hole or an electron with opposite spin is introduced at the origin in the spin-polarized lowest Landau level. The energy differences are largest for $g \to 0$ and vanish as g becomes very large. Physically, what happens when the Zeeman energy is increased by increasing g is that the region of the skyrmion in which the spin is varying shrinks in size, since it will cost increasingly more energy not to align the spin density with the external field. In the limit of zero size, the skyrmions become identical to quasihole and quasiparticle excitations.

The fact that skyrmions have lower energy than quasiparticle and quasihole excitations near $\nu = 1$ has experimental consequences. If skyrmions did not exist, or had higher energies than quasiparticles and quasiholes, the ground state would remain spin-polarized as the filling factor is reduced below unity. Electrons would simply be removed from the lowest Landau level (or new, empty states added) and the remaining ones would all be spin-polarized. Also, as the filling factor is increased above unity, the spin polarization would decrease at the rate at which new spin-reversed electrons are added, assuming still that the spin splitting of Landau levels is smaller than the cyclotron energy. However, it is experimentally observed that the ground-state polarization is rapidly destroyed when the filling factor is either increased or decreased away from unity. The reason for this is that it is cheaper for the system to add skyrmion particles or skyrmion holes than quasiparticles or quasiholes. The net spin of skyrmions, obtained by integrating the spin density, is about $7\hbar/2$, compared with the $\hbar/2$ found for quasiparticles and quasiholes. This means that considerable spin is added with each skyrmion, and there is a resulting rapid destruction of the ground-state polarization.

10.8 Composite fermions

We have so far, within the context of Laughlin's wavefunction, only discussed the 'primary' fractional quantum Hall states with filling factors $\nu = 1/p$, with p odd. In Section 10.5 we indicated that one may try to construct Laughlin-type wavefunctions for quasiparticles or quasiholes, which could conceivably form a strongly correlated liquid on top of the underlying basic fractional quantum Hall state. Then this correlated "Laughlin liquid" of quasiparticles or quasiholes would admit fractionally charged excitations, which in turn could condense and form a new correlated state, and so on. The problem with this picture, apart from its rapidly increasing and open-ended complexity, is that it does not adequately explain the sequence of energy gaps observed in fractional quantum Hall states. One would expect qualitatively

10.8 Composite fermions

that the stability of states, as reflected in the magnitude of the energy gaps, should rapidly decrease as one climbs to higher levels in this hierarchical picture. This is generally not the case. For example, the states $v = 1/3$ and $v = 2/5$ are in general more stable than the $v = 1/5$ state. Also, the quasiparticles and quasiholes in one hierarchical level have a finite size, and one would rapidly reach a situation in which the number of quasiparticles or quasiholes needed to condense at one level to form a new one would be so great that these elementary excitations would overlap substantially, in which case the notion of elementary quasiparticles and quasiholes ceases to be meaningful. Finally, the Laughlin wavefunction provides no connection between the integer quantum Hall states and the fractional ones, treating the two as fundamentally very different.

Let us go back and consider the sequence of fractions at which the fractional quantum Hall effect is observed. We can group the sequences in the following manner:

$$v = \frac{n}{2n+1} = \frac{1}{3}, \frac{2}{5}, \frac{3}{7}, \frac{4}{9}, \ldots$$

$$v = \frac{n}{2n-1} = \frac{2}{3}, \frac{3}{5}, \frac{4}{7}, \frac{5}{9}, \ldots$$

$$v = \frac{n}{4n+1} = \frac{1}{5}, \frac{2}{9}, \frac{3}{13}, \ldots$$

$$v = \frac{n}{4n-1} = \frac{2}{7}, \frac{3}{11}, \ldots$$

$$v = 1 - \frac{n}{4n+1} = \frac{4}{5}, \frac{7}{9}, \ldots$$

By inspection, we see that we can in general write these fractions as

$$v = \frac{n}{2pn \pm 1}, \tag{10.8.1}$$

and

$$v = 1 - \frac{n}{2pn \pm 1}. \tag{10.8.2}$$

For given p and n, the fractions in Eqs. (10.8.1) and (10.8.2) are related by an electron–hole symmetry that is exact if we ignore the presence of any other than the lowest Landau level. In a real system, the energy gap – the cyclotron energy – separating different Landau levels is finite, with the consequence that the real wavefunction will have some admixture of higher Landau levels. However, this admixture is small enough that the electron–hole symmetry is almost exactly preserved.

The composite fermion picture, originated by Jainendra Jain, provides a simple and natural picture within which these sequences can be viewed. It also attempts to tie together the physics of the integer and fractional quantum Hall effects. In addition, it provides a powerful computational method with which to study general fractional quantum Hall ground states, as well as their quasiparticle and collective excitations. While the origins of the composite fermion picture are empirical, numerical calculations within this picture have consistently proven to be remarkably accurate when compared with other, more direct, schemes as well as with experiments, and have provided strong support for this approach. The basic principle is to replace the strongly interacting electrons by some other particles, which are chosen to be weakly interacting. That is, we try to construct some composite particle such that the fractional quantum Hall ground states are well described by a gas of such noninteracting particles. This is reminiscent of Fermi liquid theory, in which a system of rather strongly interacting electrons can be described as a weakly interacting system of quasiparticles. In Fermi liquid theory the interaction between electrons leads to quasiparticles of different effective mass and with a modified energy dispersion relation. Similarly, we would here like to find new particles such that the strong interactions of the true electrons have been transformed into kinetic energy of these new particles. The question then is to identify the kind of particle that would be a good candidate for a similar description of fractional quantum Hall states. Jain observed that we can write the Laughlin wavefunction for, say, $n = 1/3$, in the following way:

$$\Psi_3 = \prod_{i<j}(z_i - z_j)^3 \prod_k \exp\left(-\frac{|z_k|^2}{4\ell_B^2}\right)$$

$$= \prod_{i<j}(z_i - z_j)^2 \prod_{i<j}(z_i - z_j) \prod_k \exp\left(-\frac{|z_k|^2}{4\ell_B^2}\right).$$

That is, we can think of the Laughlin wavefunction for the $\nu = 1/3$ state as obtained by starting with the filled Landau level of $\nu = 1$ and multiplying that unique wavefunction by the factors $\prod_{i<j}(z_i - z_j)^2$. But, according to our earlier discussion, this factor has the same effect as attaching two flux quanta to each electron (as seen by the other electrons). So the $\nu = 1/3$ wavefunction can be thought of as a filled $\nu = 1$ Landau level for particles that consist of

10.8 Composite fermions

electrons with two flux quanta attached to each electron. According to the statistics that we worked out earlier for the two-quasiparticle wavefunctions, each flux quantum that we attach to each electron will add a phase of π to the exchange phase of the electron. So if we add two flux quanta to each electron, the composite particles consisting of an electron plus two flux quanta must be fermions. These are the composite fermions. For a general fractional quantum Hall state $\nu = n/(2np \pm 1)$, we form the appropriate composite fermions by attaching $2p$ flux quanta to each electron.

We can now write down a general recipe for constructing composite fermion states. Start with a fractional quantum Hall state at some filling factor $\nu = n/(2np \pm 1)$. Then form composite fermions by letting each electron in this state gobble up $2p$ flux quanta of the total magnetic flux penetrating the system. The effective magnetic field B^* experienced by the composite fermions is then the field that is left over, and is given by

$$B^* = B - 2p\Phi_0\rho,$$

where ρ is the density of the electrons in the fractional quantum Hall state. Note that B^* can actually be negative, in which case the effective field acting on the composite fermions would point in the direction opposite to that acting on the electrons. The original filling factor of the electrons was

$$\nu = \frac{\rho\Phi_0}{B},$$

and the filling factor ν^* of the composite fermions is similarly given by

$$\nu^* = \frac{\rho\Phi_0}{|B^*|}.$$

Combining these two equations we can relate ν and ν^*:

$$\nu = \frac{\nu^*}{2p\nu^* \pm 1},$$

where we choose the minus sign in the denominator if B^* is antiparallel to B.

In this way, we can quickly construct the equivalent composite fermion states of all fractional quantum Hall states. For example, as we already mentioned, $\nu = 1/3$ maps to $\nu^* = 1$ when $p = 1$. The next fraction in this sequence, $\nu = 2/5$, maps to $\nu^* = 2$. The particle–hole conjugate $\nu = 2/3$ of $\nu = 1/3$ maps to $\nu^* = -1$. Similarly, the Laughlin state at $\nu = 1/5$ maps to $\nu^* = 1$ by attaching four flux quanta to each electron so that $p = 2$, and so on.

Approximate wavefunctions for fractional filling factor ground states and elementary excited states can then be generated by inverting this mapping. One then starts with a composite fermion ground state or elementary excited state at an integer filling factor v^* and maps this onto a fractional filling factor by applying factors of $\prod_{i<j}(z_i - z_j)^p$ and by finally projecting the resulting wavefunctions onto the lowest Landau level.

As we have already stated, it turns out that the wavefunctions constructed in this way are excellent approximations to exact wavefunctions, provided we are careful and project them onto the lowest Landau level. Some of the reasons for this are that the composite fermion wavefunctions are all uniquely determined, by which we mean that there is no variational freedom left to tinker with the composite fermion wavefunction once the mapping has been done. This can then also be extended to composite fermion excited states, which can be mapped to excitations in fractional quantum Hall states. So long as the composite fermion mapping provides a unique one-to-one correspondence between the states, the composite fermion prescription is remarkably accurate.

Another reason for its success lies in how the motions of the electrons are correlated in fractional quantum Hall states. According to Laughlin's wavefunction, the electrons like to attach extra powers in the relative coordinates of electrons so that the wavefunction vanishes as quickly as possible when two particles are brought together. This corresponds to attaching vortices, or flux quanta, to the electrons. It is a good approximation to assume that these flux quanta are rigidly bound to the electrons and that the electrons only experience a residual average field. The effective interactions between composite fermions are then given by the difference between the average residual field and the actual field experienced by the composite fermions. The actual field consists of the *real* applied field minus the field of the flux quanta at the locations of the composite fermions. As one can easily imagine, this actual field is rather complicated and has all kinds of unpleasant singularities stemming from the singular nature of the flux quanta attached to the electrons. Nature is kind to us in that these interactions are rather weak.

Problems

10.1 Consider a square sample with edges at $x = \pm a$, $y = \pm a$. Electrodes are attached to the left and right edges to make the electrostatic potential $\phi = -V$ at $x = -a$ and $\phi = V$ at $x = a$. Assume that $\sigma_{xx} = \sigma_{yy}$ is very small, but nonzero, and that $\sigma_{xy} = -\sigma_{yx}$ is constant. A steady-state current I flows through the sample.

(a) By using the continuity equation and Maxwell's equation show that $\partial E_x/\partial x + \partial E_y/\partial y = 0$.
(b) Show that the current flows in through one corner and out of the diagonally opposite one.
(c) Show that the Hall voltage is equal to the source-to-drain voltage. (This problem is due to Rendell and Girvin.)

10.2 A complete solution to Problem 10.1 in the limit where σ_{xx} and σ_{yy} are vanishingly small is found by solving Laplace's equation, $\nabla^2 \phi = 0$. The electrostatic potential is found to be of the form

$$\phi = \sum_{n=0}^{\infty} b_n[\cos(k_n x)\sinh(k_n y) + \cos(k_n y)\sinh(k_n x)],$$

where $k_n = (2n+1)\pi/2a$. Solve for b_n and hence find $E_y(x=0, y=0)$ and $E_y(x=0, y=a)$.

10.3 Consider a noninteracting two-dimensional electron gas in a strong magnetic field $\mathbf{B} = B\hat{\mathbf{z}}$ and in the presence of a uniform electric field $\mathbf{E} = E\hat{\mathbf{x}}$. Far from the edges of the system, there is a short-ranged elastic scatterer. Impose periodic boundary conditions along the y-direction. Show that far away from the scatterer, the only effect it has on extended states is to give rise to a phase shift $\delta_n(k)$, with the states enumerated according to their Landau level index and asymptotic wavenumber. Compare the allowed values of k with and without a scatterer using the phase shift $\delta_n(k)$. What has happened to the number of extended states if the phase shift decreases by $2\pi M$, where M is an integer, as we go through all the allowed states k for a given n? The total number of states must be the same with and without the scatterer, so what has happened to the rest of the states? Show that the phase shift is related to the transit time for a scattering state to traverse the system from one end to the other. What happens to this transit time for states that experience a nonzero phase shift?

10.4 Consider the two lowest Landau levels $n=0$ and $n=1$ in a non-interacting system with an applied electric field $\mathbf{E} = E\hat{\mathbf{x}}$. Ignore edge effects. Now suppose that the electrons can interact with acoustic phonons having a dispersion relation $\hbar\omega(\mathbf{q}) = \hbar v_s|\mathbf{q}|$, with $\mathbf{q}$ the wavevector of a phonon and v_s the speed of sound. What conditions must the electric field satisfy in order to make phonon emission by the electrons energetically permissible? The transition rate for phonon emission can be approximated using Fermi's Golden Rule. Assume that the electron-phonon interaction potential has the form $V(q) \propto \sqrt{q}$ and write down an expression for the transition rate. Estimate how large the electric field would have to be for phonon emission to be of consequence.

10.5 Skyrmion solutions were first discovered as solutions to models of classical magnetic systems, so-called nonlinear sigma models. In a dimensionless form, the magnetic field corresponding to these skyrmions can be written

$$\mathbf{B}(\mathbf{r}) = \frac{(4\lambda x, \pm 4\lambda y, r^2 - 4\lambda^2)}{r^2 + 4\lambda^2}.$$

Here $r^2 = x^2 + y^2$ and λ is a parameter having the dimension of a length that sets the scale of the size of the skyrmion. Consider the exchange Hamiltonian $\mathcal{H}_0 = -\mathbf{B}(\mathbf{r}) \cdot \mathbf{s}(\mathbf{r})$. Substitute for s_x and s_y using spin lowering and raising operators, and use the complex coordinate $z = x + iy$ to rewrite the exchange Hamiltonian in terms of spin operators. Show that this Hamiltonian must have eigenstates of the form of the skyrmion states (10.7.1) and (10.7.2). [Hint: Use the fact that z and z^* act as angular momentum raising and lowering operators.]

Chapter 11

The Kondo effect and heavy fermions

11.1 Metals and magnetic impurities

We argued in Section 8.4 that, at temperatures much lower than the Debye temperature, the resistivity of a metal should be given by an expression of the form

$$\rho(T) = \rho_0 + AT^5.$$

Here the T^5 behavior comes from electron–phonon scattering, while the impurities in the metal give rise to the constant term ρ_0. This assumed that the scattering caused by the impurities was elastic. If the scattering is inelastic, then a variety of interesting phenomena may occur. In this chapter, we discuss some of the processes that can occur when a metal is doped with magnetic impurities, whose spin states introduce extra degrees of freedom into the scattering problem. Interaction between the spins of the conduction electrons and those of the impurities then provides a mechanism for inelastic scattering of the conduction electrons.

The magnetic impurities that one can find in a metal fall into three classes. There are the transition metals, such as manganese and iron, the rare earths, such as cerium, and the actinides, of which the most important is uranium. The magnetic character of these classes of elements originates in the fact they have partially filled inner shells. For the transition metals it is the 3d shell, for the rare earths the 4f shell, and for the actinides the 5f shell that remains only partially filled, even though the outer valence states (4s, 5s, and 6s, respectively) also contain electrons. This circumstance arises because the centrifugal force experienced by an electron in a state of higher orbital angular momentum, like a 3d state in a transition metal, for example, causes its wavefunction to vanish at the nucleus, and so it resides in a region where the Coulomb potential of the nucleus is partially screened. Although the wavefunction of the 4s state extends much further from the nucleus than that of the more localized 3d state, it suffers no centrifugal force. Its wavefunction has a nonvanishing amplitude at the nucleus, where the potential is unscreened, and its energy is correspondingly lowered below that of the 3d state. In addition, the antisymmetry of the several-electron wavefunction lowers the mutual Coulomb interaction of electrons with parallel spin, resulting in one of Hund's rules, which stipulates that, other things being equal, electron spins in atoms prefer parallel alignment.

In Section 4.4 we studied band structure by using the method of tight binding. There we saw that when we assembled a periodic array of atoms to form a crystal, the width of the resulting energy bands was proportional to an integral involving the overlap of atomic wavefunctions on adjacent sites. The electrons in the partially filled d shells of the transition metals have strongly localized orbitals, and this results in rather flat energy bands. The 4s conduction band, on the other hand, arises from the considerable overlap of wavefunctions on neighboring atoms, and consequently spans an energy range that includes the energies of 3d states. In the electronic density of states one then sees a tall narrow d band superimposed on the low flat s band, as in Fig. 11.1.1. The actual electron Bloch states are mixtures of s- and d-like components, particularly in the region of overlap.

In this chapter, we begin by exploring the properties of a metal containing a dilute concentration of magnetic impurities, and then look at the consequences of increasing the impurity concentration. In addition to the inelastic scattering that the impurity sites provide for the conduction electrons, there are two different and competing phenomena that appear as the impurity concentration is increased. One is the conduction-electron-mediated interaction between different magnetic impurities. This is the so-called RKKY (or Ruderman–Kittel–Kasuya–Yosida) interaction, and occurs because a magnetic moment on one impurity site polarizes the conduction electrons, which propagate the polarization to another impurity site, in a manner somewhat

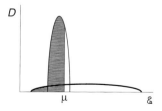

Figure 11.1.1. In the density of states of a transition metal the narrow peak composed of localized d states overlaps the much broader peak of the s states.

analogous to the phonon-mediated electron–electron interaction responsible for BCS superconductivity. The important facts for us are that the interaction strength falls off as the cube of the separation distance between impurities, and that the interaction tends to order systems magnetically. One complication (which fortunately is not important to us here) is that the sign of the interaction, which determines whether it tends to order moments ferromagnetically or antiferromagnetically, oscillates with distance. The second relevant phenomenon is a result of the increasing overlap of the wavefunctions of electrons bound to adjacent impurity sites as the concentration is increased. This overlap leads to the formation of something akin to Bloch states, which then occupy impurity bands.

The low-concentration limit of a metal in which the conduction electrons interact with impurity atoms, each of which has two localized electron spin states, up and down, is known as the Kondo limit. Here, the impurities are far enough apart that the RKKY interaction can be neglected, as can the overlap between different impurity states. The impurities can then be treated as independent, and we can study these systems by considering a single magnetic impurity in a nonmagnetic metal. In the opposite limit of high concentration, we really have a crystal consisting of both species of atoms (host plus impurity) with conduction bands formed from electrons of both species, and strong RKKY interactions. As we may expect, such systems can display very complicated magnetic and transport behavior, and are described in terms of the theory of *heavy fermions*. It turns out that some of the important ingredients in the physics of heavy fermions stem from the properties of a single impurity in a sea of conduction electrons, and so we start our examination in the dilute, or Kondo, limit.

11.2 The resistance minimum and the Kondo effect

While the resistivity of many pure metals does appear to vary as T^5 at low temperatures, adding a small amount of magnetic impurities can yield a very different behavior. When small amounts of iron, chromium, manganese, molybdenum, rhenium, or osmium are added to copper, silver, gold, magnesium, or zinc, for example, the resistivity generally exhibits a *minimum*. The temperature $T_{\min}$ at which this occurs is usually quite low, and does not appear to be related to the Fermi temperature $T_F = \mathcal{E}_F/k$ or to the Debye temperature Θ, and seems to vary with impurity concentration c roughly as $c^{1/5}$, while the depth of the minimum, $\rho(0) - \rho(T_{\min})$, is proportional to c. Since $\rho(0)$ itself is proportional to c, the relative depth of the minimum is roughly independent of c, and usually of the order of one tenth of $\rho(0)$.

The explanation for this effect was provided by Kondo in 1964. He realized that when magnetic impurities are present the conduction electrons may suffer a change of spin as they scatter, and that higher orders of perturbation theory than the first have to be treated very carefully, since the commutation relations of not only the annihilation and creation operators of the conduction electrons but also of the spin raising and lowering operators have to be taken into account. In other words, the Exclusion Principle must be used in calculating any scattering process that passes through an intermediate state when the internal degrees of freedom of the scatterer are involved. To show this, we follow Kondo's calculation, which starts with the part of the perturbing potential containing the magnetic interaction, $\mathbf{s} \cdot \mathbf{S}$, where $\mathbf{s}$ is the spin of the conduction electron, and $\mathbf{S}$ the spin of the localized electron in the d or f shell of the impurity. While the localized spin can have a total spin different from 1/2, we here for simplicity consider localized spin-1/2 states. We implicitly use a high-temperature theory in that we assume that the spins of the localized and conduction electrons are uncorrelated and independently have equal probability of initially being either up or down. While this will directly reveal the onset of the resistance minimum as the temperature is lowered, it will fail as the temperature is decreased further, since at the lowest temperatures conduction and localized spins form bound singlet pairs.

Following the notation of Section 3.10 we define spin raising and lowering operators s^+, s^-, S^+, and S^- for the Bloch and localized electrons, respectively, and find

$$\mathbf{s} \cdot \mathbf{S} = \tfrac{1}{2}\left(s^+ S^- + s^- S^+\right) + s_z S_z.$$

In the notation of second quantization for the Bloch states we can then write the perturbation as

$$\mathcal{H}_1 = \sum_{\mathbf{k},\mathbf{k}',\sigma,\sigma'} (V_{\mathbf{k}\mathbf{k}'}/\hbar^2)\, c^\dagger_{\mathbf{k}'\sigma'} c_{\mathbf{k}\sigma}\, \langle \sigma' | [\tfrac{1}{2}(s^+ S^- + s^- S^+) + s_z S_z] | \sigma \rangle.$$

In this expression σ and σ' refer to the spin states of the Bloch electrons and $V_{\mathbf{k}\mathbf{k}'}$ is the matrix element between Bloch conduction-band states of the spatial part $V(\mathbf{r})$ of the perturbing spin-dependent interaction due to an impurity at the origin,

$$V_{\mathbf{k}\mathbf{k}'} = \int \psi^*_{\mathbf{k}}(\mathbf{r}) V(\mathbf{r}) \psi_{\mathbf{k}'}\, \mathbf{dr}.$$

11.2 The resistance minimum and the Kondo effect

Because $V(\mathbf{r})$ extends over only about one unit cell, while the wavefunction extends over the whole crystal, this quantity is of order N^{-1}, with N the number of unit cells in the crystal. The only nonvanishing matrix element of s^+ is $\langle \uparrow |s^+| \downarrow \rangle = \hbar$, and for s^- only $\langle \downarrow |s^-| \uparrow \rangle = \hbar$ survives, while s_z has elements $\langle \uparrow |s_z| \uparrow \rangle = \hbar/2$ and $\langle \downarrow |s_z| \downarrow \rangle = -\hbar/2$. Thus

$$\mathcal{H}_1 = \frac{1}{2\hbar} \sum_{\mathbf{k},\mathbf{k}'} V_{\mathbf{k}\mathbf{k}'} [c^\dagger_{\mathbf{k}'\uparrow} c_{\mathbf{k}\downarrow} S^- + c^\dagger_{\mathbf{k}'\downarrow} c_{\mathbf{k}\uparrow} S^+ + (c^\dagger_{\mathbf{k}'\uparrow} c_{\mathbf{k}\uparrow} - c^\dagger_{\mathbf{k}'\downarrow} c_{\mathbf{k}\downarrow}) S_z]. \quad (11.2.1)$$

Diagrams illustrating the type of scattering caused by each of these terms would then be of the form shown in Fig. 11.2.1. The scattering probability is proportional to the square of the modulus of the elements of the T-matrix for this perturbation, as indicated in Section 4.7. In the Born approximation the T-matrix is replaced by $\mathcal{H}_1$ itself and then the scattering probability is found from terms with precisely two annihilation operators and two creation operators, and turns out to be composed of terms of the form

$$Q(\mathbf{k}, \mathbf{k}') n_{\mathbf{k}\downarrow} (1 - n_{\mathbf{k}'\uparrow})$$

corresponding to the process of Fig. 11.2.1(a) and other terms corresponding to the other processes. The scattering is still elastic, since we are not assuming the energy of the impurity to depend on its spin direction after the Bloch electron has scattered and departed. The various occupation numbers $n_\mathbf{k}$ can then be averaged to give the $f_\mathbf{k}$ that enter the Boltzmann equation as in Section 8.2, and one finds the resistivity still to be independent of the temperature.

The interesting effects occur when we consider the second-order terms in the T-matrix. Let us for simplicity look at those processes in which the net result is that an electron in the state $\mathbf{k} \uparrow$ is scattered into the state $\mathbf{k}' \uparrow$. While

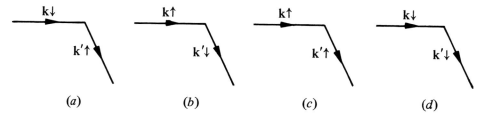

Figure 11.2.1. In the Born approximation a magnetic impurity can scatter an electron in these four different ways.

in first order only the process shown in Fig. 11.2.1(c) contributes to this, there are other possibilities in second order. We recall that

$$T = \mathcal{H}_1 + \mathcal{H}_1 \frac{1}{\mathcal{E} - \mathcal{H}_0} T \approx \mathcal{H}_1 + \mathcal{H}_1 \frac{1}{\mathcal{E} - \mathcal{H}_0} \mathcal{H}_1$$

to second order. Of the sixteen types of second-order term that we find when we substitute expression (11.2.1) into this, we examine only those two involving products of S^- with S^+. These are sums of the form

$$\left(\frac{1}{2\hbar}\right)^2 \sum_{k_1 k_2 k_3 k_4} V_{k_1 k_2} V_{k_3 k_4} c^\dagger_{k_1 \uparrow} c_{k_2 \downarrow} S^- \frac{1}{\mathcal{E} - \mathcal{H}_0} c^\dagger_{k_3 \downarrow} c_{k_4 \uparrow} S^+ \quad (11.2.2)$$

and

$$\left(\frac{1}{2\hbar}\right)^2 \sum_{k_1 k_2 k_3 k_4} V_{k_1 k_2} V_{k_3 k_4} c^\dagger_{k_3 \downarrow} c_{k_4 \uparrow} S^+ \frac{1}{\mathcal{E} - \mathcal{H}_0} c^\dagger_{k_1 \uparrow} c_{k_2 \downarrow} S^-. \quad (11.2.3)$$

For these to have the net effect only of scattering $\mathbf{k} \uparrow$ into $\mathbf{k}' \uparrow$ we must always have $\mathbf{k}_2 = \mathbf{k}_3$, $\mathbf{k}_4 = \mathbf{k}$, and $\mathbf{k}_1 = \mathbf{k}'$. In terms of diagrams we can picture these processes as in Fig. 11.2.2. The diagram (a) represents expression (11.2.2), in which an electron is first scattered from $\mathbf{k} \uparrow$ to the virtual state $\mathbf{k}_2 \downarrow$, and then finally to the state $\mathbf{k}' \uparrow$. In diagram (b), however, the first thing that happens is the creation of an electron–hole pair. That is, an electron already in state $\mathbf{k}_2 \downarrow$ is scattered into $\mathbf{k}' \uparrow$. The incoming electron in state $\mathbf{k} \uparrow$ then drops down into this vacancy in a process that we can depict as the annihilation of an electron–hole pair. The energy of the intermediate state differs from that of the initial state by $\mathcal{E}_{\mathbf{k}'} - \mathcal{E}_{\mathbf{k}_2}$, which is just the negative of the energy difference, $\mathcal{E}_{\mathbf{k}_2} - \mathcal{E}_{\mathbf{k}}$, of the process of Fig. 11.2.2(a). We can thus

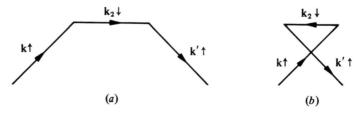

Figure 11.2.2. These two second-order processes both contribute to the scattering amplitude for a conduction electron. Because of the change of spin, the occupancy $n_{\mathbf{k}_2}$ of the intermediate state does not cancel from the total scattering amplitude.

11.2 The resistance minimum and the Kondo effect

add the contributions of expressions (11.2.2) and (11.2.3) and use the anti-commutation relations of the c's and $c^\dagger$'s to find

$$\left(\frac{1}{2\hbar}\right)^2 \sum_{\mathbf{k}_2} V_{\mathbf{k'}\mathbf{k}_2} V_{\mathbf{k}_2 \mathbf{k}} c^\dagger_{\mathbf{k'}\uparrow} c_{\mathbf{k}\uparrow} \frac{1}{\mathcal{E}_\mathbf{k} - \mathcal{E}_{\mathbf{k}_2}} [(1 - n_{\mathbf{k}_2\downarrow})S^- S^+ + n_{\mathbf{k}_2\downarrow} S^+ S^-]. \tag{11.2.4}$$

If S^+ and S^- were not operators but simply numbers they would commute, and the terms in $n_{\mathbf{k}_2}$ would cancel. We would then be back in the situation of having the type of scattering that leads to a temperature-independent resistivity. However, as we may verify from the definitions in Eqs. (3.10.12),

$$S^- S^+ - S^+ S^- = -2\hbar S_z.$$

There is thus a contribution to the scattering matrix that is proportional to

$$c^\dagger_{\mathbf{k'}\uparrow} c_{\mathbf{k}\uparrow} S_z \sum_{\mathbf{k}_2} V_{\mathbf{k'}\mathbf{k}_2} V_{\mathbf{k}_2 \mathbf{k}} \frac{\langle n_{\mathbf{k}_2\downarrow} \rangle}{\mathcal{E}_\mathbf{k} - \mathcal{E}_{\mathbf{k}_2}}. \tag{11.2.5}$$

The presence of the term in $n_{\mathbf{k}_2}$ has the consequence that the scattering probability becomes strongly temperature-dependent. When we form $|T_{\mathbf{k}\mathbf{k'}}|^2$ we shall find contributions of the form

$$P_K(\mathbf{k}\uparrow, \mathbf{k'}\uparrow) \propto \langle n_{\mathbf{k}\uparrow}(1 - n_{\mathbf{k'}\uparrow}) \rangle g(\mathbf{k'}, \mathbf{k})$$

where $g(\mathbf{k'}, \mathbf{k})$ is the sum over $\mathbf{k}_2$ in expression (11.2.5), and the subscript K refers to the contribution responsible for the Kondo effect. Because the thermal average of the expectation value of $n_{\mathbf{k}_2\downarrow}$ is given by the Fermi–Dirac function, we see that $g(\mathbf{k'}, \mathbf{k})$ depends on the temperature.

The total probability $P(\mathbf{k}\uparrow, \mathbf{k'}\uparrow)$ of a scattering event occurring in which the net effect is that an electron is transferred from $\mathbf{k}\uparrow$ to $\mathbf{k'}\uparrow$ can be written as

$$P(\mathbf{k}\uparrow, \mathbf{k'}\uparrow) = f_{\mathbf{k}\uparrow}(1 - f_{\mathbf{k'}\uparrow}) Q(\mathbf{k}\uparrow, \mathbf{k'}\uparrow).$$

Here $Q(\mathbf{k}\uparrow, \mathbf{k'}\uparrow)$ is composed of two parts. One is independent of the temperature and is due to first-order processes plus those contributions from higher-order processes that do not involve the occupation numbers of the intermediate states. The second part contains contributions from the occupation numbers of the intermediate states, and in second order has the temperature dependence of $g(\mathbf{k}, \mathbf{k'})$. The qualitative nature of this temperature dependence may be seen by making a few approximations. We first assume that the matrix elements $V_{\mathbf{k}'\mathbf{k}_2}$ and $V_{\mathbf{k}_2\mathbf{k}}$ vary slowly over the range of energies over which we have to integrate $\mathbf{k}_2$, so that we can replace each of them by a constant V_0. We then change from a sum over $\mathbf{k}_2$ to an integration over energies $\mathcal{E}_2$ by introducing the density of states $D(\mathcal{E})$, and then approximate this by its value when $\mathcal{E}_{\mathbf{k}_2}$ is equal to the chemical potential μ. We specialize to the value of $g(\mathbf{k}, \mathbf{k}')$ when $\mathcal{E}_{\mathbf{k}}$ is also equal to μ to find

$$g(\mathbf{k}, \mathbf{k}') \approx V_0^2 D(\mu) \int_{-W}^{W} \frac{f(\hat{\mathcal{E}}) d\hat{\mathcal{E}}}{\hat{\mathcal{E}}},$$

where W is an energy characteristic of the width of the band. After inserting the form of the Fermi–Dirac function f and defining $x \equiv \hat{\mathcal{E}}/2kT$ we have

$$g(\mathbf{k}, \mathbf{k}') \approx -V_0^2 D(\mu) \int_0^{W/2kT} \frac{\tanh x}{x} dx.$$

The integral is one we have met before in deriving Eq. (7.5.10) for the critical temperature of a BCS superconductor, and so we find the result

$$g(\mathbf{k}, \mathbf{k}') \approx -V_0^2 D(\mu) \ln(1.14 \, W/kT).$$

The low-temperature resistivity then takes the form

$$\rho(T) = \rho_0 + \rho_1 \left(\frac{T}{\Theta}\right)^5 - \rho_2 \ln\left(\frac{kT}{W}\right). \tag{11.2.6}$$

The resistance has its minimum when the derivative $d\rho/dT$ vanishes. Thus

$$5\rho_1 \frac{T_{\min}^4}{\Theta^5} - \frac{\rho_2}{T_{\min}} = 0$$

and

$$T_{\min}^5 = \frac{\rho_2 \Theta^5}{5\rho_1}.$$

For low impurity concentrations, the scattering events on different impurities are independent and their contributions add incoherently. The sum over scattering events on all impurities is then proportional to the concentration c of impurities, so we have thus shown that

$$T_{\min} \propto c^{1/5}.$$

The fact that Eq. (11.2.6) erroneously predicts an infinite resistance as T approaches zero is a consequence of the inadequacy of considering only second-order terms in the T-matrix, and of our implicit high-temperature assumption that the energy of the local spin is independent of its orientation. At lower temperatures the local spins form bound collective states with the conduction-band electrons. In order to flip the local spin, this binding must be broken by the thermal energy of excited conduction electrons. As the temperature is reduced this becomes more and more unlikely, and the local spin becomes "frozen out." In fact, the singularity we discovered signals this formation of a bound state.

11.3 Low-temperature limit of the Kondo problem

It is clear that perturbation theory, especially when limited to second-order calculations, is not going to be useful if we want to learn the nature of the low-temperature behavior of a system in the Kondo limit. Instead, we will try to use our intuition to guess a reasonable variational wavefunction that describes a collective bound state, and see whether we can make this state have a lower energy than a state with an independent localized spin and a Fermi sea. We start with an impurity atom embedded in a host metal. In the low-concentration limit, we can ignore interactions between impurities, and so for simplicity consider a single local spin-$\frac{1}{2}$ state coupled antiferromagnetically to the sea of conduction electrons. We would like to construct some kind of trial wavefunction that describes the formation of a bound state of conduction electrons and the local moment. One natural possibility for the case of antiferromagnetic coupling would be to have excitations above the Fermi surface combine with the local moment to form a spin singlet, which would be an antisymmetric combination of spin-up and spin-down states of the local spin and states above the Fermi surface. Another possibility would be to form spin singlets of states below the Fermi surface – holes – and local spins. Simple variational wavefunctions that describe these possibilities were first written down by Yosida. We will here follow Mahan's treatment of the problem.

The trial wavefunctions we consider are of two types. The first is

$$|\Psi_a\rangle = \sum_{|\mathbf{k}|>k_F} a_\mathbf{k} [\alpha^\dagger c^\dagger_{\mathbf{k}\downarrow} - \beta^\dagger c^\dagger_{\mathbf{k}\uparrow}] |F\rangle.$$

Here, $\alpha^\dagger$ and $\beta^\dagger$ create up- and down-spin states of the local spin, $|F\rangle$ is the filled Fermi sea, and the $a_\mathbf{k}$ are coefficients that we will have to determine. We see that this state is a spin singlet: it is an antisymmetric combination of a local up-spin plus a delocalized down-spin above the Fermi surface with a local down-spin and a delocalized up-spin that is also above the Fermi surface. The second type of state is formed with annihilation operators for conduction electrons below the Fermi surface, and is written as

$$|\Psi_b\rangle = \sum_{|\mathbf{k}|<k_F} b_\mathbf{k}[\alpha^\dagger c_{\mathbf{k}\uparrow} + \beta^\dagger c_{\mathbf{k}\downarrow}]|F\rangle.$$

This state combines local spins with holes. It too is an antisymmetric combination of antiparallel local spin and electron spin, and thus a spin singlet, but now the electrons are below the Fermi surface. The two states $|\Psi_a\rangle$ and $|\Psi_b\rangle$ are orthogonal to one another, since they contain different numbers of electrons.

Next we have to evaluate the expectation value of the Kondo Hamiltonian in the states $|\Psi_a\rangle$ and $|\Psi_b\rangle$, and minimize these expressions with respect to the coefficients $a_\mathbf{k}$ and $b_\mathbf{k}$. Note that the states are not normalized, so we have to divide the expectation values by the respective norms, which also depend on $a_\mathbf{k}$ and $b_\mathbf{k}$, and minimize these entire expressions. This is a rather lengthy operation, and so we instead take a short cut by making another approximation. The states $|\Psi_a\rangle$ and $|\Psi_b\rangle$ are the simplest spin-singlet states we can think of that consist of electrons above the Fermi sea, or holes below it, combined with local spins. However, when the interaction term in the Kondo Hamiltonian acts on either of these states, other, more complicated terms are created. Let us look at this in some detail. The interaction term is

$$\mathcal{H}_1 = -\frac{J}{\hbar}\sum_{\mathbf{k},\mathbf{p}}\{(c^\dagger_{\mathbf{k}\uparrow}c_{\mathbf{p}\uparrow} - c^\dagger_{\mathbf{k}\downarrow}c_{\mathbf{p}\downarrow})S_z + c^\dagger_{\mathbf{k}\uparrow}c_{\mathbf{p}\downarrow}S^- + c^\dagger_{\mathbf{k}\downarrow}c_{\mathbf{p}\uparrow}S^+\}.$$

Here we have taken the coupling J to be independent of momentum transfer and to be negative, which means that we are considering antiferromagnetic coupling. We then let this interaction act on our state $|\Psi_a\rangle$.

$$\mathcal{H}_1|\Psi_a\rangle = -\frac{J}{\hbar}\sum_{\mathbf{p},\mathbf{q}}\{[c^\dagger_{\mathbf{p}\uparrow}c_{\mathbf{q}\uparrow} - c^\dagger_{\mathbf{p}\downarrow}c_{\mathbf{q}\downarrow}]S_z + c^\dagger_{\mathbf{p}\uparrow}c_{\mathbf{q}\downarrow}S^- + c^\dagger_{\mathbf{p}\downarrow}c_{\mathbf{q}\uparrow}S^+\}$$
$$\times \sum_{|\mathbf{k}|>k_F} a_\mathbf{k}[\alpha^\dagger c^\dagger_{\mathbf{k}\downarrow} - \beta^\dagger c^\dagger_{\mathbf{k}\uparrow}]|F\rangle.$$

11.3 Low-temperature limit of the Kondo problem

The simplest terms in this expression are the ones for which $\mathbf{q} = \mathbf{k}$, and are

$$-\frac{J}{\hbar}\sum_{\mathbf{p},\mathbf{k}} a_{\mathbf{k}}\{[c^{\dagger}_{\mathbf{p}\uparrow}c_{\mathbf{k}\uparrow} - c^{\dagger}_{\mathbf{p}\downarrow}c_{\mathbf{k}\downarrow}]S_z[\alpha^{\dagger}c^{\dagger}_{\mathbf{k}\downarrow} - \beta^{\dagger}c^{\dagger}_{\mathbf{k}\uparrow}]|F\rangle$$

$$+ c^{\dagger}_{\mathbf{p}\uparrow}c_{\mathbf{k}\downarrow}S^{-}[\alpha^{\dagger}c^{\dagger}_{\mathbf{k}\downarrow} - \beta^{\dagger}c^{\dagger}_{\mathbf{k}\uparrow}]|F\rangle + c^{\dagger}_{\mathbf{p}\downarrow}c_{\mathbf{k}\uparrow}S^{+}[\alpha^{\dagger}c^{\dagger}_{\mathbf{k}\downarrow} - \beta^{\dagger}c^{\dagger}_{\mathbf{k}\uparrow}]|F\rangle\}.$$

The operators S_z, S^{-} and S^{+} all act only on the local spin, and $S_z\alpha^{\dagger} = \hbar\alpha^{\dagger}/2$, $S_z\beta^{\dagger} = -\hbar\beta^{\dagger}/2$, $S^{-}\alpha^{\dagger} = \hbar\beta^{\dagger}$, $S^{-}\beta^{\dagger} = 0$, $S^{+}\alpha^{\dagger} = 0$, and $S^{+}\beta^{\dagger} = \hbar\alpha^{\dagger}$, so we obtain

$$-J\sum_{\mathbf{p},\mathbf{k}} a_{\mathbf{k}}\left\{\frac{1}{2}[c^{\dagger}_{\mathbf{p}\uparrow}c_{\mathbf{k}\uparrow} - c^{\dagger}_{\mathbf{p}\downarrow}c_{\mathbf{k}\downarrow}][\alpha^{\dagger}c^{\dagger}_{\mathbf{k}\downarrow} + \beta^{\dagger}c^{\dagger}_{\mathbf{k}\uparrow}] + c^{\dagger}_{\mathbf{p}\uparrow}c_{\mathbf{k}\downarrow}\beta^{\dagger}c^{\dagger}_{\mathbf{k}\downarrow} - c^{\dagger}_{\mathbf{p}\downarrow}c_{\mathbf{k}\uparrow}\alpha^{\dagger}c^{\dagger}_{\mathbf{k}\uparrow}\right\}|F\rangle.$$

Since $|\mathbf{k}| > k_F$, the operator combination $c_{\mathbf{k}\uparrow}c^{\dagger}_{\mathbf{k}\uparrow}$ gives unity when it acts on the filled Fermi sphere, and the presence of the operator $c^{\dagger}_{\mathbf{p}}$ requires $|\mathbf{p}| > k_F$. Applying the same argument to the other terms, and collecting up the surviving components, we see that we are left with something similar to what we started with, namely singlet states made up of conduction electrons and local-moment electrons. This is good news, as we would like Ψ_a to be an eigenstate of $\mathcal{H}_1$, and the terms we have just examined satisfy that wish. The bad news is that there are also many other terms bearing less resemblance to $|\Psi_a\rangle$. For example, terms with $\mathbf{p}, \mathbf{q} \neq \mathbf{k}$ will lead to expressions like

$$c^{\dagger}_{\mathbf{p}\uparrow}c_{\mathbf{q}\downarrow}\beta^{\dagger}c^{\dagger}_{\mathbf{k}\downarrow}|F\rangle.$$

This is a term which, in addition to the electron at momentum $\hbar\mathbf{k}$ above the Fermi surface, has a particle–hole pair consisting of an electron with momentum $\hbar\mathbf{p}$ above the Fermi surface and a hole of momentum $\hbar\mathbf{q}$ below the Fermi surface. This is inconvenient, since it will lead to some cumbersome algebra when we vary $a_{\mathbf{k}}$ to find a minimum in energy. Our Ansatz state started out just with electrons above the Fermi sea, but the action of the Hamiltonian on this state generates various kinds of other electrons and holes. Another way to say this is that we would like to keep our state in a restricted part of the Hilbert space consisting only of electron-plus-local-moment singlets, but acting with the Hamiltonian on this state takes us to a bigger part of the Hilbert space. We avoid this difficulty with a simple remedy: we insist on staying in our restricted Hilbert space by throwing out all the parts of $\mathcal{H}|\Psi_a\rangle$ that do not consist of terms of the form $(\alpha^{\dagger}c^{\dagger}_{\mathbf{k}\downarrow} - \beta^{\dagger}c^{\dagger}_{\mathbf{k}\uparrow})|F\rangle$. Technically, we can do this by using the projection operator

$$\sum_{|\mathbf{k}|>k_F} [\alpha^\dagger c^\dagger_{\mathbf{k}\downarrow} - \beta^\dagger c^\dagger_{\mathbf{k}\uparrow}]|F\rangle\langle F| \sum_{|\mathbf{k}|>k_F} [\alpha c_{\mathbf{k}\downarrow} - \beta c_{\mathbf{k}\uparrow}].$$

The eigenvalue equation we wish to solve is

$$(\mathcal{H}_0 + \mathcal{H}_1 - \mathcal{E}_0 - \delta\mathcal{E}_a)|\Psi_a\rangle = 0.$$

Here $\mathcal{E}_0$ is the ground-state energy of the unperturbed Hamiltonian, $\mathcal{H}_0 = \sum_{\mathbf{k},\sigma} \mathcal{E}_\mathbf{k} c^\dagger_{\mathbf{k},\sigma} c_{\mathbf{k},\sigma}$, and $\delta\mathcal{E}_a$ is the shift in energy due to the perturbation $\mathcal{H}_1$. We wish to determine $a_\mathbf{k}$ such that this shift is as negative as possible. Let us first look at the simplest terms:

$$(\mathcal{H}_0 - \mathcal{E}_0 - \delta\mathcal{E}_a)|\Psi_a\rangle = (\mathcal{H}_0 - \mathcal{E}_0 - \delta\mathcal{E}_a) \sum_{|\mathbf{k}|>k_F} a_\mathbf{k}[\alpha^\dagger c^\dagger_{\mathbf{k}\downarrow} - \beta^\dagger c^\dagger_{\mathbf{k}\uparrow}]|F\rangle$$

$$= \sum_{|\mathbf{k}|>k_F} a_\mathbf{k}[\mathcal{E}_\mathbf{k} - \delta\mathcal{E}_a][\alpha^\dagger c^\dagger_{\mathbf{k}\downarrow} - \beta^\dagger c^\dagger_{\mathbf{k}\uparrow}]|F\rangle.$$

This term does not contain any parts outside our restricted Hilbert space and so we need not worry about projections, and the contribution to the equation for $a_\mathbf{k}$ is

$$\sum_{|\mathbf{k}|>k_F} a_\mathbf{k}[\mathcal{E}_\mathbf{k} - \delta\mathcal{E}_a].$$

We have to do a little more work for the terms generated by $\mathcal{H}_1|\Psi_a\rangle$. First, we look at the piece of $\mathcal{H}_1$ that involves S_z. This is

$$\sum_{\mathbf{p},\mathbf{q}} S_z[c^\dagger_{\mathbf{p}\uparrow}c_{\mathbf{q}\uparrow} - c^\dagger_{\mathbf{p}\downarrow}c_{\mathbf{q}\downarrow}] \sum_{|\mathbf{k}|>k_F} a_\mathbf{k}[\alpha^\dagger c^\dagger_{\mathbf{k}\downarrow} - \beta^\dagger c^\dagger_{\mathbf{k}\uparrow}]|F\rangle$$

$$= \frac{\hbar}{2} \sum_{\mathbf{p},\mathbf{q}} [c^\dagger_{\mathbf{p}\uparrow}c_{\mathbf{q}\uparrow} - c^\dagger_{\mathbf{p}\downarrow}c_{\mathbf{q}\downarrow}] \sum_{|\mathbf{k}|>k_F} a_\mathbf{k}[\alpha^\dagger c^\dagger_{\mathbf{k}\downarrow} + \beta^\dagger c^\dagger_{\mathbf{k}\uparrow}]|F\rangle, \quad (11.3.1)$$

where we have used $S_z\alpha^\dagger = \hbar\alpha^\dagger/2$, and $S_z\beta^\dagger = -\hbar\beta^\dagger/2$. Let us look at the term in $\alpha^\dagger$ in Eq. (11.3.1). Terms that survive the projection and give nonzero contribution must contain only one electron excitation above the filled Fermi sea $|F\rangle$. This means that the product of three electron operators in (11.3.1) must reduce to one by having the other two form a simple number operator. This can happen if $\mathbf{p} = \mathbf{q}$ or if $\mathbf{q} = \mathbf{k}$. For the former case, we have

$$\sum_\mathbf{p} [c^\dagger_{\mathbf{p}\uparrow}c_{\mathbf{p}\uparrow} - c^\dagger_{\mathbf{p}\downarrow}c_{\mathbf{p}\downarrow}] \sum_{|\mathbf{k}|>k_F} a_\mathbf{k}\alpha^\dagger c^\dagger_{\mathbf{k}\downarrow}|F\rangle,$$

11.3 Low-temperature limit of the Kondo problem

and $\sum_{\mathbf{p}}[c_{\mathbf{p}\uparrow}^{\dagger}c_{\mathbf{p}\uparrow} - c_{\mathbf{p}\downarrow}^{\dagger}c_{\mathbf{p}\downarrow}]$ just counts the difference between the number of up- and down-spins after we have applied $c_{\mathbf{k}\downarrow}^{\dagger}$ to the filled Fermi sea, and is thus equal to -1. We then obtain for the terms in $\alpha^{\dagger}$ in Eq. (11.3.1)

$$-\frac{\hbar}{2}\sum_{|\mathbf{k}|>k_F} a_{\mathbf{k}}\alpha^{\dagger}c_{\mathbf{k}\downarrow}^{\dagger}|F\rangle - \frac{\hbar}{2}\sum_{\mathbf{p},|\mathbf{k}|>k_F} a_{\mathbf{k}}\alpha^{\dagger}c_{\mathbf{p}\downarrow}^{\dagger}|F\rangle. \qquad (11.3.2)$$

The terms in $\beta^{\dagger}$ in Eq. (11.3.1) are

$$-S_z[c_{\mathbf{p}\uparrow}^{\dagger}c_{\mathbf{q}\uparrow} - c_{\mathbf{p}\downarrow}^{\dagger}c_{\mathbf{q}\downarrow}]\sum_{|\mathbf{k}|>k_F} a_{\mathbf{k}}\beta^{\dagger}c_{\mathbf{k}\uparrow}^{\dagger}|F\rangle = \frac{\hbar}{2}[c_{\mathbf{p}\uparrow}^{\dagger}c_{\mathbf{q}\uparrow} - c_{\mathbf{p}\downarrow}^{\dagger}c_{\mathbf{q}\downarrow}]\sum_{|\mathbf{k}|>k_F} a_{\mathbf{k}}\beta^{\dagger}c_{\mathbf{k}\uparrow}^{\dagger}|F\rangle.$$

Again we must demand that either $\mathbf{p} = \mathbf{q}$ or $\mathbf{q} = \mathbf{k}$, and we are left with only

$$\frac{\hbar}{2}\sum_{|\mathbf{k}|>k_F} a_{\mathbf{k}}\beta^{\dagger}c_{\mathbf{p}\uparrow}^{\dagger}|F\rangle + \frac{\hbar}{2}\sum_{\mathbf{p},|\mathbf{k}|>k_F} a_{\mathbf{k}}\beta^{\dagger}c_{\mathbf{p}\uparrow}^{\dagger}|F\rangle. \qquad (11.3.3)$$

Continuing with the terms generated by the operators S^- and S^+, we find that

$$\sum_{\mathbf{p},\mathbf{q}} c_{\mathbf{p}\uparrow}^{\dagger}c_{\mathbf{q}\downarrow}S^- \sum_{|\mathbf{k}|>k_F} a_{\mathbf{k}}\alpha^{\dagger}c_{\mathbf{k}\downarrow}^{\dagger}|F\rangle - \sum_{\mathbf{p},\mathbf{q}} c_{\mathbf{p}\downarrow}^{\dagger}c_{\mathbf{q}\uparrow}S^+ \sum_{|\mathbf{k}|>k_F} a_{\mathbf{k}}\beta^{\dagger}c_{\mathbf{k}\uparrow}^{\dagger}|F\rangle$$

reduces to

$$\hbar\beta^{\dagger}\sum_{\mathbf{p},|\mathbf{k}|>k_F} a_{\mathbf{k}}c_{\mathbf{p}\uparrow}^{\dagger}|F\rangle - \hbar\alpha^{\dagger}\sum_{\mathbf{p},|\mathbf{k}|>k_F} a_{\mathbf{k}}c_{\mathbf{p}\downarrow}^{\dagger}|F\rangle. \qquad (11.3.4)$$

We now combine expressions (11.3.2), (11.3.3), and (11.3.4). In (11.3.2) and (11.3.3) there are single summations over $\mathbf{k}$, while all the other terms are double summations over $\mathbf{p}$ and $\mathbf{k}$. We drop the single summations, since they will be smaller by a factor of order N than the double ones. The result is

$$\frac{3J}{2}\sum_{|\mathbf{k}|>k_F} a_{\mathbf{k}} \sum_{|\mathbf{p}|>k_F} [\alpha^{\dagger}c_{\mathbf{p}\downarrow}^{\dagger} - \beta^{\dagger}c_{\mathbf{p}\uparrow}^{\dagger}]|F\rangle.$$

The eigenvalue equation for $\delta\mathcal{E}_a$ then becomes

$$a_{\mathbf{k}}(\mathcal{E}_{\mathbf{k}} - \delta\mathcal{E}_a) + \frac{3J}{2}\sum_{|\mathbf{p}|>k_F} a_{\mathbf{p}} = 0.$$

We avoid having to solve for the eigenvectors $a_{\mathbf{k}}$ by the trick of dividing by $\mathcal{E}_{\mathbf{k}} - \delta\mathcal{E}_a$ and summing over $|\mathbf{k}| > k_F$. The sum of the $a_{\mathbf{k}}$ then cancels to leave us with

$$1 = -\frac{3J}{2} \sum_{|\mathbf{k}|>k_F} \frac{1}{\mathcal{E}_{\mathbf{k}} - \delta\mathcal{E}_a}.$$

We change from a sum over $\mathbf{k}$ to an integration in which the energy $\hat{\mathcal{E}}$ measured relative to the Fermi energy runs from zero to a value W related to the bandwidth, and approximate the density of states by its value D at the Fermi surface. We then have

$$1 = -\frac{3JD}{2} \int_0^W \frac{d\hat{\mathcal{E}}}{\hat{\mathcal{E}} - \delta\mathcal{E}_a} = \frac{3JD}{2} \ln\left|\frac{\delta\mathcal{E}_a}{W - \delta\mathcal{E}_a}\right|.$$

Remembering that J is negative, we then find the solution with $\delta\mathcal{E}_a < 0$ to be

$$\delta\mathcal{E}_a = -\frac{W}{e^{2/(3|J|D)} - 1}. \tag{11.3.5}$$

This expression reminds us of the condensation energy of a BCS superconductor given in Eq. (7.4.2), which was proportional to $e^{-2/VD}$, with V the attractive electron–electron interaction. When $|J|D$ is small, $\delta\mathcal{E}_a$ is similarly proportional to $e^{-2/(3|J|D)}$, showing that the coupling constant $|J|$ in this case enters nonperturbatively into the problem. The function $e^{-2/(3|J|D)}$ cannot be expanded in a power series in $|J|$, and is said to have an essential singularity at $|J| = 0$. For bands more than half-filled, the state $|\Psi_a\rangle$, which consists of bound spin-singlet pairs of conduction electrons and local spins, has a lower energy than $\mathcal{E}_0$, and is a better candidate for the ground state. The state $|\Psi_b\rangle$ may also be examined by means of a similar approach, the main difference in the analysis being that $|\mathbf{k}| < k_F$. One finds that the state $|\Psi_b\rangle$ should be the ground state for bands that are less than half full.

The definition of the *Kondo temperature* T_K is

$$kT_K \equiv W \exp\left(-\frac{1}{2D|J|}\right).$$

This energy is similar to the small-$|J|$ limit of our expression (11.3.5) for the energy reduction $-\delta\mathcal{E}_a$, except for the factor of $\frac{1}{2}$ instead of $\frac{2}{3}$ in the exponent. This difference comes from the fact that those who originally defined it did so in terms of a triplet state, which is obtained if the coupling J is ferromagnetic

($J > 0$), rather than the singlet state that we have considered. We again notice an analogy with the BCS theory of superconductivity, in which we find a similar expression for the critical temperature T_c. This is not an accident. Both T_K and T_c define temperatures below which perturbation theory fails. In the BCS case, T_c signals the onset of the formation of bound Cooper pairs and a new ground state with an energy gap. The Kondo effect is a little more subtle. Here T_K defines a temperature at which the energy contributions from second-order perturbation theory become important. This happens when the local spin on a single impurity starts to become frozen out at an energy set by the Kondo coupling J and the density of states at the Fermi energy.

In summary, we have seen that the internal dynamics of the local spins interacting with the sea of conduction electrons become important at low temperatures. The net effect of this interaction is very much like a resonant state appearing at temperatures $\sim T_K$. In fact, Wilkins has noted that the Kondo effect is very well described by a density-of-states expression that adds a resonant state at the Fermi energy for each impurity with a local moment. He writes this expression as

$$D_K(\mathcal{E}) = D(\mathcal{E}) + \frac{c}{\pi} \frac{\gamma}{(\mathcal{E} - \mathcal{E}_F)^2 + \gamma^2} = D(\mathcal{E}) + \delta D(\mathcal{E}), \tag{11.3.6}$$

where $\gamma = 1.6 \, kT_K$. This expression adds a Lorentzian peak of weight unity at the Fermi energy for each impurity atom with a local moment. From this expression one can calculate, for example, the change in specific heat and the change in electrical resistance. All the many-body physics has then resulted in a simple change in the density of states at the Fermi surface consisting of the addition of a resonant state for each impurity. At high temperatures, the sharp resonances are unimportant in the presence of thermal smearing at the Fermi surface. As the temperature is decreased, the sharp resonances become more important in the scattering of the conduction electrons. At still lower temperatures, there is insufficient energy to flip the local spins, which become frozen in fixed orientations.

11.4 Heavy fermions

We treated the Kondo problem by first considering the effects of a single impurity, and then simply multiplying the expected effect by the number of impurities present. We were thus assuming that the magnetic impurities were sufficiently dilutely dispersed in the metallic host material that we did not have to consider interactions between them. The starting point for the theory of heavy fermions, in contrast, is a regular lattice, typically consisting of a basis of a rare earth or actinide and a metal. Examples are UPt$_3$ and CeAl$_3$. In this type of compound, it is possible for the electrons to form Bloch states, and display metallic behavior in the sense that the resistance diminishes to quite a small value as the temperature approaches absolute zero. This is quite distinct from the effects of the dilute magnetic impurities in the Kondo problem, which lead to a minimum in the resistance. On the other hand, the magnetic elements experience RKKY interactions, which can lead to the formation of nontrivial magnetic ground states. In addition, some of the heavy-fermion materials, UPt$_3$ being an example, have superconducting ground states that are *not* the usual BCS type of superconductor. In view of this rich diversity of interesting properties, it is perhaps useful to start by pointing out what heavy-fermion materials do have in common, and why the fermions are said to be "heavy." Let us first consider the electrical resistivity. In typical transition metals, this has a temperature dependence given by $\rho(T) = \rho(0) + AT^2$ at low temperatures, where A is proportional to the effective mass of the electrons. In metallic heavy-fermion systems at sufficiently low temperatures, the resistivity still has this simple behavior, but the constant A can be as much as seven orders of magnitude larger than for transition metals! Similarly, the specific heat of normal metals at low temperatures is of the form $C(T) = \gamma T + BT^3$, where γT is the electronic contribution, and the term in T^3 is due to phonons. For heavy fermions, the electronic contribution γT is perhaps two to three orders of magnitude greater than in normal metals, and is so large that the phonon contribution can often be ignored. Finally, there is the magnetic susceptibility χ, which for heavy fermions at low temperatures is enhanced by several orders of magnitude over that of conventional metals. All these quantities, A, γ, and χ, are proportional to the effective mass m^* of the electrons for conventional metals. Their enhanced values lead us to the interpretation that we are indeed dealing with "heavy fermions." That the concept of a large effective mass is useful for these compounds is demonstrated by the *Wilson ratio*, R. This is the ratio of the zero-temperature limit of the magnetic susceptibility (in units of the moment per atom, $g_J^2 J(J+1)\mu_B^2$) to the zero-temperature derivative of the specific heat (in units of $\pi^2 k^2$), and is

$$R = \frac{\chi(0)/g_J^2 J(J+1)\mu_B^2}{\gamma(0)/\pi^2 k^2}.$$

The fact that the Wilson ratio is approximately unity for all the nonmagnetic heavy fermions suggests that, whatever the mechanism responsible for the anomalous behavior, there is a consistent pattern that can be interpreted in terms of a large effective mass. We take this argument back one further step by recalling that the electronic density of states is itself proportional to the effective mass in the case of a single band, and decide that our task should be to examine the likely magnitude of $D(\mathcal{E}_F)$ in these systems.

Our starting point will be the *Anderson Hamiltonian*, which is constructed from the following ingredients. First, there is a sea of conduction electrons formed from the conduction-band states of the host material (Al, Pt, or Zn, for example) and from the s and p electrons of a dopant such as U or Ce. These delocalized conduction electrons will in general have some dispersion relation $\mathcal{E}_{\mathbf{k},n}$ for Bloch states labeled by a wavevector $\mathbf{k}$ in the first Brillouin zone and a band index n. This is a distracting complication, and so we assume that the bands are of free-electron form. Secondly, there are the localized d or f states of the dopant atoms. To be specific we assume that we are dealing with f electrons, and name their dispersion relation $\mathcal{E}_f(\mathbf{k}, n)$. However, since d and f electrons are very closely bound to the core of the dopant atoms, these states do not overlap significantly. The band formed by them is consequently very flat, and their energies can be taken to be a constant, $\mathcal{E}_f$ (not to be confused with the Fermi energy $\mathcal{E}_F$). The next ingredient is one that leads to strong correlations. It is the *on-site* repulsion term. This is due to the localized nature of the d and f orbitals. If there are several d or f states filled on the same atom (which is possible because of their relatively high spin degeneracy), the Coulomb interaction between them contributes strongly to the energy. This interaction can then be written

$$U \sum_{i,v,v'} n_{i,v} n_{i,v'},$$

where the summation runs over all dopant sites i, and v is an enumeration of the degenerate multiplet of local states. The on-site repulsion is given by U. This kind of localized Coulomb repulsion is sometimes called a *Hubbard* term, because it is a central piece in another model of strongly correlated electrons, the Hubbard Hamiltonian. Lastly, there is an interaction between the delocalized conduction electrons and the local states. The idea here is that a conduction electron with a certain spin can hop onto a local site, or vice versa. This is in contrast to the Kondo Hamiltonian, where local states and conduction electrons could exchange spin, but were not allowed to transform into one another. We take this local interaction to be a very short-ranged potential centered at the sites i of the dopants.

Putting it all together, we then have the periodic Anderson model,

$$\mathcal{H}_A = \sum_{\mathbf{k},\sigma} \mathcal{E}_\mathbf{k} c^\dagger_{\mathbf{k},\sigma} c_{\mathbf{k},\sigma} + \mathcal{E}_f \sum_{i,\nu} f^\dagger_{i,\nu} f_{i,\nu} + U \sum_{i,\nu',\nu} f^\dagger_{i,\nu} f_{i,\nu} f^\dagger_{i,\nu'} f_{i,\nu'}$$
$$+ \sum_{\mathbf{k},i,\nu} V(c^\dagger_{\mathbf{k},\nu} f_{i,\nu} + f^\dagger_{i,\nu} c_{\mathbf{k},\nu}). \tag{11.4.1}$$

Here $f_{i,\nu}$ annihilates an f electron of quantum number ν at site i. In the last term, which is the hybridization term, we have effectively taken the interaction V to be a delta-function in real space, making the Fourier transform independent of $\mathbf{k}$. This is a reasonable approximation provided the range of the interaction is much shorter than the Fermi wavelength. We have also taken the confusing step of writing the hybridization term as coupling the local states with conduction electrons labeled by a quantum number ν rather than by a simple spin σ. The reason for this is that ν usually denotes an enumeration of the symmetries of the degenerate local states, which depend on angular momentum, spin, and spin–orbit coupling. The conduction electron states have to be decomposed into the same symmetries by combining the Bloch states into angular momentum states. This is a tedious element of general practical calculations, and we here acknowledge that it may be necessary by using the notation ν instead of σ. Fortunately, for local states having only spin-up and spin-down degeneracies, no decomposition is required, and the coupling to the conduction electrons is just through the spin channels σ.

Equation (11.4.1) is a very rich Hamiltonian, which is capable of describing a variety of physical situations, depending on the values of U and V, and of the degeneracy of the local states. In particular, this Hamiltonian contains the Kondo Hamiltonian. By this we mean that by making a clever transformation, one can show that there are Kondo-like interactions included in the periodic Anderson model. This procedure is called the *Schrieffer–Wolff* transformation. One starts with the single-impurity Anderson Hamiltonian, in which there is just one impurity site with its degenerate d or f orbitals. One then seeks a canonical transformation that eliminates the hybridization terms between the conduction electrons and the impurity states. This is very similar in spirit to the canonical transformation we applied when studying the Fröhlich and Nakajima Hamiltonians in Sections 6.5 and 6.6. There we eliminated the term linear in electron–phonon coupling, while in the present case we eliminate the terms linear in the coupling between the conduction electrons and the impurity states.

11.4 Heavy fermions

We start by considering the single-impurity Hamiltonian

$$\mathcal{H} = \mathcal{H}_0 + \mathcal{H}_1 \tag{11.4.2}$$

where

$$\mathcal{H}_0 = \sum_{\mathbf{k},\sigma} \mathcal{E}_{\mathbf{k}} c^\dagger_{\mathbf{k},\sigma} c_{\mathbf{k},\sigma} + \mathcal{E}_f \sum_\sigma f^\dagger_\sigma f_\sigma + U n_\uparrow n_\downarrow$$

and

$$\mathcal{H}_1 = \sum_{\mathbf{k},\sigma} V_{\mathbf{k}} (c^\dagger_{\mathbf{k},\sigma} f_\sigma + f^\dagger_\sigma c_{\mathbf{k},\sigma}).$$

For simplicity we are considering only spin-up spin-down degeneracy, but are allowing the hybridization term $V_{\mathbf{k}}$ to depend on $\mathbf{k}$. Were it not for the Hubbard term, $U n_\uparrow n_\downarrow$, the whole Hamiltonian $\mathcal{H}$ could easily be diagonalized, since it would then be just a quadratic form in annihilation and creation operators. We might thus be tempted to perform the diagonalization and then treat the Hubbard term as a perturbation. However, we have seen in Chapter 2 that the strength of the Coulomb interaction makes this a difficult task. Instead, we follow the procedure of Section 6.5, and look for a unitary transformation to eliminate the interaction terms $V_{\mathbf{k}}$ to first order. Just as in Eq. (6.5.3), we seek a unitary operator s that will transform $\mathcal{H}$ into a new Hamiltonian $\mathcal{H}' = e^{-s} \mathcal{H} e^s$ from which first-order terms in $\mathcal{H}_1$ have been eliminated. For this to happen, we again require

$$\mathcal{H}_1 + [\mathcal{H}_0, s] = 0,$$

which then to second order in $V_{\mathbf{k}}$ leaves us with

$$\mathcal{H}' = \mathcal{H}_0 + \tfrac{1}{2}[\mathcal{H}_1, s].$$

From our experience with the electron–phonon interaction, where the elimination of the first-order terms led to the appearance of an effective electron–electron interaction term, we are prepared for some interesting consequences in the present case.

To proceed, we try an operator s of the form

$$s = \sum_{\mathbf{k},\sigma} a_{\mathbf{k}\sigma} V_{\mathbf{k}} (f^\dagger_\sigma c_{\mathbf{k}\sigma} - c^\dagger_{\mathbf{k}\sigma} f_\sigma), \tag{11.4.3}$$

with the coefficients $a_{\mathbf{k},\sigma}$ to be determined. With this form of s, we find that

$$[\mathcal{H}_0, s] = \sum_{\mathbf{k},\sigma} a_{\mathbf{k}\sigma} V_{\mathbf{k}} (\mathcal{E}_f - \mathcal{E}_{\mathbf{k}})(f^\dagger_\sigma c_{\mathbf{k}\sigma} + c^\dagger_{\mathbf{k}\sigma} f_\sigma) + U \sum_{\mathbf{k},\sigma} a_{\mathbf{k}\sigma} V_{\mathbf{k}} (c^\dagger_{\mathbf{k}\sigma} f_\sigma + f^\dagger_\sigma c_{\mathbf{k}\sigma}) n_{-\sigma}.$$

This poses a little problem, since the presence of four operators in the last term means that this expression cannot be equated with $-\mathcal{H}_1$, which contains only two. We solve this by making the mean-field approximation of replacing $n_{-\sigma}$ by its expectation value $\langle n_{-\sigma}\rangle$. The solution for the coefficients is then

$$a_{\mathbf{k}\sigma} = (\mathcal{E}_{\mathbf{k}} - \mathcal{E}_f - U\langle n_{-\sigma}\rangle)^{-1}.$$

Because $\langle n_{-\sigma}\rangle$ can only take on the values 0 or 1, we can rewrite this solution as

$$a_{\mathbf{k}\sigma} = \frac{1 - n_{-\sigma}}{\mathcal{E}_{\mathbf{k}} - \mathcal{E}_f} + \frac{n_{-\sigma}}{\mathcal{E}_{\mathbf{k}} - \mathcal{E}_f - U}. \tag{11.4.4}$$

Note that we have quietly removed the angular brackets from $\langle n_{-\sigma}\rangle$ and restored it to the status of being an operator. This is important in the next step we take, which is to calculate the effective interaction term in our transformed Hamiltonian $\mathcal{H}'$. It is

$$\frac{1}{2}[\mathcal{H}_1, s] = -\frac{1}{2}\sum_{\mathbf{k},\mathbf{k}'\sigma,\sigma'} [a_{\mathbf{k}\sigma}V_{\mathbf{k}}(f^\dagger_\sigma c_{\mathbf{k}\sigma} - c^\dagger_{\mathbf{k}\sigma}f_\sigma), V_{\mathbf{k}'}(c^\dagger_{\mathbf{k}'\sigma'}f_{\sigma'} + f^\dagger_{\sigma'}c_{\mathbf{k}'\sigma'})]. \tag{11.4.5}$$

The presence of the operator $n_{-\sigma}$ in the expression for $a_{\mathbf{k}\sigma}$ means that we are commuting a product of four operators with a product of two operators. Two operators disappear in the process, and this leaves us with a product of four operators, the most important of which are of the form $f^\dagger_\sigma c_{\mathbf{k}\sigma} c^\dagger_{\mathbf{k}'-\sigma} f_{-\sigma}$ and their Hermitian conjugates. These terms have the form of an interaction between a conduction electron and a local state, in which the spins of both are reversed and in which the conduction electron is scattered: this is precisely the form of the Kondo interaction. In addition to these terms, the unitary transformation also yields other terms not included in the Kondo Hamiltonian. The Anderson Hamiltonian is therefore richer in the sense that it contains more physics. This also makes it more difficult to solve. While the Kondo problem can be solved exactly using so-called *Bethe Ansatz* techniques, no exact solutions are known to the Anderson Hamiltonian.

Given the fact that the single-impurity Anderson Hamiltonian contains a Kondo-like term, it was to be expected that in the dilute limit, where one can consider just a single dopant atom, there would appear in the Anderson Hamiltonian a Kondo resonance at the Fermi energy. What is more surprising is that spectroscopic measurements find such a resonance in heavy-fermion materials such as $CeAl_3$, in spite of the fact that the Ce atoms are not at all dilute in the Al host. The reason for this is that the measurements are typically performed at temperatures too high for the coherence expected for a regular lattice of Ce and Al atoms to develop. As in the Kondo model, the appearance of this resonance depends nonperturbatively on an effective coupling constant.

One approach to studying the single-impurity Anderson Hamiltonian is very similar to the way in which we analyzed the low-temperature behavior of the Kondo Hamiltonian. One builds up approximate eigenstates by including several collective states. Each collective state consists of one or more electron–hole pair excitations from the Fermi sea, plus possible combinations of occupations of the local states. Because more states with greater complexities are included than in the low-temperature treatment of the Kondo problem, the algebra is more complicated, and so we do not present the details here. The picture that emerges, and which is fairly typical of heavy fermions, is that the density of states contains a number of interesting features. The first of these is a sharp Kondo-like peak at the Fermi energy, which is expected because the Kondo physics is contained in the Anderson Hamiltonian. In addition, there are two broader peaks, one below the Fermi surface at $\mathcal{E}_f$, and the other above the Fermi surface at approximately $\mathcal{E}_f + U$. The Kondo peak is very sharp with a small spectral weight, while the spectral weights of the broader peaks depend on the degeneracy of the local state. Experimental measurements of the density of states by means of spectroscopies that probe both the filled states below the Fermi surface and the empty states above it confirm the accuracy of this generic picture.

Problems

11.1 Calculate the energy $\delta \mathcal{E}_b$ of the state $|\Psi_b\rangle$ for spin-$\frac{1}{2}$ antiferromagnetic coupling. Follow the steps of the calculation of $\delta \mathcal{E}_a$, but note the difference that the state $|\Psi_b\rangle$ involves electron states below the Fermi surface, which will change some signs and the limits of the final integral over energies. Show that $|\Psi_b\rangle$ has a lower energy than $|\Psi_a\rangle$ if the conduction band is less than half full.

11.2 A commonly encountered Kondo system consists of Fe impurities in Cu. Assume a concentration of 1% and antiferromagnetic coupling of the order of 1 eV. Estimate the Kondo temperature (you will need to guess the density of states and conduction band width for Cu). How large a change in the specific heat (cf. Eq. (11.3.6)) would you expect due to the presence of the Fe impurities at low temperatures?

11.3 In a traditional BCS s-wave superconductor, electrons at the Fermi surface with opposite momenta and spin are bound in singlets. Suppose that such a superconductor, which in its normal state has a conduction band that is more than half-filled, is now doped with magnetic impurities with spin-1/2 local moments. Form an expression for the ratio of the Kondo temperature T_K to the BCS critical temperature T_c. How do you think this system will behave as it is cooled down if $T_c > T_K$? What if $T_c < T_K$?

11.4 In our attempt to diagonalize the Hamiltonian (11.4.2) we defined in Eq. (11.4.3) an operator s in terms of coefficients $a_{\mathbf{k}\sigma}$. These coefficients were assumed to be numbers and not operators, and so they commuted with $\mathcal{H}$. However, in Eq. (11.4.4) we rather inconsistently gave them the character of operators, and this led to the Kondo interaction when we formed $[\mathcal{H}_1, s]$. What would we have found if we had assumed from the start that $a_{\mathbf{k}\sigma}$ was linear in n_σ?

11.5 What is the coefficient of the Kondo operator $f_\sigma^\dagger c_{\mathbf{k}\sigma} c_{\mathbf{k}'-\sigma}^\dagger f_{-\sigma}$ found from the commutator $[\mathcal{H}_1, s]$ in Eq. (11.4.5)?

11.6 The Kondo effect is the result of the scattering impurity having an internal degree of freedom, namely its spin. Something similar happens when electrons are scattered from a moving impurity, the internal degree of freedom in this case being the vibrational motion of the impurity. For the purposes of this exercise we consider a substitutional impurity of mass closely equal to that of the host material, so that its motion can be described in terms of phonon operators. The electron scattering matrix element $V_\mathbf{K}$, with $\mathbf{K} = \mathbf{k}' - \mathbf{k}$, is phase shifted to $V_\mathbf{K} e^{i\mathbf{K} \cdot \mathbf{y}}$ by the displacement $\mathbf{y}$ of the impurity. To first order in $\mathbf{y}$ the scattering perturbation is then $V_\mathbf{K}(1 + i\mathbf{K} \cdot \sum_\mathbf{q} \mathbf{y}_\mathbf{q} N^{-1/2})$. Show that in second order, processes like those shown in Fig. P11.1 do not exactly cancel each other because of the energy of the virtual phonons involved. (This makes an observable contribution to the Peltier coefficient in dilute alloys at low temperatures, and is known as the Nielsen–Taylor effect.)

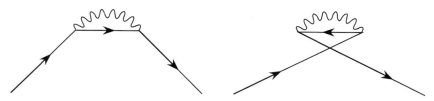

Figure P11.1. These two scattering processes do not exactly cancel.

Bibliography

Chapter 1. Semiclassical introduction

Standard introductory texts on quantum mechanics and solid state physics are

1. *Quantum Physics*, 2nd edition, by S. Gasiorowicz (Wiley, New York, 1996)
2. *Introduction to Solid State Physics*, 7th edition, by C. Kittel (Wiley, New York, 1996).

Some classic references for the topic of elementary excitations are

3. *Concepts in Solids*, by P. W. Anderson (reprinted by World Scientific, Singapore, 1998)
4. *Elementary Excitations in Solids*, by D. Pines (reprinted by Perseus, Reading, Mass., 1999).

The concept of the soliton is described in

5. *Solitons: An Introduction*, by P. G. Drazin and R. S. Johnson (Cambridge University Press, Cambridge, 1989).

Chapter 2. Second quantization and the electron gas

A detailed but accessible discussion of the electron gas and Hartree–Fock theory is given in

1. *Many-Particle Theory*, by E. K. U. Gross, E. Runge and O. Heinonen (IOP Publishing, Bristol, 1991)

This book also introduces the concepts of Feynman diagrams and Fermi liquid theory. The primary reference for the nuts and bolts of many-particle physics is

2. *Many-Particle Physics*, 3rd edition, by G. D. Mahan (Plenum, New York, 2000).

This work is particularly useful for learning how to do calculations using Feynman diagrams. Its first chapter treats second quantization with many examples from condensed matter physics, while its Chapter 5 treats electron interactions.

A classic reference for the interacting electron gas is

3. *The Theory of Quantum Liquids*, by D. Pines and P. Nozières (reprinted by Perseus Books, Cambridge, Mass., 1999).

Chapter 3. Boson systems

Lattice vibrations are thoroughly discussed in

1. *Dynamics of Perfect Crystals*, by G. Venkataraman, L. A. Feldkamp, and V. C. Sahni (M.I.T. Press, Cambridge, Mass., 1975)

and are also included in the reprinted classic

2. *Electrons and Phonons*, by J. M. Ziman (Oxford University Press, Oxford, 2000).

One of the most comprehensive references on the subject of liquid helium is

3. *The Physics of Liquid and Solid Helium*, edited by J. B. Ketterson and K. Benneman (Wiley, New York, 1978).

For the subject of magnons, and a clear introduction to magnetism in general, we recommend

4. *The Theory of Magnetism I, Statics and Dynamics*, by D. C. Mattis (Springer-Verlag, New York, 1988).

The reduction of the Dirac equation by means of the Foldy–Wouthuysen transformation may be found in

5. *Relativistic Quantum Mechanics*, by J. D. Bjorken and S. D. Drell (McGraw-Hill, New York, 1964)

Chapter 4. One-electron theory

A reference textbook on solid state physics that contains good pedagogical chapters on one-electron theory is

1. *Solid State Physics*, by N. W. Ashcroft and N. D. Mermin (Holt, Reinhart and Winston, New York, 1976).

More detail is to be found in

2. *Electronic Structure and the Properties of Solids*, by W. A. Harrison (Dover, New York, 1989)

and

3. *Elementary Electronic Structure*, by W. A. Harrison (World Scientific, Singapore, 1999)

with specialized descriptions of particular methods in

4. *The LMTO Method: Muffin-Tin Orbitals and Electronic Structure*, by H. L. Skriver (Springer-Verlag, New York, 1984)
5. *Planewaves, Pseudopotentials and the LAPW Method*, by D. J. Singh (Kluwer Academic, Boston, 1994)
6. *Electronic Structure and Optical Properties of Semiconductors*, 2nd edition, by M. L. Cohen and J. R. Chelikowsky (Springer-Verlag, New York, 1989).

A very readable text on quasicrystals is

7. *Quasicrystals: A Primer*, by C. Janot (Clarendon Press, Oxford, 1992).

Chapter 5. Density functional theory

The standard references are

1. *Density Functional Theory. An Approach to the Quantum Many-Body Problem*, by R. M. Dreizler and E. K. U. Gross (Springer-Verlag, New York, 1990)
2. *Density-Functional Theory of Atoms and Molecules* by R. G. Parr and W. Yang (Oxford University Press, Oxford, 1989).

The book by Dreizler and Gross is couched in the language of condensed matter physics, while the one by Parr and Yang is more suitable for quantum chemists. A classic review article is

3. General density functional theory, by W. Kohn and P. Vashishta, in *Theory of the Inhomogeneous Electron Gas*, edited by S. Lundqvist and N. H. March (Plenum Press, New York, 1983).

The volume

4. *Electronic Density Functional Theory. Recent Progress and New Directions*, edited by J. F. Dobson, G. Vignale, and M. P. Das (Plenum Press, New York 1998)

has a clear exposition of GGA approximations by K. Burke, J. P. Perdew and Y. Wang, as well as a good article on TDDFT by M. Petersilka, U. J. Gossmann and E. K. U. Gross, and on time-dependent current DFT by G. Vignale and W. Kohn. In addition, this volume also contains an article on ensemble DFT by O. Heinonen, M. I. Lubin, and M. D. Johnson.

There is also a review article on TDDFT by E. K. U. Gross, J. F. Dobson, and M. Petersilka in

5. *Density Functional Theory II*, Vol. **181** of *Topics in Current Chemistry*, edited by R. F. Nalewajski (Springer-Verlag, New York, 1996), p. 81.

A very understandable review article on TDDFT is

6. A guided tour of time-dependent density functional theory, by K. Burke and E. K. U. Gross, in *Density Functionals: Theory and Applications*, edited by D. Joubert (Springer-Verlag, New York, 1998).

The original papers laying out the foundation for density functional theory and the Kohn–Sham formalism are

7. P. Hohenberg and W. Kohn, *Phys. Rev.* **136**, B864 (1964)
8. W. Kohn and L. J. Sham, *Phys. Rev.* **140**, A1133 (1965).

Chapter 6. Electron–phonon interactions

A good source on this topic is Reference 2 of Chapter 3, which is further developed in

1. The electron–phonon interaction, by L. J. Sham and J. M. Ziman, in *Solid State Physics*, edited by F. Seitz and D. Turnbull, Vol. 15, p. 221 (Academic Press, New York, 1963)

and

2. *The Electron–Phonon Interaction in Metals*, by G. Grimvall (Elsevier, New York, 1981).

There is a wealth of material on electron–phonon interactions in

3. *Polarons in Ionic Crystals and Polar Semiconductors*, edited by J. T. Devreese (North-Holland, Amsterdam, 1972)

while a good source for discussion of Peierls distortions is

4. *Density Waves in Solids*, by G. Grüner (Addison-Wesley, Reading, Mass., 1994).

Chapter 7. Superconductivity

The classic reference is

1. *Theory of Superconductivity*, by J. R. Schrieffer (reprinted by Perseus, Reading, Mass., 1983).

An excellent introductory volume that treats both type I and type II superconductors is

2. *Introduction to Superconductivity*, 2nd edition, by M. Tinkham (McGraw-Hill, New York, 1996).

We can also recommend

3. *Superconductivity of Metals and Alloys*, by P. G. de Gennes (reprinted by Perseus Books, Cambridge, Mass., 1999).

A very complete review of BCS superconductivity is contained in the two volumes of

4. *Superconductivity*, edited by R. D. Parks (Marcel Dekker, 1969).

The original BCS paper, which is truly a landmark in physics, is

5. J. Bardeen, L. N. Cooper, and J. R. Schrieffer, *Phys. Rev.* **108**, 1175 (1957).

Chapter 8. Semiclassical theory of conductivity in metals

A classic text on the subject is again Reference 2 of Chapter 3. A more recent treatment is

1. *Quantum Kinetics in Transport and Optics of Semiconductors*, by H. Haug and A.-P. Jauho (Springer-Verlag, New York, 1998).

A reference for two-dimensional systems is

2. *Transport in Nanostructures*, by D. K. Ferry and S. M. Goodnick (Cambridge University Press, Cambridge, 1997).

A review article that focuses on Monte Carlo solutions of Boltzmann equations in semiconductors is

3. C. Jacoboni and L. Reggiani, *Rev. Mod. Phys.* **55**, 645 (1983).

An introduction to thermoelectricity is

4. *Thermoelectricity in Metals and Alloys*, by R. D. Barnard (Taylor and Francis, London, 1972).

Chapter 9. Mesoscopic physics

Introductory-level discussions of the Landauer–Büttiker approach are given in

1. *Electronic Transport in Mesoscopic Systems*, by S. Datta (Cambridge University Press, Cambridge, 1995).

A more advanced book is Reference 2 of Chapter 8. One of the best review articles on the Landauer–Büttiker approach is

2. The quantum Hall effect in open conductors, by M. Büttiker in *Nanostructured Systems*, edited by M. Reed (Semiconductors and Semimetals, Vol. **35**) (Academic Press, Boston, 1992).

In addition to discussing the multi-terminal Landauer–Büttiker approach, this article also describes in great detail its applications to the integer quantum Hall effect.

A good review of weak localization can be found in

3. Theory of coherent quantum transport, by A. D. Stone in *Physics of Nanostructures*, edited by J. H. Davies and A. R. Long (IOP Publishing, Bristol, 1992).

For more discussion of noise in mesoscopic systems, we recommend in particular

4. R. Landauer and Th. Martin, *Physica* **175**, 167 (1991)
5. M. Büttiker, *Physica* **175**, 199 (1991).

We also recommend the original literature on the Landauer–Büttiker formalism, namely

6. R. Landauer, *IBM J. Res. Dev.* **1**, 223 (1957)
7. R. Landauer, *Phil. Mag.* **21**, 863 (1970)
8. M. Büttiker, *Phys. Rev. Lett.* **57**, 1761 (1986).

The classic work on weak localization is

9. E. Abrahams, P. W. Anderson, D. C. Licciardello, and T. V. Ramakrishnan, *Phys. Rev. Lett.* **42**, 673 (1979).

Chapter 10. The quantum Hall effect

The standard reference is

1. *The Quantum Hall Effect*, 2nd edition, edited by R. E. Prange and S. M. Girvin (Springer-Verlag, New York, 1990)

while the book

2. *The Quantum Hall Effects, Integral and Fractional*, 2nd edition, by T. Chakraborty and P. Pietiläinen (Springer-Verlag, New York, 1995)

gives a particularly detailed discussion of the fractional quantum Hall effect and gives technical descriptions of numerical calculations. For composite fermions there is the volume

3. *Composite Fermions: A Unified View of the Quantum Hall Regime*, edited by O. Heinonen (World Scientific, Singapore, 1998).

For skyrmions we recommend

4. H. A. Fertig, L. Brey, R. Cote, and A. H. MacDonald, *Phys. Rev. B* **50**, 11018 (1994).

A volume that contains many useful articles, and particularly the one by S. M. Girvin and A. H. MacDonald, is

5. *Perspectives in Quantum Hall Effects: Novel Quantum Liquids in Low-Dimensional Semiconductor Structures*, edited by S. Das Sarma and A. Pinczuk (Wiley, New York, 1996).

The original literature reporting the discoveries of the integer quantum Hall effect and the fractional quantum Hall effect, and Laughlin's explanation of the fractional quantum Hall effect are

6. K. von Klitzing, G. Dorda, and M. Pepper, *Phys. Rev. Lett.* **45**, 494 (1980).
7. D. C. Tsui, H. L. Störmer, and A. C. Gossard, *Phys. Rev. Lett.* **48**, 1559 (1982).
8. R. B. Laughlin, *Phys. Rev. Lett.* **50**, 1395 (1983).

Chapter 11. The Kondo effect and heavy fermions

Reference 2 of Chapter 2 has a good chapter on spin fluctuations. It includes the Kondo problem and heavy fermions. A very complete account of these topics is

1. *The Kondo Problem to Heavy Fermions*, by A. C. Hewson (Cambridge University Press, Cambridge, 1997).

Some review articles on this topic are

2. D. M. Newns and N. Read, *Adv. Phys.* **36**, 799 (1987).
3. P. Fulde, J. Keller, and G. Zwicknagl, in *Solid State Physics*, Vol. **41**, p. 1, edited by H. Ehrenreich and D. Turnbull (Academic Press, San Diego, 1988).
4. N. Grewe and F. Steglich in *Handbook on the Physics and Chemistry of the Rare Earths*, Vol. **14** (Elsevier, Amsterdam, 1990).

Index

absolute thermoelectric power, 311
acoustic mode, 104
adiabatic approximation, 125
adiabatic local density approximation, 205
Aharonov–Bohm phase, 335, 348, 371
Aharonov–Casher effect, 332
Anderson Hamiltonian, 399
anharmonic oscillator, 81
annihilation operator
    for bosons, 78
    for electrons, 30, 35
    for magnons, 119
    for phonons, 96, 101
anticommutation, 37
anyon, 365

ballistic limit, 319
band index, 130
band structure, 128
basis vector, 99
BCS Hamiltonian, 235
Berry's phase, 371
Bloch electron, 125
    velocity of, 160
Bloch's theorem, 127
Bogoliubov theory of helium, 88
Bogoliubov–Valatin transformation, 237
Boltzmann equation, 285, 287
Born approximation, 172
Born–Oppenheimer approximation, 125
Bose–Einstein condensation, 87
Bose–Einstein distribution (*see* boson, distribution function)
boson, 31, 78
    distribution function, 86
boundary conditions, periodic, 28, 97, 267
Bravais lattice, 99
Brillouin–Wigner perturbation theory, 52, 138
Brillouin zone, 100, 130

canonical transformation, 224
channel, 320

chemical potential, 85, 204, 272
closed orbits, 168
coherence length, 198, 235
collective excitations, 3
commutator, 60, 79
composite fermion, 358, 379
condensate, 236
condensation energy, 244
conductance quantization, 319
conduction band, 132
conductivity
    electrical, 288, 299
    thermal, 304
contact resistance, 324
Cooper pair, 225
core state, 143
correlation energy, 49, 192
correlation potential, 192
Coulomb blockade, 336
creation operator
    for bosons, 78
    for electrons, 30, 35
    for magnons, 119
    for phonons, 96
Curie temperature, 121
current density functional theory, 205
current density operator, 161, 256
cyclotron frequency, 168, 345
cyclotron mass, 168, 221

dangerous diagrams, 90
Darwin term, 154
Debye model, 105
Debye temperature, 106
density functional theory, 183
    time-dependent, 200
density of states
    electron, 131
    phonon, 104
density operator, 59
dielectric constant, 69, 76
dimerization, 218
Dirac equation, 114, 153

Dirac notation, 27
Drude formula, 335

edge states, 353, 355
effective mass, 141, 180
Einstein model, 107, 117
elastic mean free path, 315
electrical conductivity (*see* conductivity, electrical)
electron pairs, 84, 225, 236
electron–phonon interaction, 17, 210–231, 299
electronic specific heat, 221, 251
elementary excitations, 1
energy current, 304
energy gap, 233, 243
ensemble density functional theory, 206
equation of continuity, 256, 286
exchange-correlation energy functional, 189
exchange-correlation hole, 194
exchange-correlation kernel, 205
exchange-correlation potential, 190
exchange energy, 47, 223
exchange hole, 194
exchange phase, 364
exchange scattering, 47
excitation
    collective, 3
    elementary, 1
    quasiparticle, 3
exciton, 180
Exclusion Principle, 3, 15, 18, 143, 288, 301
extended zone scheme, 131

Fermi–Dirac distribution (*see* fermion, distribution function)
Fermi energy, 132
fermion, 32
    distribution function, 86
    field operator, 57
Fermi surface, 15, 133
ferrimagnetism, 121
ferromagnetism, 10, 71, 117
Feynman–Bijl formula, 369
Fibonacci chain, 176
filling factor, 345
flux quantization, 265
Fock space, 33
Foldy–Wouthuysen transformation, 114, 153
fractional statistics, 364
Friedel sum rule, 174
Fröhlich Hamiltonian, 210
frozen phonon calculation, 197
Fuchs–Sondheimer theory, 294
functional, 184
    derivative, 203

generalized gradient approximation, 198
Ginzburg–Landau equations, 271, 277
golden mean, 176
grand canonical ensemble, 85
Green's-function method, 152
ground state, superconducting, 243
group velocity, 6, 109, 161

Hall coefficient, 297
Hall effect, 20, 297
Hall field, 20
harmonic approximation, 96
harmonic oscillator, 80
Hartree–Fock approximation, 48, 71, 75, 125
heat–current density, 304
heavy fermion, 385, 398
Heisenberg model, 118
Helmholtz energy, 113, 275, 282
heterojunction, 318
Hohenberg–Kohn theorem, 182, 184
hole-like behavior, 142
hole state, 16, 68
hole surface, 135
Holstein–Primakoff transformation, 116, 118
Hubbard term, 399
Hund's rules, 384

incompressibility, 343
inelastic scattering, 299
insulator, 132
intermediate state of superconductor, 272
inverse-effective-mass tensor, 141, 168, 180
Ising model, 118
isotope effect, 235, 251

Jahn–Teller effect, 279
jellium, 43
Johnson noise, 330
Jones zone, 156, 180
Josephson effect, 265, 270

Kohler's rule, 296
Kohn anomaly, 215
Kohn–Sham formulation, 187
Kohn–Sham orbitals, 190
Korringa–Kohn–Rostoker method, 152
Kondo effect, 385
Kondo limit, 385
Kondo temperature, 396
Kronecker delta, 27

Lagrangian, 112
Landau gauge, 344
Landau level, 345
Landauer–Büttiker formalism, 324, 356
lattice vibration (*see* phonon)
Laughlin wavefunction, 358
linear chain, 4, 7, 94
Liouville equation, 286
liquid helium, 88
local density approximation, 191
local spin density approximation, 197
London equation, 255
longitudinally polarized phonon, 101
Lorentz force, 111, 167
Lorenz number, 306

magnetic breakdown, 170
magnetic length, 345
magnetic moments, 113

magnetoresistance, 296, 314
magneto-roton, 368
magnon, 10, 117 (see also spin wave)
magnon interactions, 121
magnon–phonon interactions, 123
mass enhancement, 221
mass–velocity term, 153, 181
Matthiessen's rule, 303
mean field, 117
mean free path, 292
Meissner effect, 233, 254, 271
mesoscopic system, 316
metallic ferromagnets, 71
mixed state, 235, 274
mobility gap, 351
Mössbauer effect, 22
Mott insulator, 280
muffin tin potential, 152

Nakajima Hamiltonian, 226
nearly-free-electron approximation, 136
Nielsen–Taylor effect, 404
N-process, 109
number operator
    for bosons, 79
    for fermions, 37

occupation number representation, 32
Onsager relations, 311
optical mode, 103, 122
OPW (see orthogonalized plane wave)
OPW method, 150
orbit, 168
    periodic open, 169
orthogonalized plane wave, 146, 153

Padé approximants, 191
paramagnon, 74
Pauli Exclusion Principle (see Exclusion Principle)
Pauli spin susceptibility, 221
Peierls transition, 218
Peltier effect, 308
penetration depth, 233, 255, 258
Penrose tile, 174
Perdew–Burke–Ernzerhof theory, 200
Perdew–Zunger parametrization, 192
periodic Anderson model, 400
periodic boundary conditions, 28, 97, 267
periodic open orbit, 169
perturbation theory, 51
phase breaking length, 316
phase breaking time, 334
phase shift, 152
phonon, 7, 93–110
phonon drag, 312
phonon interactions, 23, 107
plasma frequency
    electron, 14, 64
    ion, 12
plasma oscillation, 11, 64
plasmon, 12, 14
polarization

electron density, 187
phonon, 7, 101
polaron, 219
Poisson noise, 330
positron, 16
probable occupation number, 285
pseudoboson, 80
pseudogap region, 280
pseudopotential, 148

quantized flux, 269, 346
quantum Hall effect, 19, 342
quasicrystal, 174
quasihole, 361
quasiparticle, 3, 17, 363
quenching, 114

random phase approximation, 60, 64
Rayleigh–Schrödinger perturbation theory, 52
reciprocal lattice, 99
reduced zone scheme, 131
reflection probability, 322
relativistic effect, 153
relaxation time, 292
    anisotropic, 252
remapped free-electron model, 135, 150
repeated zone scheme, 136
reservoir, 320
resistance minimum, 385
resistance quantum, 336
resistivity (see conductivity)
rigid-ion approximation, 210
RKKY interaction, 384
Runge–Gross theorem, 201

sandwichium, 139
scattering by impurities, 170
scattering matrix, 320
Schrieffer–Wolff transformation, 400
Schrödinger equation, 26
screening, 13, 65
second quantization, 34, 39, 78
Seebeck effect, 311
self-consistent Kohn–Sham scheme, 190
shot noise, 330
simple metal, 146
single-mode approximation, 369
singular value decomposition, 328
skymion, 370
Slater determinant, 32
soliton, 7, 218
Sommerfeld gas, 46
sound velocity in metals, 19, 229
specific heat
    due to phonons, 104
    electronic, 221, 251
spin, 48, 114
spin-orbit coupling, 153, 157
spin raising and lowering operators, 115
spin wave, 11, 71 (see also magnon)
strong-coupling superconductor, 243
structure factor, 156

superconductivity, 232–284
superfluid, 92
symmetric gauge, 344

thermal conductivity, 304
thermoelectric effects, 308
thermoelectric power, absolute, 311
thermopower, 311
tight-binding approximation, 144
T-matrix, 172
Toda chain, 8
trace of operator, 84
transition matrix, 172
transition metals, 150, 302
transition temperature of superconductor, 247
transmission probability, 322
transverse-even voltage, 297
transversely polarized phonon, 101
tunneling, 258–265
type I superconductor, 233, 275
type II superconductor, 233, 275

ultrasonic attenuation, 252
Umklapp scattering, 109

uncertainty principle, 53, 213, 271
U-process (*see* Umklapp scattering)
universal conductance fluctuation, 335

vacuum state, 30
valence band, 132
vector mean free path, 293
Voronoy polyhedron, 178
vortex rings, 93
v-representability, 207

weak-coupling superconductor, 243
weak localization, 332
Wiedemann–Franz law, 306
Wigner crystal, 191
Wilson ratio, 398
winding number, 370

Yukawa potential, 64, 70

Zener breakdown, 167